# Cell Polarity

# Frontiers in Molecular Biology

SERIES EDITORS

**B. D. Hames**

*Department of Biochemistry*
*and Molecular Biology*
*University of Leeds, Leeds LS2 9JT, UK*

**D. M. Glover**

*Cancer Research Laboratories,*
*Department of Anatomy and Physiology,*
*University of Dundee, Dundee DD1 4HN, UK*

## TITLES IN THE SERIES

# Cell Polarity

EDITED BY

## David G. Drubin

*Professor of Genetics*
*University of California, Berkeley, USA*

**OXFORD**
UNIVERSITY PRESS

*This book has been printed digitally in order to ensure its continuing availability*

# OXFORD
UNIVERSITY PRESS

Great Clarendon Street, Oxford OX2 6DP

Oxford University Press is a department of the University of Oxford.
It furthers the University's objective of excellence in research, scholarship,
and education by publishing worldwide in

Oxford  New York

Auckland  Bangkok  Buenos Aires  Cape Town  Chennai
Dar es Salaam  Delhi  Hong Kong  Istanbul  Karachi  Kolkata
Kuala Lumpur  Madrid  Melbourne  Mexico City  Mumbai  Nairobi
São Paulo  Shanghai  Singapore  Taipei  Tokyo  Toronto
with an associated company in Berlin

Oxford is a registered trade mark of Oxford University Press
in the UK and in certain other countries

Published in the United States by Oxford University Press Inc., New York

© Oxford University Press, 2000

The moral rights of the author have been asserted
Database right Oxford University Press (maker)
First published  2000
Reprinted 2002

A catalogue record for this book is available from the British Library

Library of Congress Cataloging in Publication Data
(Data available)
ISBN 0 19 963803 9 (Hbk)
ISBN 0 19 963802 0 (Pbk)

# Preface

The topic of cell polarity touches on essentially every aspect of cell and developmental biology and is fundamentally important for differentiation, proliferation, morphogenesis, and function of single-celled and metazoan organisms alike. Proteins, nucleic acids, macromolecular assemblies, and organelles are not dispersed homogeneously throughout the cytoplasm and along the plasma membrane. Rather, cells are composed of distinct microenvironments. Be the cell a bacterium, yeast, ciliate, plant, or animal cell, be it a nerve cell, epithelial cell, oocyte, or zygote, these microenvironments are themselves disposed asymmetrically within the cell, providing properties tailored to the needs of each cell type. Thus, epithelial cells have apical and basolateral domains, *S. cerevisiae* cells have mothers and daughters, nerve cells have dendrites and axons, and many cells segregate cell-fate determinants asymmetrically prior to division.

In this volume, principles of cell-polarity development are presented. These principles were revealed using the full panoply of genetic, cell biological, and biochemical approaches to study cell polarity in a wide variety of cell types. The comprehensive nature of this volume provides an opportunity to identify themes and to highlight variation in mechanisms used to generate cell polarity.

A general scheme for cell-polarity development has been proposed that, despite differences in specific molecular mechanisms, applies to many cell types and provides a framework for discussions (1). According to this scheme, a hierarchy of events begins with an intrinsic or extrinsic spatial cue acting at the cell's surface to orient the polarity of the cell. Next, receptors and signalling proteins mark and interpret the cue. Components including plasma membrane proteins, cell-wall constituents, and cytoskeletal elements then function to reinforce the asymmetry defined by the cue. Finally, the asymmetry marked by the cue is propagated into the cytoplasm and along the plasma membrane. In eukaryotes, propagation of this asymmetry is a function often attributed to the cytoskeleton which, via interactions with cargo-bearing cytoskeletal motor proteins, can recruit organelles and RNAs to the site of the cue, and can generate distinct plasma membrane domains.

Each of the first six chapters of this volume describes cell-polarity development in a different organism or cell type from which important insights into the process have been gleaned. Bacteria specify sites of DNA replication and cell division, they create distinct membrane domains, and they specify cell fates, all by mechanisms that function independent of a cytoskeleton. Powerful genetic studies on diverse bacterial species have yielded deep insights into intrinsic molecular mechanisms for cell-polarity development. Roles for transcription regulation and proteolytic processing, and an intimate linkage between cell division and cell-fate determination, have been revealed. Budding and fission yeast have also proved to be extremely tractable organisms for genetic studies of cell-polarity development, which in these yeasts can occur in response to intrinsic or extrinsic cues. The yeast studies are largely responsible for revealing that polarity development proceeds in a hierarchical series of steps, and for the identification of many key proteins involved in development of eukaryote cell polarity. In both yeasts, the axis of cell polarity changes dynamically as a function of cell-cycle stage. The fact that both cell-polarity development and cell-cycle regulation have been studied intensely in these yeasts has provided an exceptional opportunity to investigate mechanisms by which the cell-cycle machinery induces changes in cell polarity. Comparison between pathways of cell-polarity development in these two yeasts has been particularly valuable. While fission and budding yeast share with each other and with more complex eukaryotes some mechanisms for generating polarity (i.e. dependence on cortical actin structures under control of small GTPases), polarity development in these yeasts shows striking differences, such as independence (budding yeast) or dependence (fission yeast) on microtubules.

Global spatial patterning in ciliates has fascinating characteristics. Ciliates lack any bilateral symmetry and spatial patterning is inherited epigenetically. Studies on ciliates are distinctly non-molecular, but distinctly rich in description, such that the rules for pattern generation have been established. Ciliate patterning provides an interesting comparison with polarity development mechanisms used by bacteria and yeast. Patterning in ciliates is similar to generation of bud site selection patterns in budding yeast, in that both are epigenetic phenomena dependent on intrinsic cortical cues.

In epithelial cells, details of the hierarchy of steps for cell-polarity development are being elucidated by elegant studies that have benefited from the conceptual template and knowledge of key protein players provided by genetic studies in yeast. An emerging principle is that epithelial cell-polarity development is not an 'inside-out' phenomenon, but is rather an 'outside-in' phenomenon, in which cell adhesion provides the spatial cue. Since comparisons of many cell types lead to the conclusion that cell polarity most often develops around a cortical cue, 'outside-in' development seems to be the predominant mode of polarity development. A second important principle is that mechanisms for sorting proteins and vesicles once thought unique to epithelia, are in fact commonly employed in diverse metazoan cell types, including neurons.

Algae and vascular plants share with bacteria and yeasts, but not with animal cells, constraints imposed by a cell wall. Thus, cell migrations are not an option. Changes in cell shape, and generation of new cell layers during vascular plant development, depend on cell-wall modification and specification of division planes. As in yeast cells, localized tip expansion is a mechanism that can underlie cell elongation, and morphogenetic changes are co-ordinated with growth and division. An interesting feature of plants and algae is that light and gravity provide extrinsic cues for polarity development. Identification of the sensors that interpret these cues is an important goal for future studies. Much of what is known about polarity development in plants was learned by physiological approaches that, for example, highlighted roles for intracellular gradients of calcium ions. It is now important to determine whether calcium ions play similar roles in other organisms. Thanks to the relatively recent application of genetics to polarity development in *Arabidopsis* and maize, proteins playing important roles in cell-polarity development in these plants are now being identified and progress is being accelerated.

The elaborate polarity development processes that allow a single cell to develop into a multicellular organism have been studied intensively in *Xenopus laevis*. Processes that begin early in oogenesis establish the animal and vegetal poles, an important step for development of the three primary germ layers. The spatial cues that establish animal/vegetal polarity have yet to be identified. One of the earliest signs of polarity in the oocyte is the accumulation of Vg1 mRNA at the vegetal pole via a cytoskeleton-independent step, a microtubule-dependent step, and finally, an actin-dependent step. From the radial symmetry of the oocyte, the bilateral symmetry and, ultimately, left-right asymmetry of the embryo must develop. Sperm entry provides a cue for development of the dorsal–ventral axis, creating a second polarity axis and giving rise to bilateral symmetry. A cortical rotation aligns microtubule arrays so that they may mediate polarized organelle transport. This rotation is proposed to affect the juxtaposition of certain molecules of the Wnt signalling pathway, such that the signals that mark dorsal-ventral polarity are localized spatially.

Chapters 7-9 focus on general mechanisms for creating polarity and attempt to synthesize information gleaned from studies on different organisms. In Chapter 7, a theoretical framework is presented for understanding the remarkable ability of cells to interpret chemoattractant gradients. Cells must have a mechanism for amplifying spatial information from extremely shallow extracellular chemoattractant gradients into pronounced cellular asymmetry. Evidence from studies of neutrophils and *Dictyostelium* are used to support a model involving global inhibition and local enhancement mechanisms in signal processing. In these two cell types, the read-out of this signalling is appropriately oriented cell motility. As localized assembly of cortical actin filaments drives motility, actin can be viewed as an ultimate signalling target. Regulation of actin assembly by coupling heterotrimeric G proteins to small GTPases of the Rho family appears to be a widely conserved pathway. Many elements of this pathway are shared between migrating single-celled organisms, migrating metazoan cells, and non-motile yeast cells that orient and extend mating projections up pheromone gradients.

In Chapter 8, mechanisms of asymmetrical cell division are explored in three genetically tractable organisms—budding yeast, fruit flies, and nematodes—that span vast phylogenetic distances. In all cases, cell division is co-ordinated with segregation of cell-fate determinants. However, on a molecular level, other than sharing a reliance on the actin cytoskeleton, the trend is quite clearly toward diversity, not conservation in mechanism. This is true even when one compares different cell-fate decisions within one organism. This trend toward diversity in mechanism extends to bacteria, described in Chapter 1, in which cell-fate decisions are linked intimately to cell division, just as they are in yeast, fruit flies, and nematodes. Insights into cell-fate determination during asymmetrical divisions in plants await future analysis.

Spatial targeting of the secretory pathway is important for controlling cell shape and for generating an asymmetrical arrangement of cell-surface molecules, as well as for transporting molecules across cell layers. As described in Chapter 9, genetic and biochemical analyses in yeast and other cell types have identified two targeting steps: a transport step (often mediated by actin filaments or microtubules) and a docking/fusion step at the plasma membrane. In sharp contrast to the situation for inheritance of cell-fate determinants, the molecular machinery for the two steps involved in spatial targeting of the secretory pathway appear to be highly conserved from yeast to vertebrates.

The final chapter makes the important point that cell polarity in metazoan cells ultimately must be understood in the context of tissues. Epithelial appendages such as hair, nails, teeth, and glands are characterized by pronounced asymmetry. In skin epithelia, as with other tissues, a progression of events regulated spatially within the tissue specifies cell morphology and gene expression patterns. Cell fate and the specific form in which cell polarity is manifested are dictated by interactions between ectoderm and mesoderm and are controlled by a combination of long-range and short-range interactions in the form of gradients of morphogens and interactions with proximal cells.

In summary, broad features of cell-polarity development such as marking a cortical site with a molecular cue, reinforcing and interpreting the cue, and propagating the asymmetry marked by the cue to the cytoplasm and along the plasma membrane, seem quite general. The coupling of cell division and asymmetrical inheritance of cell-fate determinants also seems to be a feature common to many cell types. As to specific molecular mechanisms, there are examples of striking conservation, including the targeting of secretory vesicles to specific sites on the plasma membrane, and the coupling of trimeric G proteins to small GTPases of the Rho family for actin regulation in response to diffusible molecular cues. There are also examples of striking divergence in molecular mechanism, as in the sorting of cell-fate determinants. The richness of phenomena and underlying mechanisms emphasizes the importance of generating cell polarity.

This volume should prove useful to research scientists seeking to broaden their perspectives of cell-polarity development as well as to graduate students and advanced undergraduates seeking their first exposure to the field.

*Berkeley, California*                                                                 D.G.D.
May 2000

# Reference

1. Drubin, D.G. and Nelson, W.J. (1996) Origins of cell polarity. *Cell*, **84**, 335–344.

# Contents

The colour plates are located between pp. 22–23

## 3  Cell polarity in ciliates    78

JOSEPH FRANKEL

## 4  Spatial cues for cellular asymmetry in polarized epithelia    106

W. JAMES NELSON, CHARLES YEAMAN, AND KENT K. GRINDSTAFF

## 5   Cell polarity in algae and vascular plants    141

JOHN E. FOWLER

## 8 Genetic analysis of intrinsically asymmetrical cell division

240

FABIO PIANO AND KENNETH KEMPHUES

## 9 Polarized exocytosis: targeting vesicles to specific domains on the plasma membrane   269

PATRICK BRENNWALD AND JOAN ADAMO

## 10 Morphogenesis of skin epithelia 285

PIERRE A. COULOMBE AND KEVIN McGOWAN

# Contributors

JOAN ADAMO
Department of Cell Biology, Weill Medical College, Cornell University, New York, NY 10021, USA.

JÜRG BÄHLER
Imperial Cancer Research Fund, Cell Cycle Laboratory, 44 Lincoln's Inn Fields, London WC2A 3PX, UK.

HENRY R. BOURNE
Department of Cellular and Molecular Pharmacology, University of California, 543 Parnassus Avenue, San Francisco, California 94143-0459, USA.

PATRICK BRENNWALD
Department of Cell Biology, Weill Medical College, Cornell University, New York, NY 10021, USA.

YVES V. BRUN
Department of Biology, Jordan Hall 142, Indiana University, 1001 E 3rd Street, Bloomington, IN 47405-3700, USA.

PIERRE A. COULOMBE
Departments of Biological Chemistry and Dermatology, The Johns Hopkins University School of Medicine, Baltimore, Maryland 21205, USA.

PETER N. DEVREOTES
Department of Biological Chemistry, Johns Hopkins University School of Medicine, Baltimore, MD 21205, USA.

JOHN E. FOWLER
Dept. of Botany and Plant Pathology, 2082 Cordley Hall, Oregon State University, Corvallis, OR 97331, USA.

JOSEPH FRANKEL
Department of Biological Sciences, University of Iowa, Iowa City, IA 52242, USA.

KENT K. GRINDSTAFF
Department of Molecular and Cellular Physiology, Beckman Center for Molecular and Genetic Medicine, Stanford University School of Medicine, Stanford, CA 94305-5345, USA.

ANTJE HOFMEISTER
Department of Plant and Microbial Biology, University of California, 111 Koshland Hall, Berkeley, CA 94720-3102, USA.

KENNETH KEMPHUES
Department of Molecular Biology and Genetics, Cornell University, 107 Biotech Building, Ithaca, NY 14853, USA.

CAROLYN A. LARABELL
Life Sciences Division, Lawrence Berkeley National Laboratory, Berkeley, CA 94720, USA.

KEVIN MCGOWAN
Departments of Biological Chemistry and Dermatology, The Johns Hopkins University School of Medicine, Baltimore, Maryland 21205, USA.

W. JAMES NELSON
Department of Molecular and Cellular Physiology, Beckman Center for Molecular and Genetic Medicine, Stanford University School of Medicine, Stanford, CA 94305-5345, USA.

CAROLE A. PARENT
Department of Biological Chemistry, Johns Hopkins University School of Medicine, Baltimore, MD 21205, USA.

MATTHIAS PETER
Swiss Institute for Experimental Cancer Research (ISREC), Ch. des Boveresses 155, 1066 Epalinges/VD, Switzerland.

FABIO PIANO
Department of Molecular Biology and Genetics, Cornell University, 107 Biotech Building, Ithaca, NY 14853, USA.

GUY SERVANT
Department of Cellular and Molecular Pharmacology, University of California, San Francisco, California 94143-0450, USA.

ORION D. WEINER
Department of Biochemistry and Biophysics, University of California, San Francisco, California 94143-0554, USA.

CHARLES YEAMAN
Department of Molecular and Cellular Physiology, Beckman Center for Molecular and Genetic Medicine, Stanford University School of Medicine, Stanford, CA 94305-5345, USA.

# Abbreviations

| | |
|---|---|
| AF | after fertilization |
| AGP | arabinogalactan protein |
| APC | adenomatous polyposis coli tumor suppressor protein |
| βARK | β-adrenergic receptor kinase |
| ASC | axis-stabilizing complex |
| BAPTA | 1,2-bis(2-aminophenoxy)ethane-$N,N,N',N'$-tetraacetic acid |
| BCC | basal cell carcinoma |
| BFA | brefeldin A |
| BMPs | bone morphogenetic proteins |
| cAMP | cyclic adenosine monophosphate |
| CB | cytochalasin B |
| CD | cytochalasin D |
| CDPK | calmodulin-independent protein kinase |
| cGMP | cyclic guanosine monophosphate |
| CNS | central nervous system |
| CRAC | cytosolic regulator of adenylyl cyclase |
| CRIB | Cdc42–Rac interactive binding |
| CVPs | contractile vacuole pores |
| Cyp | cytoproct |
| CZ | clear zone |
| DHP | dihydropyridine |
| 6-DMAP | 6-dimethylaminopurine |
| EcadGFP | E-cadherin tagged with green fluorescent proteins |
| ECM | extracellular matrix |
| EGF | epidermal growth factor |
| EGTA | ethylene glycol-bis(β-aminoethyl ether) $N,N,N',N'$-tetraacetic acid |
| ER | endoplasmic reticulum |
| ERK | extracellular signal-regulated kinase (or MAPK) |
| ES | external sense organs |
| FGF | fibroblast growth factor |
| FH | formin homology |
| FMLP | fMet–Leu–Phe |
| GAP | GTPase-activating protein |
| GBD | G-protein-binding domain |
| GBP | GSK3-binding protein |
| GDI | GDP-dissociation inhibitor |
| GEF | guanine nucleotide exchange factor |
| GFP | green fluorescent protein |

| | |
|---|---|
| GMC | guard mother cell (Chapter 5) |
| GMC | ganglion mother cell (Chapter 8) |
| GPCR | G-protein-coupled receptor |
| GPI | glycosylphosphoinositol |
| GRK | G-protein receptor kinase |
| GSK3 | glycogen synthase kinase 3 |
| GTP | guanosine triphosphate |
| GV | germinal vesicle |
| HA | influenza haemagglutinin |
| HCF | hexacyanoferrate |
| HIV | human immunodeficiency virus |
| IAA | indole-3-acetic acid |
| IL | interleukin |
| IP$^3$ | inositol trisphosphate |
| IQGAP | IQ motif and GTPase Activating Protein motif |
| LatA | latrunculin A |
| LatB | latrunculin B |
| LDL | low density lipoprotein |
| lef-1 | lymphoid enhancer factor 1 |
| LH | left-handed |
| MAPK | mitogen-activated protein kinase or ERK |
| MDCK | Madin–Darby canine kidney cells |
| MEK | MAPK kinase or ERK kinase |
| MEKK | MEK kinase |
| MES | maternal effect sterility |
| METRO | messenger transport organizer |
| MHC | major histocompatibility complex |
| MHCK | myosin heavy chain kinase |
| MTOC | microtubule organizing centre |
| MLCK | myosin light chain kinase |
| NB | neuroblast |
| NBP | NPA-binding protein |
| N-CAM | neural cell-adhesion molecule |
| NETO | new end take off |
| NPA | 1-$N$-naphthylphthalamic acid |
| NSF | $N$-ethylmaleimide-sensitive fusion protein |
| NT | normalized time of the first cell cycle |
| NT-3 | neurotrophin-3 |
| OA | oral apparatus |
| OP | oral primordium |
| ORS | outer root sheath |
| PAK | p21-activated kinase |
| PEG | polyethylene glycol |
| PH | pleckstrin homology (domain) |

| | |
|---|---|
| PI | phosphoinositide |
| pIgA-R | poly-IgA receptor |
| PI(4)P | phosphatidylinositol 4-phosphate |
| PI(3,4)P$^2$ | phosphatidylinositol (3,4)-bisphosphate |
| PI(4,5)P$^2$ | phosphatidylinositol (4,5)-bisphosphate |
| PI(3,4,5)P$^3$ | phosphatidylinositol (3,4,5)-trisphosphate |
| PNR | procephalic neurogenic region |
| PNS | peripheral nervous system |
| Pon | partner of numb |
| PPBs | pre-prophase bands |
| PSTPIP | proline-serine-threonine phosphatase-interacting protein |
| PTB | phosphotyrosine-binding (domain) |
| PTX | pertussis toxin |
| RGS | regulators of G-protein signalling |
| RH | right-handed |
| RHOK | Rho-kinase |
| SC | subsidiary cell |
| SLO | streptolysin-O |
| SMC | subsidiary mother cell |
| Smc | structural maintenance of chromosomes |
| Spb | spindle pole body |
| SNAPs | soluble NSF attachment proteins |
| SOP | sensory organ precursor |
| SPB | spindle pole body |
| TCF | T-cell factor |
| TEM | transmission electron microscopy |
| TGF | transforming growth factor |
| TGN | *trans*-Golgi network |
| TIBA | 2,3,5-triodobenzoic acid |
| TJ | tight junction |
| UM | undulating membrane |
| URS | upstream regulatory sequence |
| UTR | untranslated region |
| VAMP | vesicle-associated membrane proteins |
| VSV G | vesicular stomatitis virus G |
| WASP | Wiskott–Aldrich syndrome protein |

# 1 | Polarity and cell fate in bacteria

ANTJE HOFMEISTER and YVES V. BRUN

## 1. Introduction

During the development of all organisms, different cell types are generated from a single precursor cell. Generation of this diversity requires that cells be able to divide into two different progeny. Formally, two mechanisms are possible (1):

1. The two progeny cells are initially identical but become different by interacting with each other, by interacting with neighbouring cells, or by responding to diffusible factors secreted from neighbouring cells.

2. Alternatively, the precursor cell can divide asymmetrically into two progeny cells that are already committed to different developmental pathways at their time of birth. Such asymmetric divisions can be generated if the progenitor cell is capable of segregating determinants preferentially into one of its two progeny cells so as to initiate a particular developmental pathway in this progeny cell but not its sister cell.

Here we describe mechanisms by which asymmetric cell divisions cause cellular diversity in the aquatic bacterium *Caulobacter crescentus* and in the soil bacterium *Bacillus subtilis*. Recent discoveries have revealed that *C. crescentus* and *B. subtilis* are organized into structural and functional domains that control their cellular morphogenesis (2, 3). Proteins are targeted to specific subcellular sites, including the incipient division plane, the septum, and the cell pole. Below, we discuss how differential protein localization in *C. crescentus* and *B. subtilis* governs:

- chromosome organization;
- DNA replication and chromosome segregation;
- cytokinesis; and
- the establishment of cell-type-specific gene expression.

## 2. Asymmetric cell division

Differentiation in *C. crescentus* and *B. subtilis* is a result of asymmetric cell division. In contrast to symmetric cell division, which yields identical progeny, asymmetric cell division generates progeny with different fates. Whereas cell division in *C. crescentus* is obligatorily asymmetric and cellular differentiation is an integral part of the cell cycle, *B. subtilis* divides asymmetrically and differentiates only during spore formation in response to environmentally induced nutrient deprivation.

Each *C. crescentus* cell division yields a sessile stalked cell and a swarmer cell that is motile by virtue of a single polar flagellum (Fig. 1A). The sessile stalked cell is capable of immediately initiating chromosomal DNA replication, assembling a flagellum at

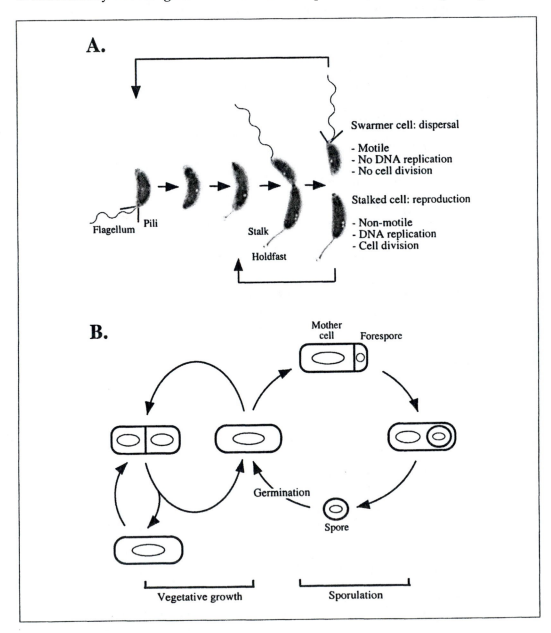

the pole opposite its stalk, and dividing into a stalked cell and a swarmer cell. By contrast, the swarmer cell is unable to replicate its chromosome and leads a foraging lifestyle. Ultimately, the swarmer cell differentiates into a stalked cell by shedding its flagellum and assembling a stalk at the newly vacated pole. At the same time, chromosomal DNA replication is reinitiated and the cell cycle is repeated.

*Bacillus subtilis* propagates by binary fission, in which a dividing cell yields two identical progeny (Fig. 1B). When deprived for nutrients, however, *B. subtilis* enters a developmental programme during which process an asymmetrically positioned septum partitions the developing cell into dissimilar-sized progeny (Fig. 1B) (4, 5). Initially, the small forespore and the large mother cell remain side by side. Later in development, the forespore becomes wholly engulfed by the mother cell, which nurtures the forespore until it develops into a mature spore that is liberated by lysis of the mother cell. The mature spore is highly resistant to environmental insults, but is capable of germinating under favourable conditions and propagating by binary fission.

# 3. Structurally and functionally distinct progeny

In *C. crescentus* structural asymmetry is intrinsic. Many of the distinct characteristics of swarmer and stalked-cell progeny are established prior to cell division, by virtue of polar gene transcription or the differential distribution of gene products to the two poles of the predivisional cell. Already the predivisional cell is asymmetric, carrying a flagellum and chemoreceptors at one pole and a stalk at the opposite pole (6–8). Although the flagellar protein components and the chemoreceptors localize to the incipient swarmer-cell pole, their polar assembly appears to be mediated by separate targeting mechanisms.

Flagellum biogenesis involves the transcription of more than 50 genes in the order required for sequential assembly of their protein products (9–12). Early flagellar genes encode proteins that compose the basal body and the hook structure, as well as regulatory proteins such as FlbD (see Section 6.2.1 below). Late flagellar genes

---

**Fig. 1** Asymmetric cell division during differentiation in *C. crescentus* and *B. subtilis*. (A) Life cycle of *C. crescentus*. The transmission electron micrographs are of cells at different stages of the cell cycle. The outline of the flagellum was enhanced for illustration purposes. The straight lines emanating from the flagellated pole of swarmer cells were drawn in to represent pili. Polar structures are indicated. The life cycle is depicted starting with the swarmer cell. Swarmer cells have a polar flagellum and are chemotactically competent. They are unable to replicate DNA and divide. After remaining in the swarmer stage of the life cycle for a fixed period, the cells differentiate into stalked cells. This differentiation involves release of the flagellum, growth of a stalk and holdfast at the same pole that had shed the flagellum, and initiation of DNA replication. The stalked cell elongates and synthesizes a flagellum at the pole opposite the stalk, thus generating an asymmetric predivisional cell. Cell division produces a swarmer cell and a stalked cell that can immediately re-enter the cell cycle. (Photos by Yves Brun.) (B) *B. subtilis* propagation and spore formation. *Bacillus subtilis* propagates by binary fission in which a medial division yields two identical progeny cells. During sporulation, however, an asymmetrically positioned septum generates a large mother cell and a small forespore. Asymmetric division is followed by engulfment of the forespore by the mother cell. Finally, the mature spore is released into the environment by lysis of the mother cell. The spore can germinate under favourable conditions and produce progeny. Thin ovals represent the chromosomes.

encode flagellins that constitute the helical filament. Already in the predivisional cell, the MS ring that anchors the flagellar basal body in the cytoplasmic membrane is asymmetrically assembled at the pole of the incipient swarmer compartment. Polar localization of the MS ring is the result of cell-cycle-dependent proteolysis and positioning of its FliF protein subunits (6). The mechanism of FliF targeting is not known, but may involve its binding to a recognition patch left at the future pole at the time of the previous cell division.

The clustering of the chemoreceptors at the pole opposite of the stalk is independent of flagellar positioning. Indeed, polar positioning of chemoreceptors is not unique to *C. crescentus* or polarly flagellated bacteria in general, but is similarly observed in *E. coli* and *B. subtilis*, which have flagella distributed over their entire cell surface (13, 14). Polar localization of the *E. coli* Tsr chemoreceptor depends on its interaction with the chemosensory transducing proteins CheA and CheW. The Tsr chemoreceptor is held in a polar complex with the cytoplasmic CheA histidine kinase through their interaction with the CheW protein (15). In the absence of CheA and CheW, or in the absence of CheW alone, Tsr is not retained in its polar position and is randomly distributed in the cytoplasmic membrane (14, 16). Similarly, in strains deleted for all chemoreceptors, CheA and CheW lose their membrane association and are randomly distributed in the cytoplasm. Therefore, CheW is responsible for the polar positioning of the chemosensory protein cluster. It is possible that the chemoreceptors are targeted directly to the cell poles and that complex formation with the chemosensory transducing proteins serves to retain them at their polar location.

During sporulation in *B. subtilis*, polarity arises as a consequence of a visibly asymmetric cell division and does not seem to exist prior to septum formation (4, 5). The polarity created by the asymmetrically positioned septum is propagated by the localization of essential regulators to the division plane and by the subsequent establishment of differential gene expression (see Section 6.2 below).

## 4. Chromosome organization

Another asymmetric characteristic is apparent in the different condensation states of the nucleoids, suggesting that the replicated chromosomes assume distinctive higher-order structures. The two chromosomes at opposite poles of the *C. crescentus* predivisional cell exhibit different condensation states and differential control of replication potential. During the *C. crescentus* cell cycle the more highly condensed nucleoid is inherited by the swarmer cell (17). Differentiation of the swarmer cell to a stalked cell later in the cell cycle is accompanied by a decondensation of the nucleoid to the condensation state of the nucleoid in the stalked cell following cell division. In sporulating *B. subtilis*, the forespore nucleoid is initially highly condensed but decondenses coincidently with later-stage forespore-specific gene expression (18). There is no evidence that the difference in chromosome condensation is involved in regulating differential gene transcription. In fact, it could merely be a consequence of the packaging requirement into the small volume of forespore cytoplasm. In this case, however, it is not apparent why one of the equally sized progeny cells in *C. crescentus* should harbour a condensed nucleoid. It is thus possible that nucleoid condensation is a means of transiently inactivating much of the genome in a specific cell type, with subsequent nucleoid decondensation allowing for the selective activation of appropriate regions of the chromosome.

Like eukaryotic cells, bacteria contain proteins that are responsible for chromosome condensation and that are crucial for faithful chromosome partitioning. Smc (structural *m*aintenance of *c*hromosomes) proteins are prevalent in bacteria, archaebacteria, and eukaryotes, and exhibit a structure similar to other eukaryotic motor proteins. In *B. subtilis* Smc is involved in chromosome condensation, contributing to the compaction of the nucleoid (19–21). It has been suggested that Smc-dependent chromosome condensation facilitates assembly of a nucleoprotein complex around the origin region and its subsequent role in sister chromosome pairing (19).

# 5. DNA replication and chromosome segregation

In all organisms the production of viable progeny depends on faithful DNA replication and chromosome segregation. The recent advent of cytological methods in conjunction with fluorescence microscopy to the study of chromosomal DNA replication and segregation in bacteria, has revealed that these processes are similar to the ones described in eukaryotes. Bacterial DNA is replicated at fixed intracellular positions (22), newly replicated origin regions of sister chromosomes are rapidly separated from each other, and the chromosomal region around the origin of replication is in a defined orientation for most of the cell cycle (23–29).

The discovery that replicative DNA polymerase localizes at discrete intracellular positions, predominantly at or near the cell centre, and is not randomly distributed has suggested a factory model of replication (22). According to this model, during the process of DNA replication DNA is threaded through a stationary replication factory. Following their duplication, the origin regions of the sister chromosomes rapidly move to, and become established at, opposite cell poles. For most of the cell cycle, the replication origin regions are then found toward the end of the highly condensed nucleoid body, oriented near the poles of the cell (23–29). These findings raise intriguing questions as to the mechanisms that mediate the dynamic assembly and maintenance of replication factories near the bacterial cell centre and replication origins near the poles in the absence of known cytoskeletal components. In addition, whereas eukaryotic cells have a conspicuous mitotic apparatus that is responsible for segregating homologous chromosomes, the nature of the motor that drives the newly duplicated replication origins toward opposite poles of the bacterial cell remains elusive. It is conceivable that the fixed replisome is a major contributor to chromosome order in the cell and that it provides the driving force for chromosome segregation (22, 30).

Cellular proteins that contribute to efficient chromosome partitioning are ParA and ParB from *C. crescentus* as well as Soj and Spo0J from *B. subtilis*. These proteins are similar to a family of plasmid-encoded proteins required for plasmid partitioning in *E. coli*. Because *parA* and *parB* homologues have not been found to be essential in other bacteria, it is remarkable that they are required for cell viability in *Caulobacter* and that their overexpression causes defects in cell division (31). The essential nature of *parA* and *parB* in *C. crescentus* suggests the exciting possibility that cellular levels and perhaps proper subcellular location of ParA and ParB proteins are involved in the co-ordination of cell division and chromosome movement. SpoOJ and ParB bind to sites located in the replication origin region, suggesting that these sites may serve a function analogous to centromeres (31, 32). Although the function of SpoOJ and other proteins of this family is yet unknown, they are thought to be involved in pairing or positioning of sister chromosomes (26, 33).

## 5.1   Control of DNA replication by CtrA

A fundamental example of asymmetry in *C. crescentus* is the differential replicative ability of the chromosomes in the two progeny cells. The stalked progeny cell immediately initiates DNA replication, whereas the swarmer progeny cell is unable to replicate its chromosome until later in the cell cycle when it metamorphoses into a stalked cell.

The response regulator, CtrA, represses the initiation of DNA replication in the swarmer portion of the predivisional cell and in the swarmer progeny cell (Fig. 2) (34). In addition, CtrA plays a pivotal role in the *C. crescentus* cell cycle by modulating the transcription of genes important for flagellar biosynthesis, DNA methylation, and cell division (35–37). CtrA activity is under tight cell-cycle control both by phos-

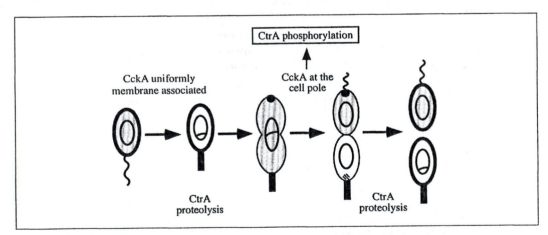

**Fig. 2** Regulation of CtrA. Shading inside the cells indicates the presence of the CtrA response regulator. The localization of the membrane-associated CckA histidine kinase that regulates CtrA activity by phosphorylation is shown throughout the cell cycle. In swarmer cells and stalked cells, CckA is distributed throughout the inner membrane, as indicated by a thick cell outline. In predivisional cells, CckA localizes at the flagellar pole, as depicted by a black dot. The observation that CckA is less abundant or absent at the stalked pole in many cells is indicated by the small dots. Polar localization of CckA is concurrent with the phosphorylation of CtrA. The ovals inside the cells represent non-replicating chromosomes, and the theta structures represent replicating chromosomes. Because of the presence of phosphorylated CtrA, the chromosome of the swarmer cell is not replicating.

phorylation and by proteolysis (38). CtrA is synthesized in the early predivisional cell when it is phosphorylated. In the late predivisional cell, CtrA is selectively proteolysed in the stalked compartment, but not in the swarmer compartment. As a result, CtrA~P is inherited by the swarmer cell where it represses DNA replication. During the swarmer to stalked-cell transition, CtrA activity is lost from the new stalked cell by dephosphorylation and its subsequent degradation by the ATP-dependent serine protease ClpXP, leading to the initiation of DNA replication (39). Similarly, because CtrA~P is absent from the stalked progeny cell, DNA replication can be initiated immediately.

Recently, the essential CckA histidine kinase has been identified that plays a crucial role in the cell cycle by regulating CtrA activity through phosphorylation (40, 41). Unlike CtrA, the CckA histidine kinase is a stable protein that appears to be present at similar levels throughout the cell cycle. However, CckA exhibits a cell-cycle-dependent dynamic spatial control, whereby it changes its localization patterns from a polar position to a uniform association with the inner membrane (Fig. 2). Polar localization of CckA may determine its histidine kinase activity. CckA localization to the cell pole occurs after the initiation of DNA replication, coincident with CtrA synthesis and phosphorylation. The CckA histidine kinase then remains at the incipient swarmer cell pole until after division, when it assumes its uniform membrane localization in both progeny cells. In the stalked progeny cell CtrA is dephosphorylated and degraded, ensuring that DNA replication is initiated. The swarmer progeny inherits CtrA~P from the predivisional cell, which prevents the initiation of DNA replication. CckA is delocalized in the swarmer cell and CtrA is no longer continuously phosphorylated. It is thus possible that the swarmer to stalked cell transition is a result of a continuous decrease of CtrA~P below a threshold level that is needed to prevent initiation of DNA replication.

## 5.2 Temporal genetic asymmetry

In all organisms, the replicated chromosomes are segregated prior to cytokinesis. During the developmental process of spore formation in *B. subtilis*, however, polar septation precedes chromosome segregation, thus creating transient genetic asymmetry (Fig. 3). In the predivisional sporulating cell, the replicated chromosomes are arranged in an extended structure (4, 42). Because of this structural organization, formation of the septum close to the cell pole bisects the chromosome destined for the forespore. Consequently, only a portion of this chromosome is initially trapped in the forespore, the remainder of the genetic material being contained in dual copies in the mother cell. The full complement of the chromosome is then translocated across the septum into the forespore. This process depends on the SpoIIIE protein, which is localized in the septum (43) and shares similarities with gene products from Streptomycetes that are implicated in conjugational transfer of plasmid DNA (44, 45).

The transient genetic asymmetry is not random. Rather it is fixed by the defined orientation of the chromosomes in the predivisional cell (28, 44). Thus only about one-third of the chromosome surrounding the origin of replication is initially present

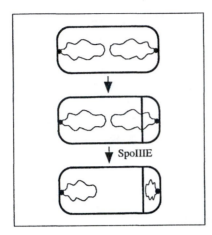

**Fig. 3** Temporal genetic asymmetry during sporulation in *B. subtilis*. Sister chromosomes that were generated during the last round of DNA replication assume an extended structure in the predivisional cell. Their origin regions, which are represented by the dots, are in a defined orientation near opposite cell poles. The sporulation septum is laid down close to one cell pole and bisects one of the chromosomes, creating transient genetic asymmetry. Subsequently, the full complement of the chromosome is translocated across the septum into the forespore. Chromosome translocation depends on the SpoIIIE protein, which is localized in the septum.

in the forespore (46). The fact that a specific segment of DNA is initially trapped in the forespore was inferred from the observation that certain *spoIIIE* mutants show a characteristic position dependence for expression of chromosomal genes whose transcription is under the control of the forespore-specific transcription factor $\sigma^F$ (44). This observation was subsequently explained by the finding that the origin regions of the two chromosomes are localized near opposite poles of the predivisional cell (23, 26, 28). As described above, movement of replication origins towards opposite cell poles is not specific to sporulation, but is a general feature of chromosome segregation during the vegetative bacterial cell cycle (see above) (23, 24, 26, 28, 29, 31). In addition, the arrangement of the chromosome within the bacterial cell seems to retain the linear order of genes (46, 47). Therefore, chromosome translocation across the septum is expected to lead to the sequential appearance of origin-distal genes in the forespore.

A recent study suggests that the transient genetic asymmetry resulting from the polar position of the sporulation septum in *B. subtilis* could be involved in the establishment of cell specificity (48, 49). Moving the gene encoding the forespore-specific transcription factor $\sigma^F$ from its normal origin-distal to an origin-proximal locus on the chromosome is sufficient to produce spores in the absence of otherwise essential activator proteins. These findings indicate that the transient genetic asymmetry can result in the differential accumulation of proteins in the two progeny cells to a degree that governs the establishment of cell fate. However, it remains to be determined whether this mechanism is normally exploited during the developmental process of spore formation. It is an exciting prospect that *B. subtilis* not only modifies cell division but also DNA segregation to establish cell-type-specific gene expression during sporulation.

## 6. Checkpoints that regulate polar development

In *C. crescentus*, different stages of development require the completion of previous stages of the replication and division cycles (50). The inhibition of DNA replication blocks flagellum synthesis (50, 51) by preventing the transcription of early flagellar genes (52). Cells that can replicate DNA but that are blocked in cell division are also affected in their progression through development. The initiation of cell division in *C. crescentus* and during *B. subtilis* sporulation plays an essential role in the establishment of differential programmes of gene expression that set up the fates of the progeny cells. In wild-type cells of *C. crescentus*, a flagellum is synthesized at the new pole during the first DNA replication cycle. Cells gain motility when flagellar rotation is initiated in late predivisional cells. In the second replication cycle, after cell separation, this same pole of the progeny swarmer cell sheds its flagellum and synthesizes a stalk and a holdfast. However, in cells inhibited at an early stage of cell division, polar development is blocked before the initiation of flagellar rotation. Thus, cells are unable to eject the flagellum and to synthesize a stalk and holdfast, producing filamentous cells that are still asymmetric, with a stalk and holdfast at one pole and a flagellum at the other pole. If, however, cell division is blocked after constriction has begun but before cell separation, long filaments with regularly spaced constrictions are formed. These cells are able to activate the development of the flagellated pole, ultimately yielding filamentous cells that bear a stalk and holdfast at both poles. Consequently, polar development is not simply coupled to progression through the DNA replication cycle or to an increase in cell mass; it is dependent on the progression of cytokinesis as well (50).

## 6.1 Regulation of FtsZ

Cytokinesis is initiated by polymerization of the essential GTPase FtsZ into a ring structure at the future division site (53). Localization of FtsZ seems to be the key event in assembly of the cell-division apparatus. The FtsZ ring is required for recruitment of other essential components of the cell-division machinery, as well as proteins that are involved in the establishment of cell fate (53–57). However, the central question of how the future division site is recognized by FtsZ remains unresolved.

In *C. crescentus*, and during vegetative growth in *B. subtilis*, FtsZ rings form exclusively at midcell in preparation for medial cell division. At the onset of sporulation in *B. subtilis*, however, the localization pattern of FtsZ changes from medial to bipolar, forming two rings near opposite poles of the cell (58). Despite the presence of two FtsZ rings, only one polar site is chosen for cell division, indicating that FtsZ ring formation is not sufficient to drive septation. The switch in FtsZ localization from a medial to a bipolar pattern is dependent on the transcription factor Spo0A, the master regulator for entry into sporulation (58). In a subsequent step that is regulated by the $\sigma^H$ transcription factor, the sporulation septum is stochastically formed at one of the polar sites.

### 6.1.1 Proteolytic control of cell-type-specific inheritance of FtsZ

The cell-type-specific distribution of FtsZ after cell division in *C. crescentus* is a combined effect of transcriptional and proteolytic regulation (Fig. 4). At the end of

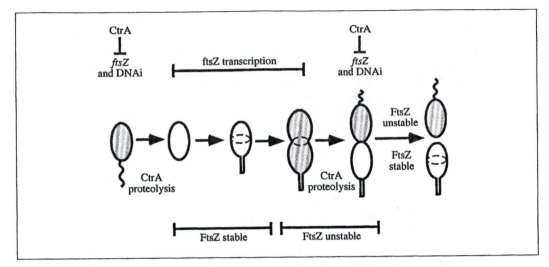

**Fig. 4** Model of FtsZ regulation in *C. crescentus*. CtrA, which is indicated by grey shading inside the cells, represses *ftsZ* transcription and the initiation of DNA replication (DNAi) in swarmer cells. During swarmer-cell differentiation, CtrA is degraded and allows for the initiation of *ftsZ* transcription and DNA replication. The FtsZ concentration increases and FtsZ polymerizes to form a ring at the site of cell division (represented by the dashed oval). During this time FtsZ is stable. The reappearance of CtrA activity in the predivisional cell inhibits *ftsZ* transcription. FtsZ depolymerizes as the cell and the FtsZ ring constricts. FtsZ is degraded rapidly, especially in the swarmer pole of the predivisional cell where it becomes depleted. Degradation of CtrA in the stalked pole of the predivisional cell allows *ftsZ* transcription and DNA replication to resume immediately. FtsZ forms a ring at the midcell in preparation for the next division.

the cell cycle, FtsZ is cleared from the incipient swarmer cell (59). As a consequence, FtsZ is only present in stalked cells after cell division. FtsZ begins accumulating during swarmer-cell differentiation, coincident with the initiation of DNA replication. The concentration of FtsZ then increases rapidly until cell constriction begins, when it is present at its highest concentration. Once cells have started to constrict, the concentration of FtsZ decreases rapidly. The decrease in FtsZ concentration at the end of the cell cycle is due to a dramatic increase in its degradation. At the beginning of the cell cycle and until cells begin to constrict, FtsZ has a half-life of 80 minutes. Concurrent with cell division, FtsZ becomes highly unstable with a half-life of 10–20 minutes (35). Thus, the half-life of FtsZ decreases from an equivalent of half a cell cycle to one-tenth of a cell cycle in a short period. The fact that the increased degradation of FtsZ is coincident with the beginning of cell constriction suggests that one of the factors controlling the proteolysis of FtsZ in *C. crescentus* may be its assembly state. Perhaps the domain of FtsZ that is recognized by a protease for degradation is inaccessible when FtsZ is assembled in the cytokinetic ring, but becomes exposed when FtsZ protomers are released into the cytoplasm by the disassembly of the ring. The fact that FtsZ is medially positioned in stalked cells, in which it is relatively stable, is consistent with the assembly model (35). An alternative model, although not necessarily exclusive, is that the synthesis or the activity of the protease responsible for FtsZ degradation is subject to cell-cycle control.

The increase of FtsZ degradation at the end of the cell cycle is not sufficient to explain its presence solely in stalked cells after cell division. Indeed, the unequal partitioning of FtsZ in the progeny cell is caused by its differential stability in the two cell types. In one experiment, *ftsZ* transcription was placed under the control of an inducible promoter on a high-copy-number plasmid. Induction of the promoter resulted in an identical rate of FtsZ synthesis in swarmer cells and in stalked cells. However, the concentration of FtsZ was still 10–20 times higher in stalked cells than in swarmer cells, indicating that FtsZ is much less stable in swarmer cells (35). Because *ftsZ* transcription is considerably decreased at the end of the cell cycle, the high rate of FtsZ degradation in the swarmer cell compartment leads to its disappearance from swarmer cells after cell division. The protease that degrades FtsZ may only be present and/or active in the swarmer compartment. Alternatively, FtsZ could be unable to polymerize at the future division site in the swarmer compartment, making it susceptible to degradation.

## 6.1.2 Transcriptional regulation of *ftsZ*

While proteolysis plays a major role in the regulation of FtsZ concentration, transcriptional control is clearly important to determine the onset of FtsZ synthesis at the beginning of the cell cycle (Fig. 4). It is also important to shut down *ftsZ* expression late in the cell cycle to allow proteolysis to take over the control of FtsZ concentration. Accordingly, *ftsZ* is not transcribed in swarmer cells but its transcription increases at the beginning of the DNA replication period (S phase) (35). At the end of S phase, when cell division is first apparent and FtsZ becomes unstable, transcription of *ftsZ* begins to decrease. Immediately after the completion of cell division, *ftsZ* transcription rapidly resumes in stalked cells but remains low in swarmer cells. The transcription rate of *ftsZ* is inversely proportional to the concentration of CtrA during the cell cycle (35). CtrA is present in swarmer cells, where it represses *ftsZ* transcription, is degraded at the same time as *ftsZ* transcription begins, and reappears when *ftsZ* transcription decreases at the end of the cell cycle (Fig. 4). Because CtrA also plays a negative role in the regulation of DNA replication (see Section 5.1, above), the degradation of CtrA during swarmer-cell differentiation co-ordinates the onset of the replication and division cycles. Late in the cell cycle when DNA replication is complete and cell division has been initiated, CtrA is synthesized and represses *ftsZ* transcription and initiation of DNA replication. Just before cell separation, CtrA is degraded in the stalked compartment. The absence of CtrA in stalked cells after cell division allows *ftsZ* transcription to resume and DNA replication to be initiated.

## 6.2 Cytokinesis and the establishment of cell fate

Asymmetric cell divisions during *B. subtilis* sporulation and in the *C. crescentus* cell cycle induce differential gene expression in the progeny cells following cytokinesis. In both organisms, proteins that are localized to the division plane are essential for the establishment of dissimilar fates in the progeny. Examples of proteins that are localized to the division plane and are involved in the establishment of cell fate include:

- the FlbE histidine kinase in *C. crescentus*;
- the SpoIIE serine phosphatase in *B. subtilis*; and
- the SpoIIGA putative aspartic protease in *B. subtilis*.

### 6.2.1 Establishment of swarmer-cell fate in *C. crescentus*

Several genes encoding external flagellar components are transcribed preferentially in the swarmer compartment of the late predivisional cell (60, 61). Compartmentalized expression of late flagellar genes is attributable to the swarmer pole-specific activation of FlbD (62), a member of the two-component family of response regulators. Activation of FlbD by phosphorylation is regulated through the action of the FlbE histidine kinase (63). At a time when flagellar biogenesis has proceeded beyond the components of the motor, phosphorylated FlbD induces the transcription of late flagellar genes and represses the transcription of early flagellar genes, the products of which are no longer needed in the swarmer compartment. Significantly, the FlbE histidine kinase is distributed asymmetrically in the predivisional cell (63) and is localized to the pole of the stalked compartment as well as to the division plane. Because FlbE itself is not an integral membrane protein, it might associate with a protein or a cellular structure involved in cell division. After formation of a barrier between the two compartments, FlbE is eventually selectively trapped in the swarmer compartment of the predivisional cell. The presence of FlbE at the stalked cell pole is not sufficient for FlbD activation. Therefore, the initial accumulation of FlbE at the division plane could govern its histidine kinase function by enabling its subsequent selective sequestration to the swarmer compartment.

### 6.2.2 Establishment of cell fate in *B. subtilis*

During spore formation in *B. subtilis* the fates of the progeny cells are determined by the differential activation of two transcription factors, $\sigma^F$ and $\sigma^E$, which establish dissimilar programmes of gene expression in the two cell types, involving additional transcription factors at later stages of development (5, 64). The $\sigma^F$ and $\sigma^E$ factors are synthesized in the predivisional cell but they are inactive in directing gene expression prior to polar septation. Instead, their activation is delayed until after asymmetric division, when $\sigma^F$ directs gene transcription in the forespore and $\sigma^E$ in the mother cell (65–67). Thus temporal as well as spatial controls govern the establishment of cell fate by restricting the activities of $\sigma^F$ and $\sigma^E$ to the right time and the right place. Although the mechanisms for compartmentalizing $\sigma^F$ and $\sigma^E$ activities are unrelated, the asymmetrically positioned sporulation septum is intimately involved in the establishment of cell fate.

The activity of $\sigma^F$ is regulated by a pathway consisting of the SpoIIAB, SpoIIAA, and SpoIIE proteins. SpoIIAB is a dual-function protein that is both an anti-sigma factor (68) and a serine kinase (69), SpoIIAA is an anti-anti-sigma factor (70, 71), and SpoIIE is a serine phosphatase (72). In the predivisional cell and in the mother cell

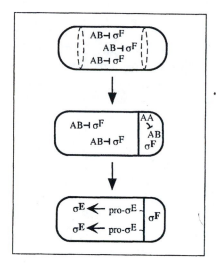

**Fig. 5** Model for the establishment of cell fate during sporulation in *B. subtilis*. In the predivisional sporulating cell FtsZ localizes in a bipolar pattern, as indicated by the dashed ovals, and recruits the SpoIIE serine phosphatase (not shown). Localized SpoIIE dephosphorylates SpoIIAA (not shown) and likely prevents it from attacking the SpoIIAB-$\sigma^F$ complex. After septum formation, SpoIIAA becomes competent to induce the release of $\sigma^F$ from the SpoIIAB'''-$\sigma^F$ complex in the forespore but not in the mother cell. Concurrent with septation, pro-$\sigma^E$ is eliminated from the forespore (not shown). Finally, $\sigma^F$-directed transcription of *spoIIR* in the forespore signals SpoIIGA-mediated pro-$\sigma^E$ processing in the mother cell (not shown). As a result, mature $\sigma^E$ is released from the septum into the cytoplasm of the mother cell, where it directs specific gene transcription.

following polar septation, the SpoIIAB protein sequesters $\sigma^F$ in an inactive protein complex (Fig. 5). In the forespore, $\sigma^F$ is disengaged from the SpoIIAB–$\sigma^F$ complex, associates with core RNA polymerase, and directs specific gene transcription. The $\sigma^F$ factor is freed for association with core RNA polymerase when the anti-anti-sigma factor, SpoIIAA, directly attacks the SpoIIAB–$\sigma^F$ complex (73). The ability of SpoIIAA to attack the SpoIIAB–$\sigma^F$ complex is governed by the phosphorylation state of SpoIIAA. When SpoIIAA is phosphorylated, it cannot bind to SpoIIAB and hence $\sigma^F$ is inactive. Only unphosphorylated SpoIIAA is capable of reacting with SpoIIAB–$\sigma^F$ to cause the release of $\sigma^F$. The SpoIIE serine phosphatase is an essential activator of $\sigma^F$ and is responsible for converting the inactive, phosphorylated form of SpoIIAA to the active, dephosphorylated form (72, 74, 75).

The activation of $\sigma^F$ is coupled to polar septation through the presence of the SpoIIE phosphatase in the septum. SpoIIE is an integral membrane protein that progresses through three patterns of subcellular localization during the early stages of sporulation (56, 76, 77). The SpoIIE phosphatase is synthesized in the predivisional cell, where it initially localizes in a bipolar pattern to the potential sites of polar septation (76, 78). After formation of the sporulation septum at one of the polar sites, SpoIIE likely becomes incorporated into the mature septum. At the same time, the ring of SpoIIE molecules persists at the other polar site. Following the activation of $\sigma^F$ in the newly formed forespore, the SpoIIE ring dissipates at the distal pole, resulting in a unipolar pattern of SpoIIE localization (56, 76, 79).

Because SpoIIE is present in the predivisional cell, regulatory mechanisms must exist to prevent the inappropriate activation of $\sigma^F$ prior to septation. Indeed, a recent study shows that SpoIIE function is regulated by two mechanisms (80). The first mechanism operates at the level of SpoIIE serine phosphatase activity and involves the FtsZ-dependent assembly of SpoIIE molecules into rings. Thus, in the presence of FtsZ, SpoIIE localizes in a bipolar pattern in the predivisional cell and dephosphorylates SpoIIAA~P. At the same time, the SpoIIE protein itself seems to prevent unphosphorylated SpoIIAA from attacking the SpoIIAB–$\sigma^F$ complex. In a second mechanism that involves the SpoIIE protein, activation of $\sigma^F$ is tied to completion of the sporulation septum, indicating that after septum formation the anti-anti-sigma factor SpoIIAA becomes competent to induce the release of $\sigma^F$ from the SpoIIAB–$\sigma^F$ complex. Thus, $\sigma^F$ activity is regulated at a step subsequent to the dephosphorylation of SpoIIAA~P. It is possible that this mechanism is involved in compartmentalizing $\sigma^F$ activity to the forespore.

The activation of $\sigma^E$ in the mother cell is regulated by two pathways. One pathway is a timing device that delays the activation of $\sigma^E$ until after polar septation by tying it to $\sigma^F$-dependent gene expression in the forespore (81). The $\sigma^E$ factor is synthesized as a membrane-bound proprotein, pro-$\sigma^E$, which carries an N-terminal extension of 27 amino acids. Proteolytic processing of pro-$\sigma^E$ to mature $\sigma^E$ is thought to be mediated by the membrane protein SpoIIGA in response to the SpoIIR signalling protein, which is produced in the forespore under the control of $\sigma^F$ (82, 83). SpoIIR is secreted from the forespore into the intermembrane space of the sporulation septum where it is believed to activate SpoIIGA for cytoplasmic pro-$\sigma^E$ processing (84).

The activation of $\sigma^E$ is associated with its progression through three distinct patterns of subcellular localization (85–87). In the predivisional cell, pro-$\sigma^E$ is associated with the cytoplasmic membrane. During polar septation, pro-$\sigma^E$ selectively accumulates at the sporulation septum that forms the interface between the forespore and the mother cell. It remains unresolved whether the change in pro-$\sigma^E$ distribution is regulated by relocalization of the existing pro-$\sigma^E$ proteins or by their specific proteolysis and *de novo* synthesis. Regardless of the precise mechanism, when polar division is complete, pro-$\sigma^E$ is poised at the septum for its subsequent proteolytic processing by the components of the intercellular signal transduction pathway. Indeed, the putative processing enzyme SpoIIGA and pro-$\sigma^E$ are co-localized at the sporulation septum. Their localizations, however, are independent of each other. Pro-$\sigma^E$ is sequestered at the septum in a *spoIIGA* deletion mutant of *B. subtilis* (85) and SpoIIGA is localized to the septum in the absence of pro-$\sigma^E$ (88). Although localization of pro-$\sigma^E$ and SpoIIGA occurs independently, the tight correlation suggests a similar mechanism. It is as yet unknown how pro-$\sigma^E$ and SpoIIGA accumulate at the septum, but their septal localization is likely a prerequisite for the proteolytic conversion of pro-$\sigma^E$ to $\sigma^E$.

Accumulation of pro-$\sigma^E$ at the sporulation septum that forms the interface between the two progeny cells culminates in the asymmetric localization of the $\sigma^E$ factor to one cell type. After proteolytic processing, mature $\sigma^E$ is freed from the septum and released into the cytoplasm of the mother cell, where it associates with core RNA polymerase and directs specific gene expression (Fig. 5) (85). When mature $\sigma^E$ is released from the septum it is absent from the forespore and chiefly or exclusively found in the mother cell (79, 89). Therefore, the sequestration of pro-$\sigma^E$ at specific subcellular sites might play an intimate role not only in its activation but also in confining $\sigma^E$ to its proper cell type. The absence of $\sigma^E$ from the forespore contributes to the establishment of mother-cell-specific gene transcription, because *spoIIIE* mutant cells of *B. subtilis* that fail to deplete pro-$\sigma^E$ from the forespore are capable of misactivating $\sigma^E$ in this cell type (79). Interestingly, the *spoIIIE* mutant cells that are deficient in eliminating pro-$\sigma^E$ from the forespore are also impaired in chromosome segregation during sporulation (44). This finding raises the possibility that the establishment of mother-cell fate is tied to chromosome segregation.

As we have seen, proteins that are targeted to the sporulation septum play an important role in the activation of $\sigma^F$ in the forespore and in the activation of $\sigma^E$ in the mother cell. Therefore, the septal localization of key regulatory proteins could be crucial in establishing different patterns of gene expression in the two progeny cells. As a consequence of their localization to the asymmetrically positioned septum, these proteins may either be unequally concentrated in, or asymmetrically segregated to, the two cell types. A recent study argues that it is not the unequal distribution of SpoIIE due to its incorporation in the asymmetrically positioned sporulation septum that governs the forespore-specific activation of $\sigma^F$ (77). Instead, it was suggested that $\sigma^F$ activation entails the exclusion of a SpoIIE-specific inhibitor from the forespore. Regardless of the precise mechanisms by which $\sigma^F$ activity and $\sigma^E$ activity are restricted to their specific cell types, the septal localization of essential activators of cell-fate determinants helps to co-ordinate cytokinesis with the establishment of cell fate.

## 7. Conclusions

The establishment of cell-type-specific gene expression is likely governed by cell-cycle events such as chromosome partitioning and cytokinesis. We have seen that in *C. crescentus* the swarmer-cell-specific activation of FlbD is coupled to cell division in a FlbE-dependent fashion. Similarly, the forespore-specific activation of $\sigma^F$ and the mother-cell-specific activation of $\sigma^E$ in *B. subtilis* are linked to septum formation in a manner that depends on the SpoIIE and SpoIIGA proteins, respectively. Thus, by coupling the initiation of cell-type-specific gene expression to cell division, the genesis of daughter cells is concurrent with the establishment of cell fate. Mechanisms of cell-fate determination in *B. subtilis* may also include the asymmetric distribution of genes, rather than the localization of proteins alone.

In all organisms, restricted protein distributions are necessary for the generation of structurally and functionally distinct progeny. Proteins can either be recruited directly to, or randomly distributed and then selectively retained at, their specific subcellular sites. All bacterial proteins analysed thus far are targeted directly to, and can be retained at, their specific sites. Direct recruitment usually relies on an intrinsic targeting signal in the protein. However, proteins with similar localization sites within the bacterial cell do not seem to contain common targeting motifs. Moreover, their localization has been shown to depend on distinct protein domains that can either be cytoplasmic, periplasmic, or transmembrane anchors. Elucidating the mechanisms that govern the recruitment of proteins within bacterial cells therefore remains one of the major challenges in our understanding of their cell cycle and differentiation.

Cell–cell interactions at the division plane could mediate the establishment of localized recognition sites that constitute different membrane domains for protein assembly. It is thus conceivable that polar multiprotein complexes, such as the flagellar structure and the chemosensory apparatus, localize to a target site that is assembled initially at the division plane by virtue of protein–protein interactions. Because the cell poles are sites of earlier cell divisions, the multiprotein complexes would assume their polar localization and could be retained by virtue of their size-limited lateral diffusion.

The finding that newly replicated origin regions are rapidly segregated to opposite cell poles, and that proteins such as the CckA histidine kinase can oscillate between distinct subcellular sites, raises further intriguing questions as to the mechanisms for DNA and protein movement in bacterial cells. Finally, the possible dependence of protein functions on their specific subcellular localization patterns awaits the development of assays for activity measurements in living cells.

# References

1. Horvitz, R. and Herskowitz, I. (1992) Mechanisms of asymmetric cell division: two Bs or not two Bs, that is the question. *Cell,* **68**, 237.
2. Losick, R. and Shapiro, L. (1999) Changing views on the nature of the bacterial cell: from biochemistry to cytology. *J. Bacteriol.,* **181**, 4143.
3. Shapiro, L. and Losick, R. (1997) Protein localization and cell fate in bacteria. *Science,* **276**, 712.
4. Piggot, P. J. and Coote, J. G. (1976) Genetic aspects of bacterial endospore formation. *Bacteriol. Rev.,* **40**, 908.
5. Stragier, P. and Losick, R. (1996) Molecular genetics of sporulation in *Bacillus subtilis. Annu. Rev. Genet.,* **30**, 297.
6. Jenal, U. and Shapiro, L. (1996) Cell cycle-controlled proteolysis of a flagellar motor protein that is asymmetrically distributed in the *Caulobacter* predivisional cell. *EMBO J.,* **15**, 2393.
7. Nathan, P., Gomes, S. L., Hahnenberger, K., Newton, A., and Shapiro, L. (1986) Differential localization of membrane receptor chemotaxis proteins in the *Caulobacter* predivisional cell. *J. Mol. Biol.,* **191**, 433.
8. Alley, M. R., Maddock, J. R., and Shapiro, L. (1992) Polar localization of a bacterial chemoreceptor. *Genes Dev.,* **6**, 825.
9. Agabian, N., Evinger, M., and Parker, G. (1979) Generation of asymmetry during development. Segregation of type-specific proteins in *Caulobacter. J. Cell Biol.,* **81**, 123.
10. Champer, R., Dingwall, A., and Shapiro, L. (1987) Cascade regulation of *Caulobacter* flagellar and chemotaxis genes. *J. Mol. Biol.,* **194**, 71.
11. Lagenaur, C., and Agabian, N. (1978) *Caulobacter* flagellar organelle: synthesis, compartmentation, and assembly. *J. Bacteriol.,* **135**, 1062.
12. Osley, M. A., Sheffery, M., and Newton, A. (1977) Regulation of flagellin synthesis in the cell cycle of *Caulobacter*: dependence on DNA replication. *Cell,* **12**, 393.
13. Kirby, J. R., Niewold, T. B., Maloy, S., and Ordal, G. W. (1999) CheB is required for behavioral responses to negative stimuli during chemotaxis in *Bacillus subtilis. Mol. Microbiol.,* **35**, 44.
14. Maddock, J. and Shapiro, L. (1993) Polar location of the chemoreceptor complex in the *Escherichia coli* cell. *Science,* **259**, 1717.

15. Gegner, J. A., Graham, D. R., Roth, A. F., and Dahlquist, F. W. (1992) Assembly of an MCP receptor, CheW, and kinase CheA complex in the bacterial chemotaxis signal transduction pathway. *Cell*, **70**, 975.

16. Shapiro, L. (1993) Protein localization and asymmetry in the bacterial cell. *Cell*, **73**, 841.

17. Evinger, M. and Agabian, N. (1979) *Caulobacter crescentus* nucleoid: analysis of sedimentation behavior and protein composition during the cell cycle. *Proc. Natl Acad. Sci., USA*, **76**, 175.

18. Setlow, B., Magill, N., Febbroriello, P., Nakhimousky, L., Koppel, D. E., and Setlow, P. (1991) Condensation of the forespore nucleoid early in sporulation of *Bacillus* species. *J. Bacteriol.*, **173**, 6270.

19. Britton, R. A., Lin, D. C., and Grossman, A. D. (1998) Characterization of a prokaryotic SMC protein involved in chromosome partitioning. *Genes Dev.*, **12**, 1254.

20. Graumann, P. L., Losick, R., and Strunnikov, A. V. (1998) Subcellular localization of *Bacillus subtilis* SMC, a protein involved in chromosome condensation and segregation. *J. Bacteriol.*, **180**, 5749.

21. Moriya, S., Tsujikawa, E., Hassan, A. K., Asai, K., Kodama, T., and Ogasawara, N. (1998) A *Bacillus subtilis* gene-encoding protein homologous to eukaryotic SMC motor protein is necessary for chromosome partition. *Mol. Microbiol.*, **29**, 179.

22. Lemon, K. P. and Grossman, A. D. (1998) Localization of bacterial DNA polymerase: evidence for a factory model of replication. *Science*, **282**, 1516.

23. Glaser, P., Sharpe, M. E., Raether, B., Perego, M., Ohlsen, K., and Errington, J. (1997) Dynamic, mitotic-like behavior of a bacterial protein required for accurate chromosome partitioning. *Genes Dev.*, **11**, 1160.

24. Gordon, G. S., Sitnikov, D., Webb, C. D., Teleman, A., Straight, A., Losick, R., Murray, A. W., and Wright, A. (1997) Chromosome and low copy plasmid segregation in *E. coli*: visual evidence for distinct mechanisms. *Cell*, **90**, 1113.

25. Lewis, P. J. and Errington, J. (1997) Direct evidence for active segregation of oriC regions of the *Bacillus subtilis* chromosome and co-localization with the SpoOJ partitioning protein. *Mol. Microbiol.*, **25**, 945.

26. Lin, D. C. H., Levin, P. A., and Grossman, A. D. (1997) Bipolar localization of a chromosome partition protein in *Bacillus subtilis*. *Proc. Natl Acad. Sci., USA*, **94**, 4721.

27. Niki, H. and Hiraga, S. (1998) Polar localization of the replication origin and terminus in *Escherichia coli* nucleoids during chromosome partitioning. *Genes Dev.*, **12**, 1036.

28. Webb, C. D., Teleman, A., Gordon, S., Straight, A., Belmont, A., Lin, D. C., Grossman, A. D., Wright, A., and Losick, R. (1997) Bipolar localization of the replication origin regions of chromosomes in vegetative and sporulating cells of *B. subtilis*. *Cell*, **88**, 667.

29. Webb, C. D., Graumann, P. L., Kahana, J. A., Teleman, A. A., Silver, P. A., and Losick, R. (1998) Use of time-lapse microscopy to visualize rapid movement of the replication origin region of the chromosome during the cell cycle in *Bacillus subtilis*. *Mol. Microbiol.*, **28**, 883.

30. Gordon, G. S. and Wright, A. (1998) DNA segregation: Putting chromosomes in their place. *Curr. Biol.*, **8**, R925.

31. Mohl, D. A. and Gober, J. W. (1997) Cell cycle-dependent polar localization of chromosome partitioning proteins in *Caulobacter crescentus*. *Cell*, **88**, 675.

32. Lin, D. C. and Grossman, A. D. (1998) Identification and characterization of a bacterial chromosome partitioning site. *Cell*, **92**, 675.

33. Niki, H. and Hiraga, S. (1997) Subcellular distribution of actively partitioning F plasmid during the cell division cycle in *E. coli*. *Cell*, **90**, 951.

34. Quon, K. C., Yang, B., Domian, I. J., Shapiro, L., and Marczynski, G. T. (1998) Negative control of bacterial DNA replication by a cell cycle regulatory protein that binds at the chromosome origin. *Proc. Natl Acad. Sci., USA*, **95**, 120.

35. Kelly, A. J., Sackett, M. J., Din, N., Quardokus, E., and Brun, Y. V. (1998) Cell cycle-dependent transcriptional and proteolytic regulation of FtsZ in *Caulobacter*. *Genes Dev.*, **12**, 880.

36. Quon, K. C., Marczynski, G. T., and Shapiro, L. (1996) Cell cycle control by an essential bacterial two-component signal transduction protein. *Cell*, **84**, 83.

37. Wright, R., Stephens, C., Zweiger, G., Shapiro, L., and Alley, M. R. (1996) *Caulobacter* Lon protease has a critical role in cell-cycle control of DNA methylation. *Genes Dev.*, **10**, 1532.

38. Domian, I. J., Quon, K. C., and Shapiro, L. (1997) Cell type-specific phosphorylation and proteolysis of a transcriptional regulator controls the G1-to-S transition in a bacterial cell cycle. *Cell*, **90**, 415.

39. Jenal, U. and Fuchs, T. (1998) An essential protease involved in bacterial cell-cycle control. *EMBO J.*, **17**, 5658.

40. Jacobs, C., Domian, I. J., Maddock, J. R., and Shapiro, L. (1999) Cell cycle-dependent polar localization of an essential bacterial histidine kinase that controls DNA replication and cell division. *Cell*, **97**, 111.

41. Stephens, C. (1999) The migrating kinase and the master regulator. *Curr. Biol.*, **9**, R493.

42. Ryter, A. (1965) Etude morphologie de la sporulation de *Bacillus subtilis*. *Ann. Inst. Pasteur (Paris)*, **108**, 40.

43. Wu, L. J. and Errington, J. (1997) Septal localisation of the SpoIIIE chromosome partitioning protein in *Bacillus subtilis*. *EMBO J.*, **16**, 2161.

44. Wu, L. J. and Errington, J. (1994) *Bacillus subtilis* SpoIIIE protein required for DNA segregation during asymmetric cell division. *Science*, **264**, 572.

45. Wu, L. J., Lewis, P. J., Allmansberger, R., Hauser, P. M., and Errington, J. (1995) A conjugation-like mechanism for prespore chromosome partitioning during sporulation in *Bacillus subtilis*. *Genes Dev.*, **9**, 1316.

46. Wu, L. J. and Errington, J. (1998) Use of asymmetric cell division and *spoIIIE* mutants to probe chromosome orientation and organization in *Bacillus subtilis*. *Mol. Microbiol.*, **27**, 777.

47. Teleman, A. A., Graumann, P. L., Lin, D. C. H., Grossman, A. D., and Losick, R. (1998) Chromosome arrangement within a bacterium. *Curr. Biol.*, **8**, 1102.

48. Frandsen, N., Barak, I., Karmazyn-Campelli, C., and Stragier, P. (1999) Transient gene asymmetry during sporulation and establishment of cell specificity in *Bacillus subtilis*. *Genes Dev.*, **13**, 394.

49. Losick, R. and Dworkin, J. (1999) Linking asymmetric division to cell fate: teaching an old microbe new tricks. *Genes Dev.*, **13**, 377.

50. Ohta, N. and Newton, A. (1996) Signal transduction in the cell cycle regulation of *Caulobacter* differentiation. *Trends Microbiol.*, **4**, 326.

51. Huguenel, E. D. and Newton, A. (1982) Localization of surface structures during procaryotic differentiation: role of cell division in *Caulobacter crescentus*. *Differentiation*, **21**, 71.

52. Stephens, C. M. and Shapiro, L. (1993) An unusual promoter controls cell-cycle regulation and dependence on DNA replication of the *Caulobacter fliLM* early flagellar operon. *Mol. Microbiol.*, **9**, 1169.

53. Lutkenhaus, J. and Addinall, S. G. (1997) Bacterial cell division and the Z ring. *Annu. Rev. Biochem.*, **66**, 93.

54. Addinall, S. G., Bi, E., and Lutkenhaus, J. (1996) FtsZ ring formation in *fts* mutants. *J. Bacteriol.*, **178**, 3877.

55. Chen, J. C., Weiss, D. S., Ghigo, J. M., and Beckwith, J. (1999) Septal localization of FtsQ, an essential cell division protein in *Escherichia coli. J. Bacteriol.*, **181**, 521.

56. Levin, P. A., Losick, R., Stragier, P., and Arigoni, F. (1997) Localization of the sporulation protein SpoIIE in *Bacillus subtilis* is dependent upon the cell division protein FtsZ. *Mol. Microbiol.*, **25**, 839.

57. Weiss, D. S., Chen, J. C., Ghigo, J. M., Boyd, D., and Beckwith, J. (1999) Localization of FtsI (PBP3) to the septal ring requires its membrane anchor, the Z ring, FtsA, FtsQ, and FtsL. *J. Bacteriol.*, **181**, 508.

58. Levin, P. A. and Losick, R. (1996) Transcription factor Spo0A switches the localization of the cell division protein FtsZ from a medial to a bipolar pattern in *Bacillus subtilis. Genes Dev.*, **10**, 478.

59. Quardokus, E., Din, N., and Brun, Y. V. (1996) Cell cycle regulation and cell type-specific localization of the FtsZ division initiation protein in *Caulobacter. Proc. Natl Acad. Sci., USA*, **93**, 6314.

60. Gober, J. W., Champer, R., Reuter, S., and Shapiro, L. (1991) Expression of positional information during cell differentiation of *Caulobacter. Cell*, **64**, 381.

61. Gober, J. W. and Shapiro, L. (1992) A developmentally regulated *Caulobacter* flagellar promoter is activated by 3' enhancer and IHF binding elements. *Mol. Biol. Cell*, **3**, 913.

62. Wingrove, J. A., Mangan, E. K., and Gober, J. W. (1993) Spatial and temporal phosphorylation of a transcriptional activator regulates pole-specific gene expression in *Caulobacter. Genes Dev.*, **7**, 1979.

63. Wingrove, J. A. and Gober, J. W. (1996) Identification of an asymmetrically localized sensor histidine kinase responsible for temporally and spatially regulated transcription. *Science*, **274**, 597.

64. Losick, R. and Stragier, P. (1992) Crisscross regulation of cell-type-specific gene expression during development in *Bacillus subtilis. Nature*, **355**, 601.

65. Driks, A. and Losick, R. (1991) Compartmentalized expression of a gene under the control of sporulation transcription factor $\sigma^E$ in *Bacillus subtilis. Proc. Natl Acad. Sci., USA*, **88**, 9934.

66. Harry, E., Pogliano, K., and Losick, R. (1995) Use of immunofluorescence to visualize cell-specific gene expression during sporulation in *Bacillus subtilis. J. Bacteriol.*, **177**, 3386.

67. Margolis, P., Driks, A., and Losick, R. (1991) Establishment of cell type by compartmentalized activation of a transcription factor. *Science*, **254**, 562.

68. Duncan, L. and Losick, R. (1993) SpoIIAB is an anti-sigma factor that binds to and inhibits transcription by regulatory protein $\sigma^F$ from *Bacillus subtitis. Proc. Natl Acad. Sci., USA*, **90**, 2325.

69. Min, K.-T., Hilditch, C. M., Diederich, B., Errington, J., and Yudkin, M. D. (1993) $\sigma^F$, the first compartment-specific transcription factor of *B. subtilis*, is regulated by an anti-sigma factor that is also a protein kinase. *Cell*, **74**, 735.

70. Alper, S., Duncan, L., and Losick, R. (1994) An adenosine nucleotide switch controlling the activity of a cell type-specific transcription factor in *B. subtilis. Cell*, **77**, 195.

71. Diederich, B., Wilkinson, J. F., Magnin, T., Najafi, S. M. A., Errington, J., and Yudkin, M. (1994) Role of interactions between SpoIIAA and SpoIIAB in regulating cell-specific transcription factor $\sigma^F$ of *B. subtilis. Genes Dev.*, **8**, 2653.

72. Duncan, L., Alper, S., Arigoni, F., Losick, R., and Stragier, P. (1995) Activation of cell-specific transcription by a serine phosphatase at the site of asymmetric division. *Science*, **270**, 641.

73. Garsin, D. A., Duncan, L., Paskowitz, D. M., and Losick, R. (1998) The kinase activity of

the antisigma factor SpoIIAB is required for activation as well as inhibition of transcription factor $\sigma^F$ during sporulation in *Bacillus subtilis*. *J. Mol. Biol.*, **284**, 569.

74. Arigoni, F., Duncan, L., Alper, S., Losick, R., and Stragier, P. (1996) SpoIIE governs the phosphorylation state of a protein regulating transcription factor $\sigma^F$ during sporulation in *Bacillus subtilis*. *Proc. Natl Acad. Sci., USA*, **93**, 3238.

75. Feucht, A., Magnin, T., Yudkin, M. D., and Errington, J. (1996) Bifunctional protein required for asymetric cell division and cell-specific transcription in *B. subtilis*. *Genes Dev.*, **10**, 794.

76. Arigoni, F., Pogliano, K., Webb, C., Stragier, P., and Losick, R. (1995) Localization of protein implicated in establishment of cell type to sites of asymmetric division. *Science*, **270**, 637.

77. Arigoni, F., Guérout-Fleury, A.-M., Barák, I., and Stragier, P. (1999) The SpoIIE phosphatase, the sporulation septum and the establishment of forespore-specific transcription in *Bacillus subtilis*: a reassessment. *Mol. Microbiol.*, **31**, 1407.

78. Barak, I., Behari, J., Olmedo, G., Guzman, P., Brown, D. P., Castro, E., Walker, D., Westpheling, J., and Youngman, P. (1996) Structure and function of the SpoIIE protein and its localization to sites of sporulation septum assembly. *Mol. Microbiol.*, **19**, 1047.

79. Pogliano, K., Hofmeister, A. E. M., and Losick, R. (1997) Disapppearance of the $\sigma^E$ transcription factor from the forespore and the SpoIIE phosphatase from the mother cell contributes to establishment of cell-specific gene expression during sporulation in *Bacillus subtilis*. *J. Bacteriol.*, **179**, 3331.

80. King, N., Dreesen, O., Pogliano, K., and Losick, R. (1999) Septation, dephosphorylation and the activation of $\sigma^F$ during sporulation in *Bacillus subtilis*. *Genes Dev.*, **13**, 1156.

81. Zhang, L., Higgins, M. L., Piggot, P. J., and Karow, M. L. (1996) Analysis of the role of prespore gene expression in the compartmentalization of mother cell-specific gene expression during sporulation of *Bacillus subtilis*. *J. Bacteriol.*, **178**, 2813.

82. Karow, L. M., Glaser, P., and Piggot, P. J. (1995) Identification of a gene, *spoIIR*, which links the activation of $\sigma^E$ to the transcriptional activity of $\sigma^F$ during sporulation in *Bacillus subtilis*. *Proc. Natl Acad. Sci., USA*, **92**, 2012.

83. Londoño-Vallejo, J.-A. and Stragier, P. (1995) Cell–cell signaling pathway activating a developmental transcription factor in *Bacillus subtilis*. *Genes Dev.*, **9**, 503.

84. Hofmeister, A. E. M., Londoño-Vallejo, A., Harry, L., Stragier, P., and Losick, R. (1995) Extracellular signal protein triggering the proteolytic activation of a developmental transcription factor in *B. subtilis*. *Cell*, **83**, 219.

85. Hofmeister, A. (1998) Activation of the proprotein transcription factor pro-$\sigma^E$ is associated with its progression through three patterns of subcellular localization during sporulation in *Bacillus subtilis*, *J. Bacteriol.*, **180**, 2426.

86. Ju, J., Luo, T., and Haldenwang, W. G. (1997) *Bacillus subtilis* pro-$\sigma^E$ fusion protein localizes to the forespore septum and fails to be processed when synthesized in the forespore. *J. Bacteriol.*, **179**, 4888.

87. Zhang, B., Hofmeister, A., and Kroos, L. (1998) The pro-sequence of pro-$\sigma^K$ promotes membrane association and inhibits RNA polymerase core binding. *J. Bacteriol.*, **180**, 2434.

88. Fawcett, P., Melnikov, A., and Youngman, P. (1998) The *Bacillus* SpoIIGA protein is targeted to sites of spore septum formation in a SpoIIE independent manner. *Mol. Micro.*, **28**, 931.

89. Hofmeister, A. and Losick, R. (1997) Establishment of cell type in a primitive organism by cell-specific elimination and proteolytic activation of a transcription factor. *Cold Spring Harbor Symp. Quant. Biol.*, **62**, 49.

# 2 | Cell polarity in yeast

JÜRG BÄHLER and MATTHIAS PETER

## 1. Introduction

Eukaryotic cells respond to intracellular and extracellular cues to direct cell growth and division. Selecting sites of polarized growth and division is crucial for the development of both unicellular and multicellular organisms. The two yeasts *Saccharomyces cerevisiae* (budding yeast) and *Schizosaccharomyces pombe* (fission yeast) have rigid cell walls. Cellular polarization targets secretion of cell wall and other materials to restricted areas in the cell. As a result, cell polarity in these yeasts directly underlies cellular morphogenesis. Both budding and fission yeast establish cell polarity at several stages for growth, cytokinesis, mating, and sporulation. Thus, they are useful model organisms for studying various aspects of cell polarity. In recent years, many regulatory and cytoskeletal components important for directing and establishing polarity have been identified, and molecular mechanisms underlying polarity development in yeast have been elucidated. Key signalling pathways that regulate polarization during the cell cycle and mating response have been described. Since many of the components important for polarized cell growth are conserved in other organisms, the basic mechanisms mediating cell polarity are likely to be universal among eukaryotes.

## 1.1 The polarized yeast cell

Cell polarity in yeast is most simply defined as an asymmetric distribution of specific proteins and organelles near a defined spatial site, thereby allowing the cell to increase its surface in an asymmetrical or polar fashion. In *S. cerevisiae*, cells in early $G_1$ phase of the cell cycle grow isotropically and insert new cell-wall material all over their surface until they reach a certain size, at which time activation of the $G_1$ cyclin-dependent kinase (Cdc28p-Clnp) or of a MAP kinase pathway activated by mating pheromones, initiates cellular polarization (1). The establishment of cell polarity then leads to asymmetrical growth, resulting in formation of either a bud or a mating projection. It is useful to distinguish polarity establishment from polarized growth: establishment of cell polarity involves the asymmetrical localization of proteins and organization of the actin cytoskeleton towards that site, while polarized growth requires secretion and surface extension. Several mutants are able to establish cell polarity but fail to exhibit polarized growth (see below).

In a polarized cell, all major cytoskeletal elements and organelles are organized around the site of cell-surface growth (Plate 1). The spindle pole body (SPB), embedded in the nuclear envelope, is positioned on the side of the nucleus near the bud (Plate 1A) or the shmoo tip (Plate 1B). Astral microtubules radiate from the SPB towards the site of growth. The nucleolus is located near the nuclear envelope and is generally positioned opposite the bud—or shmoo tip. In cells exposed to phero- mones, the nucleus becomes dumb-bell shaped, containing distinct chromosomal and nucleolar domains (2), and it is displaced toward the shmoo tip. Cortical actin patches are concentrated in both the shmoo tip and in the bud, with actin cables oriented toward them. A ring of septins is found both at the shmoo- and the mother- bud neck. Secretory vesicles are concentrated in the shmoo or the bud where the insertion of new cell-wall material occurs, leading to polarized growth. The differ- ence between a mating projection and formation of a bud is the width of the area of polarized growth: budding cells have a restricted zone of growth, while mating projections have a wide neck. However, it is not understood why cells exposed to pheromones have a wider neck. The effect may be due to a change in the organ- ization of the septins: the morphology of cells lacking the septin-associated kinase Gin4p is reminiscent of cells exposed to pheromones, and cells treated with mating pheromones no longer localize Gin4p at the neck (3–5).

Unlike *S. cerevisiae*, *S. pombe* cells grow as straight cylindrical rods of constant diameter (3–4 μm) and 7–15 μm length (6). Polarized growth occurs exclusively by elongation at the cell ends (Plate 1C). *Schizosaccharomyces pombe* cells establish polar- ized growth immediately after cell division, and they never show a phase of isotropic growth. After doubling in cell length, cell polarity is directed towards the cell centre, where a division septum is laid down to generate two equal daughter cells (Plate 1D). As in *S. cerevisiae*, the intrinsic programme of cell polarity is overriden by extrinsic signals during mating. Growth of the mating projection starts at a cell end but is no longer in a straight axis, and the growing tip is thinner than a normal cell end (Plate 1E). Upon contact with a partner cell, the nucleus migrates towards the shmoo tip in preparation for karyogamy. Both in vegetatively growing and in mating *S. pombe* cells several cytoplasmic microtubules extend along the cell length to the growing ends. Most of these microtubules do not touch the SPB. F-actin patches and secretory vesicles are concentrated at the growing ends of *S. pombe* cells, at the cell division site, and at the tip of the mating projection. The insertion of new cell-wall material occurs at all these places. F-actin cables are thought to be oriented towards the sites of polarity. Septins are also present in *S. pombe* (7, cited in 8). However, it is not clear at present whether they are involved in cell polarity. In vegetative cells, septins localize to the cell centre late during cytokinesis, and it is not known whether they show any specific localization during mating.

Cell polarity may also be important during sporulation, but although the cytology of spore morphogenesis has been described in considerable detail in budding and fission yeast (9, 6), very little is known about underlying molecular mechanisms. In

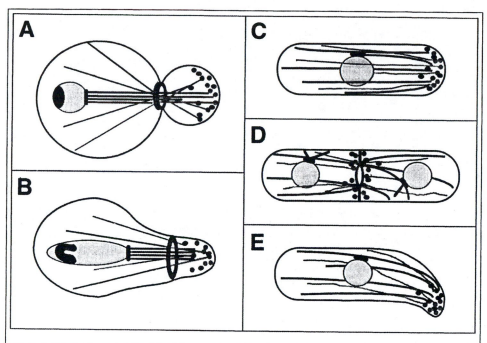

**Plate 1** Polarized yeast cells. A budding yeast cell showing polarized components in the cell cycle (A) and in response to mating pheromones (B). A septin ring (blue) is found at the shmoo or bud neck. Cortical actin patches (red circles) are concentrated at the site of polarized growth and actin cables (red) orient towards the bud or shmoo tip. The spindle pole body (green rectangle) is embedded in the nuclear envelope facing the site of polarization, and microtubules (green) radiate from the SPB into the bud or shmoo tip. Secretory vesicles (black circles) are concentrated around the site of polarized growth. The cell nucleus (grey) elongates in response to pheromones and the nucleolus (black) is positioned in the nucleus on the side opposite to the shmoo or bud tip. Fission yeast cells showing polarized components during vegetative cell growth (C), cell division (D), and in response to mating pheromones (E). Components are presented as in (A and B). Cortical actin patches and secretory vesicles are concentrated at the sites of cell polarity and actin cables orient towards these sites. Microtubuies extend along the cell length and end at the sites of cell polarity.

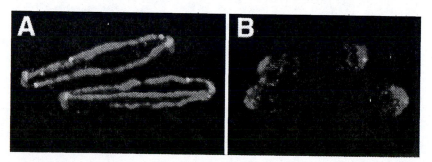

**Plate 2** Cellular localization of Tea1p and Pom1p in fission yeast. (A) Tea1p (green) is found at both cell ends and at the tips of microtubules (red). Picture courtesy of J. Mata and P. Nurse. (B) Pom1p (green) is concentrated at the new cell end, whereas F-actin (red) localizes to the old cell end before NETO.

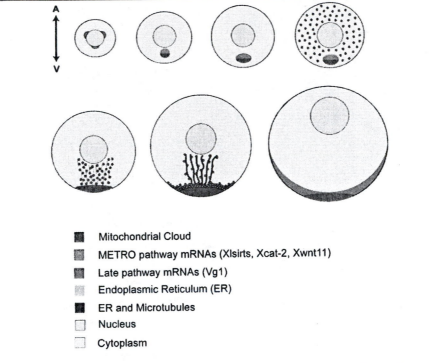

- ■ Mitochondrial Cloud
- ■ METRO pathway mRNAs (Xlsirts, Xcat-2, Xwnt11)
- ■ Late pathway mRNAs (Vg1)
- ■ Endoplasmic Reticulum (ER)
- ■ ER and Microtubules
- □ Nucleus
- □ Cytoplasm

**Plate 3** Localization of mRNA at the vegetal pole. Two pathways exist to position mRNAs at the vegetal pole: the early, or METRO, pathway and the late pathway. Those mRNAs using the METRO, including Xlsirts, Xcat2 and Xwnt11, are first seen in pre-mitochondrial aggregates around the nucleus, then the mitochondrial cloud. The RNAs move to the cortex during stage I and dock at the prospective vegetal pole during stage II. As the METRO pathway is functioning, Vg1 RNA is seen throughout the cytoplasm exclusive of the mitochondrial cloud. A subdomain of ER forms at the trailing edge of the cloud and vg1 begins accumulating at this site in a microtubule-independent fashion. Movement of Vg1 to the vegetal cortex continues by a microtubule-dependent mechanism that also relies on endoplasmic reticulum. Vg1 mRNA then anchors at the vegetal cortex by a microfilament-dependent mechanism.

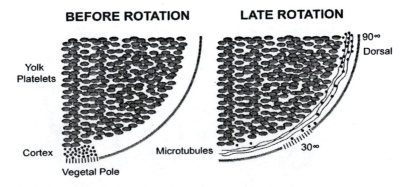

**BEFORE ROTATION**   **LATE ROTATION**

Yolk Platelets

Cortex
Vegetal Pole
Microtubules

$90\infty$
Dorsal
$30\infty$

**Plate 4** Microtubule-mediated transport during cortical rotation. During rotation the cortex moves 30° towards the plus ends of microtubules and the prospective dorsal side of the egg. At the same time, organelles located at the vegetal pole before rotation travel along the microtubules, at 30–50 μm min⁻¹, as much as 90°. These organelles are likely to attach to and detach from microtubules during this transport, leaving a gradient of dorsalizing components in the vegetal dorsal quadrant of the embryo.

*Saccharomyces cerevisiae*, at the beginning of meiosis II, a flattened membrane sac termed 'prespore wall' or 'prospore membrane' forms in close apposition to the cytoplasmic face of each SPB. These membrane sacs extend along the outer surface of the nuclear envelope to form cup-shaped structures that surround the nuclear lobes containing the separating chromosomes. During formation of the prospore membrane, vesicles normally bound to the plasma membrane are thought to be redirected to the incipient prospore membrane. With the exception of several *SEC* genes, including the sporulation-specific *SEC9* homologue *SPO20* (10), the genes that participate in the assembly of this specialized membrane bound structure remain largely unknown. It has been speculated that reorganization of the actin cytoskeleton may be required for the redistribution of vesicles, and several genes, including *CDC42*, *BNR1*, *SPA2*, and *PEA2*, are induced at the time of prospore membrane formation (11). However, the so far described *cdc24* or *cdc42* temperature-sensitive mutants do not appear to exhibit a sporulation defect (A. Neiman, personal communication). Interestingly, two sporulation-specific septins (*SPR3* and *SPR28*) have been identified that localize to the leading edge of the prospore membrane and may be part of the coated bulb-like lip structure at the leading edge which grows from its origin at the SPB around the nuclear membrane (12, cited in 13). In *S. pombe*, spore formation also starts during meiosis II and proceeds in a similar way via prospore membranes that emerge from the SPBs and engulf each of the four nuclei in the ascus (14). This process is accompanied by a polarised localization of both actin and fimbrin, which are essential for sporulation (15; J.-Q. Wu, J. Bähler, and J. R. Pringle, unpublished data). In *S. pombe* there are three sporulation-specific septins that localize in a similar way to Spr3p and Spr28p (J. Bähler, O. S. Al-Awar, J.-Q. Wu, and J. R. Pringle, unpublished data), but their function remains to be determined. In conclusion, how cell polarity is established during sporulation is poorly understood, but may involve a common set of proteins implicated in polarity establishment and polarized growth during the cell cycle or mating, as well as several sporulation-specific homologues of cell-polarity proteins.

## 1.2 Basic steps of polarity establishment

Establishment of cell polarity can be divided into three basic steps (Fig. 1) (see also Chapter 4), which can be separated genetically from each other. First, cells need to choose a site on their surface towards which they will polarize. Throughout this review, we will call this spatial site of polarization the 'landmark'. During cell division, the landmark is defined intrinsically in a cell-type-specific manner, whereas cells responding to a pheromone gradient choose a landmark that is determined by the position of the mating partner. In the absence of a landmark, cells are still able to polarize, but they do so at abnormal positions. In a second step, the landmark has to be recognized by a series of proteins, collectively named polarity establishment proteins or actin-organizing complex. In the absence of a functional polarity establishment complex, cells fail to polarize and continue to grow in an isotropic manner. In the last step, the polarity establishment proteins recruit the machinery required to

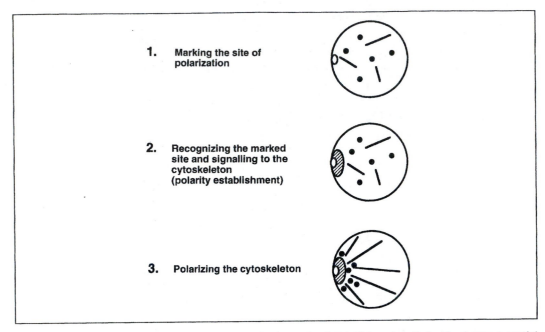

**Fig. 1** Basic steps of polarity establishment. Establishment of cell polarity requires that cells choose a spatial site (landmark) on their cell cortex (open circle): the position of this site can be defined genetically or in response to internal or external signals. Once a site of polarization has been chosen, this landmark recruits a number of proteins, collectively termed polarity establishment proteins or actin-organizing components (hatched circle). These proteins localize to the landmark and in turn organize the cytoskeleton (black).

organize and polymerize the actin cytoskeleton. The polarized cytoskeleton then targets exocytosis or secretion towards the landmark, leading to polarized growth. While this simple hierarchical scheme is useful for the general picture, there are important exceptions. For example, some genes required for exocytosis are also involved in the development of cell polarity, suggesting that polarized secretion is not only a consequence of cell polarity, but may also be involved in the establishment or maintenance of cellular polarization (see Section 2.2.2.).

In this chapter, we will discuss mechanisms by which a landmark is established during the cell cycle and in response to extracellular signals. The proteins and mechanisms required for defining landmarks seem to differ between various polarization events. We will then summarize recent insights of how the polarity establishment proteins are recruited to these landmarks, and how they organize the actin cytoskeleton. The proteins and mechanisms involved in polarity establishment seem to be conserved between the two yeasts. Finally, we will discuss how the polarity establishment proteins are regulated by the cell-cycle machinery or signalling pathways. We will only mention a few selected proteins which are responsible for organizing and polymerizing actin filaments. Clearly, many additional actin-binding proteins contribute to the dynamic changes of the cytoskeleton. For a detailed discussion on this topic, we refer the reader to some recent reviews (16–19). In the final section, we will attempt to highlight common pathways and to compare and contrast the mechanisms of polarity establishment in the two yeasts.

# 2. Cell polarity in *Saccharomyces cerevisiae*

## 2.1 Microtubule and actin-dependent processes

Because cytoskeletal elements play a role in organizing the cytoplasm and establishing distinct plasma membrane domains (16), they are well suited to position spatial landmarks. In budding yeast, the establishment of cell polarity during cell division and in response to pheromones is dependent on a functional actin cytoskeleton but does not appear to require microtubules (20, 21). Cells with a depolymerized actin cytoskeleton are unable to undergo any form of polarized growth, whereas transient actin depolymerization causes cells to abandon a bud site or mating projection and initiate growth at a second site, suggesting that actin filaments are also required for the maintenance of an axis of cell polarity (22). Experiments using the actin assembly inhibitor, latrunculin-A (LatA), revealed actin-dependent and actin-independent pathways to establish cell polarity (22). The actin-dependent pathway localizes secretory vesicles and a putative vesicle docking complex to sites of cell-surface growth, providing an explanation for the dependence of polarized surface growth on actin function. In contrast, several proteins required for the establishment of cell polarity, such as Bud6p/Aip3p, are localized in an actin-independent manner, suggesting that an anchor at the cell cortex binds these proteins.

In contrast to the involvement of the actin cytoskeleton, cells with a completely disassembled microtubule network are able to select axial and bipolar bud sites, properly localize marker proteins to these landmarks, and efficiently organize the actin cytoskeleton (20). Microtubules are necessary for the migration and proper orientation of the nucleus, SPB separation, spindle function, and nuclear division (20). Positioning of the mitotic spindle requires the interaction of microtubules with the cell cortex, but the components that mediate these interactions are still poorly understood (see Section 2.2.3). Taken together, establishment and maintenance of cell polarity and polarized growth in budding yeast requires the organization and regulation of the actin- but not the microtubule cytoskeleton.

## 2.2 Defining a landmark during cell division

In budding yeast, the position of the new bud determines the orientation of the mitotic spindle. This site is not chosen randomly, but is positioned in a genetically determined pattern with respect to the site used during previous cell divisions (23–26). Different cell types exhibit a distinct budding pattern: haploid **a** or α cells bud in an axial pattern (Fig. 2A) while **a**/α diploid cells bud in a bipolar manner (Fig. 2C). In the axial pattern, mother cells (cells that have undergone at least one budding event) position their next bud adjacent to the previous bud site, and daughter cells (cells that have yet to bud) bud adjacent to the birth scar. In the bipolar pattern,

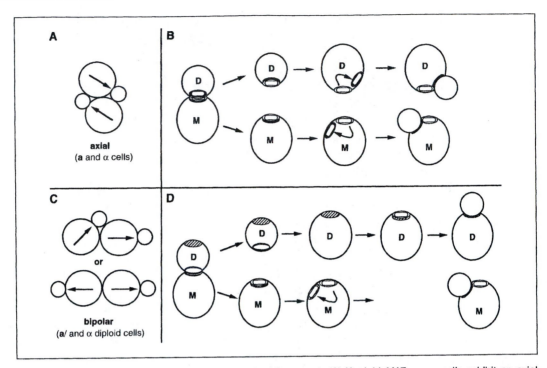

**Fig. 2** Choosing landmarks during cell division in budding yeast. (A) Haploid *MAT* **a** or α cells exhibit an axial budding pattern in which a mother and daughter cells usually bud adjacent to the previous site of cytokinesis. The arrows indicate the axis of cell division. (B) A cytokinesis tag is thought to direct the axial budding pattern. The septin ring (open ring) serves as a template that allows binding of the axial bud-site components Bud3p, Bud4p, and Axl2p/Bud10p (filled ring). This complex then directs the budding machinery adjacent to the old site (arrow) before its disassembly. M, mother cell; D, daughter cell. (C) *MAT* **a**/α cells diploid cells follow a bipolar budding pattern in which daughter cells usually bud from the pole opposite the mother (distal pole), and mother cells bud from either pole. (D) A tag remaining after cytokinesis (open circle) is used by diploid mothers (M) to position the next bud adjacent to the previous one (arrow); older mothers can also position the next bud adjacent to their birth scar (not shown). Diploid daughter cells (D) lose their cytokinesis tag in G$_1$ and select the distal tip as a bud site (shaded patch), possibly through a landmark deposited at the site of previous cell-surface growth.

daughter cells usually bud at the pole opposite to their birth scar, new mother cells bud near the previous bud site, and old mothers bud at either pole. It has been speculated that the purpose of the axial budding pattern is to facilitate diploid formation (27): because wild-type mother cells switch their mating type, the axial budding pattern juxtaposes *MAT* **a** and *MAT* α cells, thereby facilitating mating and diploid formation. Diploid cells have survival advantages. They are more resistant to environmental stresses and DNA-damaging agents. The diploid budding pattern allows cells to maximize access to nutrients; upon nutrient starvation yeast cells bud in a unipolar manner, become strikingly elongated, and the cells remain attached to each other (28). This pseudohyphal mode of growth allows chains of cells to spread across considerable distances in the search for nutrients. In addition, pseudohyphae have the capacity to invade agar. The unipolar and bipolar budding patterns of diploids are related and depend on at least some of the same proteins.

Three groups of genes are required to specify these cell-type-specific budding patterns. The first group (consisting of *RSR1/BUD1*, *BUD2*, and *BUD5* ) is required for the correct budding pattern in all cell types. Mutants in any of these genes result in a random budding pattern. A second group (consisting of *BUD3*, *BUD4*, *AXL1*, *AXL2/BUD10*, and others) is required for the correct budding pattern only in haploid cells: **a** or α cells lacking any of these products exhibit a bipolar budding pattern, and thus these proteins may be involved in choosing an axial landmark. A third group of genes (including *BNI1*, *BUD6*, *BUD7*, *BUD8*, *BUD9* and others) is required for the correct budding pattern only in **a**/α diploid cells. **a**/α cells lacking any of these proteins exhibit a random pattern, while **a** or α haploid cells lacking any of these products exhibit a normal, axial pattern. It is proposed that these proteins are involved in creating a bipolar landmark, which is thought to be located at the distal tip of daughter cells and at both ends of mother cells (see below).

## 2.2.1 The axial landmark

The observation that during axial budding new buds form adjacent to previous ones led to a model proposing that a 'tag' or landmark localizes to the site of cytokinesis and directs subsequent budding events to this site (29; Fig. 2B). This persistent positional signal thus marks the region of the division site (birth scar) on a newborn daughter cell, as well as each previous division site (bud scar) on a mother cell. What is the nature of this axial landmark? Loss of several proteins, including the septins, Bud3p, Bud4p, Axl1p, and Axl2p/Bud10p, results in a bipolar instead of axial budding pattern in haploid cells, whereas the budding pattern of diploid cells is unaffected. The septins are probably the major structural components of the filament system called the 10 nm filaments (8, 30, 31), and a purified septin complex containing Cdc3p, Cdc10p, Cdc11p and Cdc12p is able to assemble into long filaments *in vitro* (32). The septins localize as a ring at the incipient bud site, remain at the mother-bud neck during bud formation and bud growth, and are seen as a double ring structure during mitosis (Fig. 2). Thus, this ring-structure satisfies both the spatial and functional requirements to be an element of the axial landmark. Bud4p, Bud3p, and Axl2p/Bud10p co-localize with the septins throughout the cell cycle. Their localization is disrupted when *cdc12* mutant cells are shifted to the restrictive temperature, demonstrating that their localization requires an intact septin structure. Thus, these proteins are thought to use the septin ring as an assembly platform (33–36). Interestingly, Bud4p also influences the localization of the septin structure, suggesting that Bud3p, Bud4p, and Axl2p/Bud10p may in turn act as a platform for septin assembly in the next cycle (34). Taken together, Axl2p/Bud10p, Bud3p, and Bud4p may contribute to the axial landmark and are thought to direct axial budding by localizing additional bud-site selection components to the presumptive bud site (see below). Bud4p encodes a large protein with a GTPase domain, whereas *AXL2/ BUD10* encodes a transmembrane protein (35, 36) which is *o*-glycosylated by the mannosyl transferase Pmt4p (37). Interestingly, *pmt4* mutant cells exhibit a daughter-specific axial budding defect, suggesting that the activity of Axl2p/Bud10p in daughter cells requires *o*-glycosylation (37).

*AXL1* encodes a protein with an insulinase-like protease activity which is required for **a**-factor processing (38). Surprisingly, however, this protease activity is not required for its role in axial budding. It is possible that Axl1p regulates or interacts with other axial-specific components (Bud3p, Bud4p, Bud10p/Axl2p) to carry out its bud-site selection function. The localization of Axl1p remains to be determined.

As mentioned earlier, **a** or α cells lacking genes required for the haploid-specific budding pattern (Bud4p, Bud3p, or Axl2p/Bud10p) exhibit a bipolar pattern characteristic of diploid cells. Thus, the simplest hypothesis to explain this cell-type-specific difference is that one or more of these products is downregulated in diploid cells. However, Bud3p, Bud4p, and Axl2/Bud10p are found at similar levels in all cell types, arguing that these proteins are not responsible for controlling the diploid-specific budding pattern. On the other hand, transcription of the *AXL1* gene is repressed in **a**/α diploid cells, most likely by the action of the homeodomain protein **a**1-α2 (39). Moreover, **a**/α diploid cells that express *AXL1* ectopically, exhibit an altered budding pattern that can be viewed as partially axial, suggesting that Axl1p is at least one of the components responsible for triggering axial budding in haploid cells, perhaps by preventing use of the bipolar landmark.

### 2.2.2   The bipolar landmark

Diploid cells use a different landmark for budding than haploid cells (Fig. 2C). Mutations in several genes, including *ACT1*, *SPA2*, *PEA2*, *RVS161*, *RVS167*, *SEC3*, *SEC4*, *SEC9*, *BNI1*, *BUD6/AIP3*, *BUD7*, *BUD8*, and *BUD9*, affect bud-site selection in diploid cells but not haploid cells. Spa2p and Pea2p associate tightly with each other (40), and may be part of a large 12S multiprotein complex (called the polarisome) containing Bud6p/Aip3p, Spa2p, and Bni1p as well as the signalling proteins Ste11p, Ste7p, and Mkk1p and Mkk2p (41). Spa2p, Pea2p, Bud6p/Aip3p, and Bni1p all co-localize at the presumptive bud site, and the localization of Pea2p at the bud tip is dependent on its binding to Spa2p (40, 42). Presumably, these components help to mark and/or recognize a cytokinesis landmark and establish a new growth site. Bud6p/Aip3p also interacts with actin (43), providing a link between the polarisome and the actin cytoskeleton (see below).

Sec3p, Sec4p, and Sec9p are required for polarized exocytosis, but they are also involved in the development of cell polarity (44): homozygous diploids carrying mutations in these genes establish new bud sites at random positions rather than in the bipolar pattern (45). Interestingly, Sec3p localizes to the sites of exocytosis throughout the cell cycle; the localization of Sec3p is independent of a functional secretory pathway, independent of the actin and septin cytoskeletons, and, strikingly, also independent of the polarity establishment proteins Cdc24p and Cdc42p (46). Thus, Sec3p may not only target the exocytosis machinery to the site of polarization, but may also be part of the bipolar landmark, or influence its activity. It will be important to determine how Sec3p interacts with components of the bipolar landmark such as the polarisome. In particular, the identification of *SEC3* alleles which are specifically defective in secretion or bipolar budding would be useful to separate the two activities. Similarly, Rvs161p and Rvs167p are not only involved in bipolar budding, but are also required for endocytosis and cell fusion during mating (47). In this case, specific alleles of *RVS167* have been isolated which are efficient for endocytosis and have a normal actin cytoskeleton, but they are defective for cell fusion. It is not known whether the bud-site selection defect is also restored in these mutant cells.

Interestingly, diploid daughter cells harbouring specific *ACT1* alleles (21) or lacking Bud6p/Aip3p (43) place their first bud correctly at the distal poles but choose random sites for budding in subsequent cell cycles, suggesting that actin and associated proteins are involved in placing the bipolar positional marker at the division site but not at the distal tip of the daughter cell. Bud8p or Bud9p exhibit a striking bias limiting bud formation to either the axial pole (*bud8*) or the distal pole (*bud9*), suggesting that these proteins may directly mark the poles or assist in placing or recognizing the bipolar landmark. Marking of the proximal site may thus involve a cortical component such as Bud9p, while marking the pole opposite to the cytokinesis mark may be accomplished by Bud8p and growth components that have previously assembled at that site. Consistent with this result, Bud8p has been localized to the distal tip of daughter cells (H. Harkins, L. Schenkman, and J. Pringle, personal communication). Early recognition of the cortical landmark may involve the polarisome, actin, and other components that localize to the neck region at cytokinesis.

### 2.2.3 Use of the landmarks as organizers for other processes

Besides their role as landmarks for polarized growth, components of the bipolar landmark may also direct the asymmetrical localization of mRNA or proteins involved in a variety of other biological processes. *ASH1* encodes a daughter-specific nuclear repressor of *HO* transcription (48, 49). Remarkably, this asymmetrical distribution of Ash1p is mediated by localizing the *ASH1* mRNA in a cap at the distal pole of daughter cells (50, 51). Polarized localization of *ASH1* mRNA requires actin and *BNI1* (52), suggesting that components of the bipolar landmark may be involved in directing or anchoring *ASH1* mRNA at the distal pole. Consistent with this notion, recent experiments indicate that *ASH1* mRNA is mislocalized in cells deleted for *BUD6*, *SPA2*, or *BUD8* (K. Irie and A. Sil, personal communication). Thus, the asymmetrical localization of the *ASH1* transcript may use the bipolar landmark as a daughter-specific anchor.

In *S. cerevisiae*, positioning of the cell nucleus and the mitotic spindle requires both cytoplasmic microtubules and actin. Interestingly, *bni1*Δ and *bud6*Δ mutant cells exhibit nuclear positioning defects and misoriented cytoplasmic microtubules (53); the spindle orientation function of Bni1p requires its ability to bind Bud6p/Aip3p, and all known alleles of *BNI1* defective for bipolar budding are also defective for spindle orientation (54), suggesting that the bipolar landmark might be involved in both spindle orientation and bud-site selection. Thus, the polarisome may be required to localize a component involved in orientation of the microtubule cytoskeleton. *KAR9* was originally identified because it is required for karyogamy during mating (55), but subsequent analysis revealed that cells lacking Kar9p are also defective in spindle orientation (56). Kar9p is found at the cell cortex in an actin- but not microtubule-dependent fashion, and genetic analysis indicates that Bni1p functions in the same process as Kar9p (53). Strikingly, the cortical localization of Kar9p requires several components of the polarisome, including Bni1p, Spa2p, Pea2p, and Bud6p/Aip3p (57), indicating that the bipolar landmark may provide a docking site for Kar9p. In turn, Kar9p may direct spindle orientation by recruiting or stabilizing cytoplasmic microtubules (57).

### 2.2.4 Recognition of the axial and bipolar landmarks

Both the axial and bipolar landmarks are thought to be recognized by the Rsr1p/Bud1p GTPase module (Fig. 3), because haploid and diploid cells deleted for any of these proteins lose their predetermined budding pattern and instead position their buds randomly. The small GTPase, Rsr1p/Bud1p, is activated by the exchange factor Bud5p and inactivated by the GTPase-activating protein (GAP) Bud2p, and this GTPase module may recruit the polarity-establishment proteins to the bud site (see below). Rsr1p/Bud1p is localized at the plasma membrane throughout the cortex (300), whereas Bud2p and Bud5p are localized at the presumptive axial or bipolar bud site (H.-O. Park, personal comunication). Thus, these results suggest that Rsr1p/Bud1p is locally activated at the axial or bipolar landmark. However, it is not known how the Rsr1p/Bud1p GTPase module is recruited to the axial or bipolar landmarks, and it will be important to identify the components that interact with the Rsr1p/Bud1p module. Interestingly, alleles of *BUD2* and *BUD5* have been identified which specifically disrupt the bipolar budding pattern (58). These mutant proteins may have lost their ability to interact with the bipolar landmark but have retained their ability to interact with the axial landmark.

## 2.3 Defining the landmark in response to an extracellular signal

During mating, cell polarity is determined by an external cue, a morphogenetic gradient, which is established by the position of an appropriate mating partner (59–61). Before we discuss recent progress on the identification and regulation of the mating landmark, we will briefly introduce the events that are triggered during mating. For a more detailed discussion on signal transduction and mating processes, we refer to previously published reviews (62–64).

Haploid cells secrete mating pheromones which are recognized by cell-surface receptors in a cell-type-specific manner: **a**-cells produce **a**-factor, which binds to the receptor of α-cells (Ste2p), while α-cells produce α-factor which binds to the receptor on **a**-cells (Ste3p). Binding of pheromones to the receptor triggers a mitogen-activated protein kinase (MAPK) signal transduction pathway, culminating in arrest of the cell cycle, changes in gene expression, and formation of a pear-shaped cell, called a 'shmoo' (Plate 1B). These responses are initiated by a cell-surface receptor coupled to a heterotrimeric guanosine triphosphate (GTP) binding protein (G protein). Activation of the pheromone receptor leads to dissociation of the G protein into Gα and Gβγ, which in turn signals to downstream effectors to induce cellular responses. Ste5p and the PAK-like protein kinase Ste20p have been shown to bind to Gβγ (65–68) and both proteins are required to activate the MAPK pathway in response to pheromones (69, 64). However, it is possible that they play additional roles in the establishment of cell polarity (see below; 64, 70). Ste5p is able to bind multiple components of the MAPK cascade, including the MEKK Ste11p, the MEK Ste7p, and the MAPK Fus3p, and is thus proposed to function as a scaffold for signal transduction during mating (71). Ste20p is thought to activate the MEKK Ste11p, which in turn activates Ste7p, which then phosphorylates and activates Fus3p. The transcriptional repressors Dig1p and Dig2p are nuclear targets of Fus3p (72, 73), and their phosphorylation leads to their dissociation from the transcription factor Ste12p, which then induces expression of multiple mating-specific genes (74). In addition, Fus3p phosphorylates Far1p, a protein that is required for arresting the cell cycle, presumably by inhibiting the cyclin-dependent kinase Cdc28p-Clnp (75).

The spatial signal emanating from the mating partner that directs polarized growth is apparently a gradient of mating pheromone. Given a choice between mating partners, yeast cells mate almost exclusively with cells producing the strongest pheromone signal (76). More recent work has directly demonstrated polarized growth of **a**-cells toward a micropipette filled with α-factor (59) (for a discussion of polarity development in response to chemical gradients in more complex cells, see Chapter 7). Thus, cells use a pheromone gradient to locate their mating partner and polarize their actin cytoskeleton towards the site of the highest pheromone concentration. What is the nature of the mating landmark that depends on the pheromone signal? It is likely that the pheromone receptors contribute to this spatial site. First, *ste2Δ* cells, in which the mating pathway has been activated downstream of the receptor, show decreased mating-partner discrimination in comparison with isogenic wild-type strains (76). Furthermore, two Ste2p truncation mutants lacking most or all of the cytoplasmic tail show severe defects in forming and orienting projections along an α-factor gradient (M. Snyder, personal communication). However, the receptor tail is also required for rapid ubiquitin-mediated internalization and subsequent degradation of the receptor, raising the possibility that these receptor tail mutants may have a cell-polarity defect due to lack of internalization (77).

Genetic experiments implicate the G protein in the regulation of cell polarity during mating (78), suggesting that an effector of Gβγ may be involved in signalling cell polarity. Cells which have the signalling pathway activated by a membrane-bound Ste5p (67), are still dependent on Gβγ for their ability to orient cell polarity during mating, indicating that Gβγ must have a target which is not involved in signal transduction (P. Pryciak, personal communication). Several lines of evidence now suggest that the polarity effector of Gβγ may be Far1p. First, specific alleles of *FAR1* (*far1-s*) have been identified which cause a specific mating defect, resulting from inability of these cells to locate their mating partner (79). Thus, Far1p is necessary for oriented cell polarity *in vivo*. Secondly, Far1p binds to Gβγ through an amino-terminal ring-finger domain (80, 81) and, importantly, cells expressing a mutant Far1p with reduced ability to interact with Gβγ are unable to polarize along a

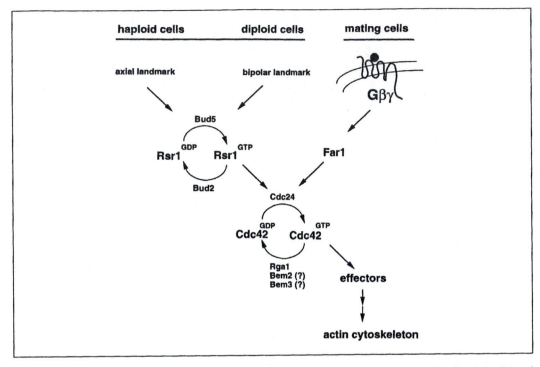

**Fig. 3** Recruitment of the Cdc42p GTPase module during budding and mating. In budding cells, the axial and bipolar landmarks are thought to localize the Rsr1p/Bud1p GTPase module. In turn, Rsr1p/Bud1p-GTP binds to Cdc24p, thereby recruiting the Cdc42p GTPase module to the landmark. During mating, these landmarks are ignored and instead the cells polarize towards a landmark established by the position of the mating partner. This landmark may include the activated receptor and Gβγ, which recruits the Cdc42p GTPase module in a manner that requires the adaptor protein Far1p. In turn, the Cdc42p module organizes the actin cytoskeleton.

pheromone gradient (79, 80). Thus, these results suggest that the landmark established during mating may be a complex composed of the receptor, Gβγ, and Far1p (Fig. 3).

It is noteworthy in this context that yeast cells are not only able to choose a site of polarization in a pheromone gradient but they are able to correct polarized growth along the pheromone gradient, suggesting that they continuously monitor their direction with respect to the source of pheromone. In addition, an intact actin cytoskeleton is required for developing a polarized organization of the pheromone receptors and other proteins, including Bem1p, Cdc24p, and Cdc42p, at the cell cortex (19, 81). These results suggest that the mating landmark is dynamic and must continuously be reinforced during oriented polarization. Once the actin cytoskeleton has been organized towards the mating partner, polarized delivery of newly synthesized receptor to the surface of highest receptor occupancy may result in dramatic amplification of cellular polarity. It is possible that clustering of the receptor may occur at early time points and amplify small differences in receptor occupancy across the cell surface. However, it is important to keep in mind that occupied receptors are internalized rapidly in response to pheromones (77), suggesting that perhaps limited activation of receptors may be important to polarize along a pheromone gradient. Clearly, a better understanding of the early events is needed to elucidate the mechanisms of how cells determine and maintain a spatial landmark in response to an extracellular signal.

Mating cells not only establish a new landmark, but they must ignore the predetermined site of polarization (the bud site) (82). In the absence of a chemotropic system, cells use the bud site as landmark during mating (79, 83, 84). At present it is not known whether active mechanisms exist which prevent use of the bud site during mating, or whether the mating landmark has simply a higher affinity for the polarity establishment complex than the bud site, thereby making the use of the bud site less favourable. It has been observed that Bud4p and Axl2p/Bud10p are down-regulated in response to pheromones (30, 36), but it is not clear whether this has any functional significance during mating. Interestingly, recent results suggest that the bud-site-selection proteins Rsr1p/Bud1p and Bud2p are required for shmoo formation in response to pheromone treatment in cells carrying mating-specific alleles of *CDC24* (*cdc24-m*) or *FAR1* (*far1-s*) (81; Butty and Peter, unpublished results). Thus, if the bud site is actively repressed during mating, these results indicate that the repression mechanism may also require functional Far1p and Cdc24p. Consistent with this notion, overexpression of a cytoplasmic Far1 mutant protein interferes with the correct budding pattern in haploid cells (85).

## 2.4  Establishment of cell polarity

The landmarks established during the cell cycle and mating reaction serve to recruit a set of proteins which are collectively called polarity establishment proteins or the actin-organizing complex (86). In turn, these proteins organize the actin cytoskeleton (see below). The polarity establishment proteins comprise a GTPase module which includes the small GTPase Cdc42p, its guanine nucleotide exchange factor (GEF) Cdc24p, and an unknown GTPase-activating enzyme (GAP) for Cdc42p (Fig. 3). Several proteins, including Rga1p, Bem2p, and Bem3p, have been proposed to function as GAPs for Cdc42p, but their functions *in vivo* remain to be determined (87–90). The activity of Cdc42p may also be regulated by Rdi1p, a GDP-dissociation inhibitor (GDI) (91). The SH3 domain containing protein, Bem1p, interacts with several proteins, including the PAK-like kinase Ste20p, Cdc24p, and Cdc42p, and is thus thought to function as an adaptor for the Cdc42p module. Cdc42p cycles between its GDP-bound and the GTP-bound state: Cdc24p stimulates GTP exchange, thereby activating Cdc42p, whereas the GAP proteins increase GTP hydrolysis, thereby inactivating Cdc42p. Cdc42p–GTP interacts with downstream effectors which, in turn, organize the actin cytoskeleton (see below). Bem1p does not appear to be required for the conversion of Cdc42p–GDP to Cdc42p–GTP (92), but may be required to properly localize the Cdc42p GTPase module. Cdc42p, Cdc24p, and Bem1p are distributed throughout the cytoplasm in early $G_1$ cells, but later localize to the presumptive bud site and to the shmoo tip in cells exposed to pheromone (81, 93; A.-C. Butty and M. Peter, unpublished results). Cdc42p, Cdc24p, and Bem1p are also found at the division site during cytokinesis, but it is not clear whether this localization is functionally significant. How are the polarity establishment proteins targeted to the landmark? As discussed earlier, it is thought that the Rsr1p/Bud1p GTPase module recruits the Cdc42p GTPase module, providing an intriguing link between two GTPase modules (94). Interestingly, *CDC24* and *CDC42* alleles which cause a random budding pattern have been isolated, suggesting that these mutant proteins may be defective in their interaction with the Rsr1p/Bud1p GTPase module (95; E. Bi, personal communication). A functional bud site requires that Rsr1p/Bud1p cycles between the GDP- and the GTP-bound form. Proteins have been identified which bind specifically to either the GDP– or GTP–Rsr1p/Bud1p (96). Importantly, Cdc24p and Bem1p have been found as effectors of Rsr1p/Bud1p; Cdc24p binds to Rsr1p/Bud1p in its GTP-bound form, and this interaction requires an intact effector domain of Rsr1p/Bud1p. In addition, overexpression of Rsr1p/Bud1p is able to suppress a *cdc24ts* mutation (97). In contrast, at least *in vitro*, Bem1p binds Rsr1p/Bud1p preferentially in its GDP-bound form and this interaction is not dependent on a functional effector domain of Rsr1p/Bud1p (96). Thus, through a GTPase-cycling reaction, Rsr1p/Bud1p is speculated to localize or activate Cdc24p and/or Bem1p at the incipient bud site (98). Rsr1p/Bud1p–GTP may be involved in recruiting Cdc24p to the bud site, while subsequent GTP hydrolysis may be important to allow activation of Cdc24p (see below). However, although recruitment of Cdc24p and Bem1p by Rsr1p/Bud1p is a likely possibility, this has not been demonstrated directly. One testable prediction of the model may be that cells expressing a Cdc24p or Bem1p mutant protein unable to interact with Rsr1p/Bud1p should exhibit a random budding pattern.

During mating, recruitment of the polarity establishment proteins to the Gβγ landmark is not dependent on the general bud-site-selection machinery (99), but requires Far1p (see Section 2.3). Interestingly, Far1p not only binds Gβγ, but its carboxy-terminal domain also interacts with the polarity establishment proteins Cdc24p, Cdc42p, and Bem1p (80, 81) (Fig. 4). Far1p binds to Cdc42p preferentially in its GTP-form and this interaction requires the presence of Bem1p (80). Thus, Far1p may function as an adaptor that links Gβγ to the polarity establishment proteins during mating. Interestingly, cells harbouring specific *CDC24* alleles (*cdc24-m*) have been identified that, like *far1-s* cells, display a mating defect, because they are unable to polarize along a pheromone gradient (84). These cells exhibit a normal budding pattern, suggesting that the Cdc24p-m mutant proteins are able to interact efficiently with Rsr1p/Bud1p. Genetic analysis suggested that Far1p and Cdc24p are functioning in the same polarization pathway during mating. Indeed, all the Cdc24p-m mutant proteins are defective for their interaction with Far1p (80, 81). Taken together, these results suggest that Rsr1p/Bud1p and Far1p bind separable domains on Cdc24p. In vegetative cells, Far1p is localized to the nucleus of $G_1$ cells (100) but some Far1p redistributes to the cytoplasm upon pheromone treatment (80). There is no apparent staining of Far1p at the shmoo tip, suggesting that complexes between Far1p and Gβγ may be transient. Nuclear Far1p is needed to arrest the cell cycle by binding to the Cdc28p-Clnp kinase, while cytoplasmic Far1p is required for cytoskeletal polarization along a pheromone gradient (85).

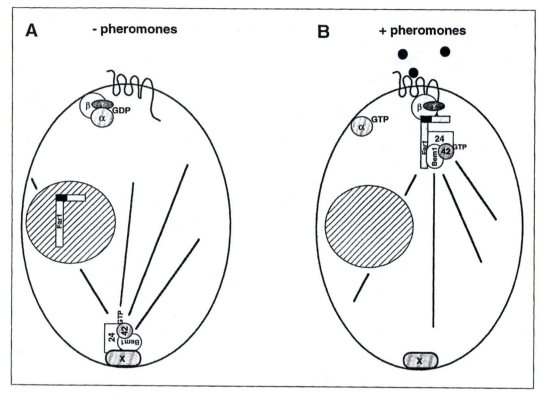

**Fig. 4** Model for the polarization in response to pheromones in budding yeast. During vegetative growth, Far1p is sequestered in the nucleus and cell polarity is established towards an internal landmark, the bud site (A). Binding of pheromones to the receptor leads to the formation of Gβγ at the plasma membrane (B). In turn, Gβγ activates the signal transduction pathway by binding to Ste5p and Ste20p, resulting in export of Far1p from the nucleus to the cytoplasm. Cytoplasmic Far1p is recruited to the site of the incoming signal by binding to Gβγ, thereby bringing the polarity establishment proteins Cdc24p, Bem1p, and Cdc42p to the site of polarization. The polarity establishment complex then locally activates Cdc42p by converting it to the GTP-bound form, which is required to polymerize the actin cytoskeleton.

Taken together, these results suggest the following model for how cells recruit the polarity establishment proteins to the mating landmark: in the absence of pheromones, Far1p is sequestered in the nucleus and cell polarity is established towards an internal landmark, the bud site (Fig. 4). Binding of pheromones to the receptor leads to dissociation of the G protein and formation of Gβγ at the plasma membrane. In turn, Gβγ activates a signal transduction pathway, resulting in export of Far1p from the nucleus to the cytoplasm (85). Cytoplasmic Far1p is than recruited to the site of the incoming signal by binding to Gβγ, thereby bringing the polarity establishment proteins Cdc24p, Bem1p, and Cdc42p to the site of polarization. The polarity establishment complex then locally activates Cdc42p by converting it to the GTP-bound form, which is required to polymerize the actin cytoskeleton (see below).

## 2.5   Signalling to the actin cytoskeleton

### 2.5.1   Targets of Cdc42p involved in cellular polarization in the $G_1$ phase of the cell cycle

Yeast cells express six different Rho-type GTPases (Cdc42p and Rho1p–Rho5p) which are thought to regulate actin-dependent processes at many different stages of the cell cycle (18, 101, 301; A. Rodal, K. G. Kozminski, and D. Drubin, personal communication). However, Cdc42p is the only Rho-type GTPase which is able to trigger cellular polarization at bud emergence (M.-P. Gulli and M. Peter, unpublished results), suggesting that the relevant target proteins required to polarize the actin cytoskeleton in the $G_1$ phase of the cell cycle must be specific effectors of Cdc42p (for a related discussion of Cdc42 regulation of the actin cytoskeleton, see Chapter 7). Five effectors have been identified in yeast which bind Cdc42p in its GTP-bound form and do not appreciably interact with other Rho-type GTPases. These proteins include the PAK-like protein kinases Cla4p, Ste20p, and Skm1p (102–104), and Gic1p and Gic2p (105, 106). The interaction occurs through a characteristic CRIB motif (Cdc42–Rac interactive binding), which is found in many Cdc42p targets from yeast to man (107). With the entire genome sequence of *S. cerevisiae* available, it is clear that these proteins represent the complete set of CRIB domain containing proteins in this organism. Ste20p, Gic1p, and Gic2p localize to sites of polarized growth in a Cdc42p-dependent manner (105, 106, 108–110). This localization requires a functional CRIB domain, suggesting that Cdc42p targets these proteins to the site of polarization. Cells deleted for both *GIC* genes exhibit severe polarization defects at bud emergence (105, 106). Likewise, cells lacking both *STE20* and *CLA4* exhibit cytoskeletal defects which resemble those of *cdc42–1* mutants, suggesting that the PAK-like kinases are also critical effectors for the organization of the cortical actin cytoskeleton at bud emergence (70). Thus, these results suggest that several effectors of Cdc42p need to co-operate in $G_1$ to polarize the actin cytoskeleton (Fig. 5). However, it is not understood how these CRIB domain containing proteins trigger cellular polarization and how they regulate assembly of the actin cytoskeleton. Using a permeabilized cell system, it has been shown that Cdc42p is required to nucleate filamentous actin (111). Interestingly, cells deleted for *STE20* or *CLA4* are defective in actin nucleation, suggesting that these PAK-like kinases may be involved in activating actin nucleation sites (70). Ste20p interacts with Bem1p, which in turn has been shown to co-immunoprecipitate with actin (112), probably as a large multi-subunit complex. However, the relevant substrates of the PAK-like kinases involved in actin nucleation and bud emergence remain to be identified. The sequence of the two Gic proteins does not provide any clues to their function, but it is clear that they are not required to regulate the activity or localization of the PAK-like kinases (105). Recent evidence suggests that the Gic proteins may link Cdc42p to the polarisome (M. Jaquenoud and M. Peter, unpublished results).

Proteins that contain a CRIB motif are currently the only effectors of Cdc42p that do not interact with other GTPases of the Rho subfamily. Thus, these proteins are likely to trigger cellular polarization at bud emergence. However, several proteins

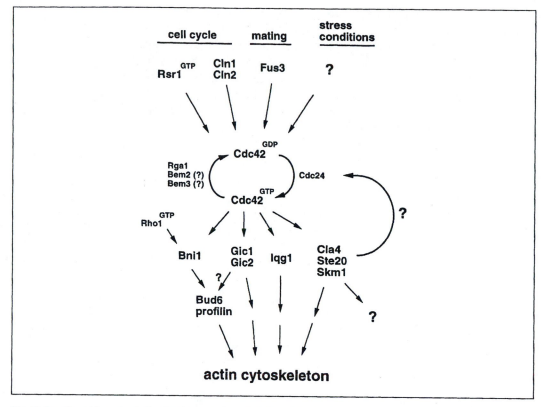

**Fig. 5** A pathway for cytoskeletal polarization mediated by Cdc42p in budding yeast. Cell cycle progression and extracellular signals lead to the conversion of GDP to GTP on Cdc42p, possibly by activation of the exchange factor Cdc24p. GTP-Cdc42p then binds and thereby localizes several effectors to sites of polarized growth, which in turn contribute to polarization of the actin cytoskeleton. Cdc42p might exert its effects on the cytoskeleton through several effectors, including Gic1p and Gic2p, the PAK-like kinases Ste20p, Cla4p and Skm1p, Bni1p, and Iqg1p. Bni1p is known to also interact with other members of the Rho-GTPase family. The PAK-like kinases may be part of a feedback loop that regulates Cdc24p. For further details see text.

without a recognizable CRIB domain have also been shown to interact with Cdc42p in a GTP-dependent manner and clearly contribute to cytoskeletal polarization at many stages of the cell cycle (Fig. 5). For example, the formins Bni1p and Bnr1p bind to several Rho-type GTPases, including Cdc42p, and cells lacking Bni1p show defects in polarized morphogenesis and in the organization of the underlying actin cytoskeleton (113–115). Bni1p is localized at the presumptive bud site and at the shmoo tip, and this asymmetrical localization requires its interaction with Rho-GTPases and Spa2p (113, 114, 116). Bni1p also interacts with profilin and Bud6p/Aip3p (114, 115); profilin is an actin-monomer-binding protein implicated in actin assembly (117) and Bud6p/Aip3p interacts with actin in the two-hybrid system (43). Thus, Bni1p may generally link Rho-GTPases to the regulation of actin polymerization, and its interaction with Cdc42p may contribute to actin polarization during bud emergence.

Iqg1p is a yeast member of the IQGAP family of proteins (118) which is clearly involved in cytokinesis (119, 120) but may also play a role during bud emergence (121). Iqg1p interacts with Cdc42p by two-hybrid analysis, but it is not known whether Iqg1p also binds other Rho-GTPases. Iqg1p is found together with Myo1p in a ring at the mother–bud junction after initiation of anaphase (119, 120), but, at least when over-expressed, Iqg1p also co-localizes with Cdc42p at the presumptive bud site (121). Like its mammalian counterparts, Iqg1p co-immunoprecipitates with actin and calmodulin, consistent with its proposed role in the regulation of the actin cytoskeleton.

Finally, Cdc42p and other Rho-GTPases have been shown to interact with Bem4p/Rom7p, Boi1p and its homologue Boi2p (122–124). Because the Boi proteins also bind to Bem1p via their SH3 domains it is possible that Bem1p bridges their interaction with Cdc42p. The function of these proteins is not known, and their sequence does not provide any clues. They could thus function as regulators or targets of Rho-GTPases, although to date no GEF, GDI, or GAP activity has been associated with any of them.

Besides these proteins, which have been shown at least by some assays to interact with Cdc42p or other Rho-type GTPases, the proteins discussed below have been proposed, or are suspected, to be targets of Cdc42p, but an interaction with any of the Rho-type GTPases has not been demonstrated. For example, Cdc42p interacts with proteins of the WASP family in many organisms (125, 126) (see also Chapter 7), but while mammalian WASP proteins contain a CRIB motif, the yeast homologue, Las17p/Bee1p, lacks this Cdc42p interaction domain (127). Indeed, two-hybrid analysis did not detect an interaction between Cdc42p and full-length Las17p/Bee1p (M. Jaquenoud and M. Peter, unpublished result), and it is thus not clear whether Las17p/Bee1p directly interacts with Cdc42p *in vivo*. In addition, Cdc42p is localized at the presumptive bud site, whereas Las17p/Bee1p is found predominantly in actin patches (128). Thus, while members of the WASP family in mammalian systems are likely candidates which may link Cdc42p to actin polymerization, there is no direct biochemical or genetic evidence to support this model in yeast. Cells lacking Las17p/Bee1p are viable but exhibit severe defects in the organization of the actin cytoskeleton and a defect in actin patches (128). Interestingly, like other WASP family members, Las17p/Bee1p has been shown to interact genetically and physically with Arp2p (129–131), a member of the Arp2/3-complex which is required for actin polymerization (132).

Msb3p and Msb4p are homologous proteins which have been isolated as high-copy suppressors of *cdc24* mutants and they may be targets of Cdc42p (133). Like Cdc42p, both Msb3p and Msb4p localize to the presumptive bud site, the bud tip, and the mother–bud neck, and this localization is Cdc42p dependent. However, it is not known whether these proteins interact directly with Cdc42p. Cells lacking both *MSB3* and *MSB4* are viable, but they are larger and rounder than normal and exhibit a disorganized actin cytoskeleton. Importantly, the localization of the septins is not affected in *msb3Δ msb4Δ* double mutants. Taken together, these data suggest that Msb3p and Msb4p may function downstream of Cdc42p specifically in a pathway leading to actin polymerization.

*ZDS1* and *ZDS2* have been isolated in a genetic screen for negative regulators of *CDC42*: overexpression of Zds1p or Zds2p appears to reduce the activity of Cdc42p (134). Zds1p and Zds2p are homologous to each other but have no detectable homology to known regulators of GTPases such as GAPs. Cells deleted for *ZDS1* and *ZDS2* grow slowly with an apparent delay in mitosis, and produce elongated cells and buds with morphological abnormalities. Zds1p is localized at the presumptive bud site and at the bud tip, suggesting that Zds1p and Zds2p may interact with Cdc42p. *ZDS1* and *ZDS2* have also been isolated as multicopy suppressors of the *cdc28–1N* mutant cells (135) and as a suppressor of genes regulating the Swe1p kinase (136). Although these different functions are confusing at first, they may be explained, at least in part, by the existence of the morphology checkpoint (137, 138): in response to an aberrant actin cytoskeleton, cells activate the tyrosine kinase, Swe1p, which in turn inhibits the mitotic Cdc28p-Clbp kinase to delay entry into mitosis. Because overexpression of *ZDS2* reduces expression of *SWE1*, we speculate that in the absence of *ZDS1* and *ZDS2*, Swe1p is upregulated and thus may prevent activation of the Cdc28p-Clbp kinase. This model predicts that deletion of *SWE1* in *zds1*Δ *zds2*Δ cells should rescue their growth and morphology defects. Taken together, the Zds proteins may provide a link between bud emergence and the regulation of entry into mitosis.

## 2.5.2  Additional functions of Cdc42p

Although Cdc42p is clearly required to trigger actin polarization at bud emergence, it may also function at other stages of the cell cycle and in signal transduction pathways (101). Besides their role in the $G_1$ phase of the cell cycle, Ste20p and Cla4p are required to phosphorylate Myo1p during cytokinesis, leading to constriction of the actin ring and cleavage of the two daughter cells in late telophase (139). Importantly, this cytokinesis function of Ste20p and Cla4p requires their ability to interact with Cdc42p (108, 109, 140), suggesting that Cdc42p is also involved in cytokinesis. In addition, Ste20p is required for signal transduction during mating, low nitrogen conditions, and in response to high osmolarity (109, 141). Interestingly, signalling in response to high osmolarity requires Ste20p with a functional CRIB domain as well as Cdc42p and Cdc24p (V. Reiser and G. Ammerer, personal communication; O'Rourke, personal communication), suggesting that Ste20p may be regulated by Cdc42p in response to high osmolarity conditions. The role of Cdc42p for signal transduction in response to pheromones is controversial: while early reports found that Cdc42p is required for signalling (142, 143), other experiments challenge this conclusion (108, 109, 144). Possibly Cdc42p and Gβγ may play partially redundant roles with respect to regulation of Ste20p during mating. Taken together, Ste20p and its regulator Cdc42p appear to play multiple roles during the cell cycle and in response to environmental conditions, suggesting that different effectors of Cdc42p may be involved in regulating diverse biological processes (Fig. 6). It will thus be important to isolate *CDC42* alleles which are specifically defective in some of these responses (95).

### 2.5.3 Regulation of the Cdc42p module

The switch from isotropic to polarized growth during the cell cycle or in response to mating pheromones is triggered by the activation of the $G_1$ cyclin-dependent kinase (Cdc28p-Clnp) or a MAPK pathway in response to pheromones (Fig. 5). The targets which are activated by these kinases are not known. Importantly, activated Cdc42p (Cdc42p-G12V) is able to induce actin polarization in the absence of the $G_1$ cyclins (M.-P. Gulli and M. Peter, unpublished results), suggesting that Cdc28p-Clnp kinase functions upstream of the Cdc42p module and may thus activate GTP exchange on Cdc42p. Cdc28p-Clnp may increase the activity of Cdc24p, inhibit the activity of the GAP, or affect the GDI for Cdc42p. Similarly, activation of cellular polarization in response to pheromones is independent of Cdc28p-Clnp kinase but requires a functional MAPK, Fus3p (F. van Drogen and M. Peter, unpublished results). Cyclin-dependent kinases and MAPKs display overlapping substrate specificity, suggesting that Fus3p and Cdc28p-Clnp may phosphorylate common targets. The Cdc42p-GTPase module may also be activated during cytokinesis and in response to stress conditions such as high osmolarity (see above), but the mechanism of this activation remains to be elucidated. Interestingly, Cdc24p is phosphorylated in a cell-cycle-dependent manner and in response to pheromones (M.-P. Gulli and M. Peter, unpublished results), but it is not known whether phosphorylation of Cdc24p regulates its activity. Thus, the simplest model would propose that phosphorylation of Cdc24p by Cdc28p-Clnp or Fus3p may activate Cdc24p and as a result trigger actin polarization. However, although hyperphosphorylation of Cdc24p correlates with Cdc28p-Clnp activity *in vivo*, available results suggest that Cdc24p is not directly phosphorylated by Cdc28p-Clnp but requires the PAK-like kinase Cla4p (M.-P. Gulli and M. Peter, unpublished results; I. Bose and D. J. Lew, personal communication). As discussed above, cells lacking both Ste20p and Cla4p are unable to polarize their actin cytoskeleton in $G_1$, consistent with a role upstream of Cdc24p (70). Interestingly, Ste20p may be a direct substrate of the Cdc28p-Clnp kinase (110, 145). Although it is not known whether Cla4p is also a target of Cdc28p-Clnp, it is interesting to note that Cla4p was isolated in a synthetic lethal screen with a strain lacking CLN1 and CLN2 (103, 140). Thus, it is possible that activation of the PAK-like kinases by Cdc28p-Clnp may trigger cellular polarization, although the kinase activity of Ste20p *in vivo* was not dependent on the $G_1$ cyclins (110, 145). In any event, because activation of Cla4p requires its ability to interact with Cdc42p-GTP (140), Cla4p may function in a feedback loop: Cla4p regulates the production of Cdc42p-GTP which in turn activates Cla4p. It is not known at present whether such a feedback loop is positive or negative in nature: it may be important to either reinforce a polarization signal at a given site or to turn off the signal once polarization has occurred. Clearly, further work is required to distinguish among these possibilities.

While the Rsr1p/Bud1p GTPase module is important to localize the Cdc42p module spatially during the cell cycle (see above), it may also play a role in regulating the activity of Cdc24p. *bud2* mutants have been found to be synthetically lethal with mutants of the genes encoding the $G_1$ cyclins Cln1p and Cln2p (146, 147).

Interestingly, this lethality is caused by increased activity of Rsr1p/Bud1p–GTP, as deletion of Rsr1p/Bud1p restores viability of the *bud2Δ cln1Δ cln2Δ* strain (146), suggesting that Rsr1p/Bud1p–GTP may inhibit the Cdc42p module. Cdc24p has been shown to be an effector of Rsr1p/Bud1p, suggesting that Rsr1p/Bud1p in its GTP form may prevent the activation of Cdc24p. Thus, the Rsr1p/Bud1p GTPase module may not only localize the Cdc42p module, but may also influence its function. As discussed earlier, a functional Rsr1p/Bud1p GTPase module requires both the GDP- and the GTP-form of Rsr1p/Bud1p (96). We thus speculate that GTP-bound Rsr1p/Bud1p is needed to recruit Cdc24p to the bud site, but that activation of Cdc24p requires conversion of Rsr1p/Bud1p to the GDP state.

While activation of the Cdc42p module is required to promote cellular polarization, its inactivation may be equally important to ensure a transient polarization signal, thereby preventing polarization towards multiple sites or at the wrong time in the cell cycle. Little is currently known about how cells inactivate the polarization signal, but at least two mechanisms can be proposed: first, as discussed above, phosphorylation of Cdc24p by Cla4p may not activate but instead inhibit the activity of Cdc24p. In this model, Cla4p would be part of a negative feedback loop leading to inhibition of Cdc24p in the presence of Cdc42p–GTP. However, even cells over-expressing active Cdc42p–GTP do not grow uniformly all over their surface but instead form one or more buds simultaneously (148). Thus, in these cells growth can still be restricted to few sites, suggesting that additional mechanisms downstream of Cdc42p must exist which prevent polarization towards multiple sites. One such mechanism may involve degradation of specific Cdc42p effectors required for actin polarization. The Cdc42p effector Gic2p is expressed in a cell-cycle-dependent manner, and its levels peak at the time of bud emergence. Interestingly, Gic2p is removed rapidly by ubiquitin-dependent degradation once cellular polarization has occurred (92). Degradation of Gic2p requires its ability to interact with Cdc42p, suggesting that the active pool of Gic2p is eliminated specifically. Interestingly, cells that are unable to degrade Gic2p exhibit morphological abnormalities consistent with polarization towards multiple sites (92). Thus, degradation of Gic2p may be important to restrict cellular polarization during the cell cycle. Alternatively, degradation of Gic2p may ensure alterations of different Cdc42p effectors, thereby allowing the dynamic establishment of a polarization site and recruitment of the actin cytoskeleton. Taken together, these results suggest that the Cdc42p module is regulated by positive and negative signals, thereby ensuring that cellular polarization is restricted both spatially and temporally.

## 2.6 Regulation of polarized growth: the role of other Rho-GTPases

As discussed earlier, a polarized organization of the actin cytoskeleton leads to polarized secretion and, as a result, polarized growth (see Section 1.2). Thus, not surprisingly, cells lacking proteins involved in exocytosis are defective for polarized growth. We will not discuss exocytosis in this chapter, and instead refer the reader to previous reviews on this topic (44, 149) and Chapter 9 of this volume. In addition, polarized growth requires stabilization of the cell wall to prevent lysis at sites of polarized secretion. The composition and regulation of cell-wall biosynthesis has been reviewed extensively (150). In this chapter we will focus on the regulation of the actin cytoskeleton after bud emergence and discuss the involvement of several components functioning downstream of Rho-GTPases. While Cdc42p is required for cellular polarization in the $G_1$ phase of the cell cycle, the five additional Rho-type GTPases (Rho1p–Rho5p) are involved in the regulation of various stages of polarized growth after bud emergence (18, 151, 301; A. Rodal, K.G. Kozminski, and D. Drubin, personal communication). It is not clear whether Cdc42p functions upstream of the Rho-GTPases or whether activation of the Rho-GTPases occurs independently of Cdc42p (18, 152). Although there is some functional redundancy between the different Rho-type GTPases, these proteins exhibit distinct functions which are not shared among the different Rho-GTPases. Rho1p localizes to sites of polarized growth and, interestingly, the SH3 domain containing protein, Sla1p, is required for its proper localization in $G_1$ cells (153). Overexpression of Rho1p in its active GTP form prevents cellular polarization, possibly by interfering with actin polymerization (M.-P. Gulli and M. Peter, unpublished results). In contrast, cells harbouring temperature-sensitive alleles of *RHO1* lyse either as small-budded cells or arrest growth with a disorganized actin cytoskeleton (154–156). Rho1p is required for the activity of the glucan synthase Fks1p both *in vitro* and *in vivo* (157, 158), providing an explanation for the lysis defect of *rho1* mutant cells. Rho1p also activates protein kinase C (159), which in turn increases expression of *FKS1* and *FKS2* (160). Interestingly, overexpression of either Pkc1p or components functioning in the cell integrity MAPK cascade downstream of Pkc1p are able to suppress the growth defect of *rho1* temperature-sensitive cells (156). Taken together, activation of Fks1p and the Pkc1p pathway by Rho1p is required for polarized growth. Interestingly, Myo2p and Exo70p have been identified recently as specific effectors of Rho3p–GTP (302), providing a direct molecular function of Rho3p on exocytosis and the regulation of the cytoskeleton. In addition, the formins Bni1p and Bnr1p (see above) and the signalling protein Skn7p have been identified as effectors of multiple Rho proteins (18, 151). These targets may be involved in organizing or maintaining the polarized organization of the actin cytoskeleton throughout the cell cycle (see above). Clearly, our knowledge of specific effectors of the Rho-GTPases is still limited, and it seems likely that many additional proteins will be discovered which function downstream of the multiple Rho-GTPases.

## 3.  Cell polarity in *Schizosaccharomyces pombe*

### 3.1  Defining landmarks for cell growth

*Schizosaccharomyces pombe* is a haplontic organism: diploid zygotes generated by mating of two cells immediately undergo meiosis to form four haploid ascospores. Ascospores germinate by swelling and then form a polarized outgrowth of a new cell

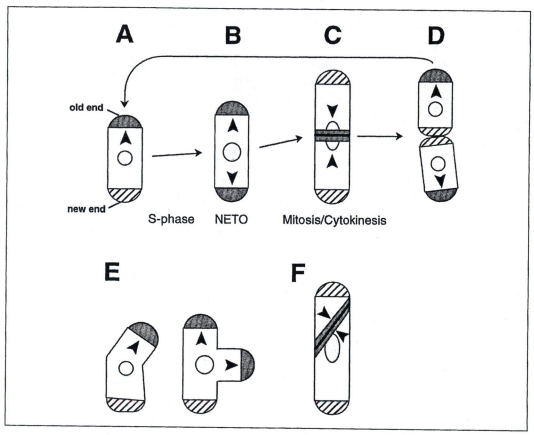

**Fig. 6** Normal and aberrant patterns of cell polarity in vegetative cells of fission yeast. (A) A newly born cell growing in a unipolar way at the old end (the end that was not generated by the previous cell division). (B) After S phase, the cell switches from unipolar to bipolar growth. This event is called NETO ('new end take off'). (C) During mitosis there is no growth at the cell ends, and polarity is directed towards the cell centre, where a symmetrical cell division takes place. (D) After cell separation, polarized growth is again directed towards the old ends of both daughter cells, and the cycle starts all over. (E) Cells that direct polarized growth towards abnormal sites due to aberrant microtubules or mutations in landmark genes. (F) Cell with a mislocalized septum due to mutations in landmark genes. The arrowheads show the direction of cell polarization. Filled and hatched areas represent active and inactive sites of cell polarity, respectively.

end, which will generate a cell of normal dimensions after division from the spore body (161). *Schizosaccharomyces pombe* cells have a regular cylindrical shape, and they show a defined pattern of polarized growth at the cell ends and symmetrical division at the cell centre (6; Fig. 6). The positions of cell growth and division are reflected by the distribution of the F-actin cytoskeleton during the cell cycle (162, 163). A newly born cell initiates growth at the end that was present in the mother cell prior to cell division ('old end'; Fig. 6A). At a stage in the early $G_2$ phase of the cell cycle, an additional growth zone is initiated at the opposite end, generated by the preceding cell division ('new end'). This transition from unipolar to bipolar growth is known as

NETO (for 'new end take off'; Fig. 6B). Throughout interphase, growth is such that the ends are opposed along the long axis of the cell, leading to cell elongation in a straight line. At the onset of mitosis, polarized growth at the cell ends ceases, and the actin cytoskeleton redistributes from the cell ends to the centre, where cell division takes place to form two equal daughter cells (Fig. 6C). After cell division, actin redistributes to the old ends of both daughter cells, from where polarized growth resumes (Fig. 6D).

### 3.1.1 Microtubules and cell polarity

Unlike in budding yeast (see Section 2.1), there is good evidence that the microtubule cytoskeleton is involved in polarized growth in fission yeast. Treatment with microtubule-inhibiting drugs (164, 165) or mutations in tubulin genes (166–170) lead to the formation of bent and/or branched cells (Fig. 6E). Similar defects in localization of cell growth have been observed in *tea2* and *ban* mutants, which have short or otherwise misorganized microtubules (171), as well as in mutants in the *alp1*, *alp11*, and *alp21* genes encoding cofactors D, B, and E, respectively, which are required for tubulin folding and proper microtubule function (172; P. Radcliffe and T. Toda, personal communication). Furthermore, overexpression or deletion of the microtubule-associated proteins Mal3p and Dis1p also produces abnormally shaped cells (173, 174).

*Schizosaccharomyces pombe* cells show an elaborate system of cytoplasmic microtubules throughout interphase. They are organized along the long axis of the cell, with some microtubules running from end to end (Plate 1C). At the onset of mitosis, cytoplasmic microtubules are replaced by spindle microtubules, and, after anaphase, new arrays of cytoplasmic microtubules are formed from organizing centres at the medial division site (Plate 1D; 175, 176). The spindle microtubules are nucleated by the spindle pole body (SPB), but essentially nothing is known about how cytoplasmic microtubules are organized and oriented in interphase cells. However, evidence is accumulating that at least some of the cytoplasmic microtubules are also nucleated from the SPB. Although the majority of cytoplasmic microtubules are not associated with the SPB, the ends of some microtubules are clearly connected to the cytoplasmic face of the SPB (177). In some mutant cells (*ban3*, *ban4*), all cytoplasmic microtubules seem to emerge from the single SPB at the nuclear periphery (171, 176). Moreover, microtubules that have been depolymerized by cold shock will regenerate from the nuclear region upon incubation in warm medium (J. Hyams, V. Snell, J. Cope, and P. Nurse, unpublished data; 178). These observations raise the possibility that all cytoplasmic microtubules are nucleated from the SPB and are then released (176). Recently, microtubule behaviour has been observed in live cells (179; K. Sawin, personal communication). Cytoplasmic microtubules seem to be generated from regions at the nuclear surface, and they extend and shorten dynamically towards both cell ends. In conclusion, the microtubules are directed from the cell centre towards the sites of polarized growth at the cell ends.

### 3.1.2 Microtubule-dependent and microtubule-independent landmarks?

What is the function of the microtubule cytoskeleton during polarized growth? Microtubules might potentially provide a system for polarized transport of proteins to the growth site(s). However, interference with microtubule function leads to cells that are still able to grow in a polarized way, but they do so at abnormal positions (Fig. 6E; see above). This function for microtubules in directing polarized growth may be mediated through transport of marker proteins to the growth sites. Consistent with this model, the two proteins Tea1p and Pom1p are delivered by microtubules to the cell ends (178, 180). In the absence of these proteins, the cells show a variety of defects in the positioning of cell polarity such as bent and branched growth (Fig. 6E), failure to switch to bipolar growth, and initiation of growth from the new end after cell division.

Tea1p shows some similarities to *Drosophila* Kelch and human Ezrin, and there are intriguing parallels between the inferred role of Tea1p in polarity development and roles for microtubules in animal polarity development (181, 182). Tea1p is located at the tips of microtubules and at the cell ends (Plate 2). Microtubules are required to deliver and maintain Tea1p at cell ends, whereas actin is not required for Tea1p localization. These observations have led to a model for Tea1p as a dynamic marker that finds and marks the ends of the cell continuously as a consequence of the property of microtubules, which explore the space of the cell and become aligned parallel to its long axis (178, 183). Constant delivery of marker proteins to the regions of both cell ends would then, on average, lead to cell growth in a straight line. Consistent with this view, both microtubule organization and localization of polarized growth are altered in cells with aberrant shape, and seem to be guided by the curvatures of the cell (184, 185). Tea1p, which is present in a large protein complex, may itself contribute to proper microtubule alignment within the cell by inhibiting further growth of microtubules at the cell ends (178). Other proteins, such as Tea2p, Tea3p, and Tip1p, have been identified that are required for correct localization of Tea1p and also affect the organization of cytoplasmic microtubules when deleted and/or overexpressed (H. Browning, D. Brunner, J. Hayles, M. Arellano, N. Peat, J. McIntosh, and P. Nurse, personal communication).

The idea of the microtubule and Tea1p system marking the cell ends dynamically and continuously is in contrast to the situation in *S. cerevisiae*, where the position of a cortical marker is determined by the position of a marker laid down at specific stages of the previous cell cycle (see Section 2.2; Fig. 2). However, there is some evidence that microtubule-independent landmarks for polarized growth also exist in fission yeast. Most cells can orient polarized growth towards an end even with a compromised microtubule cytoskeleton or in the absence of Tea1p. Tea1p is clearly not sufficient to activate cell growth, because it localizes also to new cell ends before they are growing (178). Moreover, polarity proteins such as actin, Ral3p, and even Tea1p can find the cell ends in the absence of microtubules, although inefficiently and slowly, and their localization is temporarily disrupted after treatment with the microtubule inhibitor, thiabendazole (165). Finally, disruption of microtubules with inhibitors does not seem ultimately to affect the efficiency of secretion or rates of polarized growth itself (186; K. Sawin, personal communication). This is further evidence that microtubules are not involved directly in the mechanism of polarized growth *per se* (see above). Nothing is known thus far about the nature of the possible

**Table 1** Cell polarity proteins in budding yeast, fission yeast, and Metazoa

| S. cerevisiae protein | S. pombe protein | Metazoan protein | References |
|---|---|---|---|
| **Landmark proteins** | | | |
| Bud3p | - | - | 33 |
| Bud4p | - | - | 34 |
| Axl1p | O14077 | Insulinase | 39 |
| Axl2/Bud10p | - | - | 35, 36 |
| Septins | Septins | Septins | 8 |
| Bud6/Aip3p | Z97208 | - | 43 |
| Bud7p | P87317 | - | Unpublished |
| **Bud8p and Bud9p** | - | - | Unpublished |
| Spa2p | - | - | 295 |
| Pea2p | - | - | 40 |
| Sec3p | - | - | 46 |
| Far1p | - | - | 80, 81 |
| Kel1p and Kel2p | **Tea1p** | (Kelch, Ezrin) | 178, 181, 296 |
| - | **Pom1p** | (Dyrk2 and Dyrk3) | 180, 297 |
| - | **Mid1/Dmf1p** | (Anillin) | 180, 196, 198 |
| | | | |
| **Polarity establishment and maintenance proteins** | | | |
| Bud1/Rsr1p | Ras1p | Ras-GTPases | 211 |
| Ras1p and Ras2p | | | |
| Cdc42p | Cdc42p | Cdc42 | 218 |
| Rga1p | (O74360) | (Rho-GAPs) | 88 |
| Cdc24p | Ral1/Scd1p | (P91620) | 217 |
| Bem1p | Ral3/Scd2p | - | 217 |
| Rho1-Rho4p | Rho1-Rho4p | Rho-GTPases | 18, 298 |
| Pkc1p | Pck1p and Pck2p | Protein kinases C | 299 |
| | | | |
| **Proteins signalling to the actin cytoskeleton** | | | |
| Ste20p, Cla4p, and Skm1p | Pak1/Shk1p and Pak2/Shk2p | Pak-like kinases | 238, 239, 240, 241 |
| Las17/Bee1p | O36027 | WASP | 131 |
| Iqg1p | Rng2p | IQGAP | 267 |
| Bni1p and Bnr1p | Cdc12p and Fus1p | Formins/FH proteins | 273 |
| Gic1p and Gic2p | - | - | 105, 106 |

Some representative polarity proteins of three classes are compared between yeast and metazoan cells. The complete genome of *S. cerevisiae*, more than 70 per cent of the *S. pombe* genome, and varying amounts of metazoan genomes (nearly 100 per cent for *C. elegans*) are available. In the first class, only the proteins printed in **bold** have been shown to function as landmark proteins. Note that classifying the proteins into three groups is somewhat arbitrary and some proteins, such as *S. cerevisiae* formins and Rsr1p/Bud1p, also function as landmark proteins.

For proteins that were not studied biologically the databank accession numbers are shown.

-; no homologue found thus far.

Proteins in parentheses show weak homology, or homology based on short stretches.

microtubule-independent tags in fission yeast, and whether they relate to bud-site selection strategies in budding yeast. However, most proteins required for bud-site selection do not seem to exist in *S. pombe* (Table 1).

The microtubule system may be most important for the initiation of new growth sites after cell division and during NETO. Cells that have been treated with microtubule-depolymerizing drugs will branch at high frequency, but only when the drug is added before NETO (165). Short, residual microtubules remain associated with the nucleus in the drug-treated cells, and these microtubules appear to play a role in cell branching (187; K. Sawin, personal communication). Thus, microtubules may be instrumental in triggering new growth sites, both at the ends in normal cells and at aberrant sites in drug-treated cells. Although actin can polarize in the absence of microtubules, it will return to either the old or new end (165), raising the possibility that microtubules are also involved in distinguishing the two ends of the cell. Spores and protoplasts can establish polarity in round cells (161, 188, 189) (i.e. in the absence of a pre-shaped cell end and more similar to the situation in *S. cerevisiae*). It would be interesting to determine whether microtubules and marker proteins such as Tea1p are also required to establish polarity in germinating spores or regenerating protoplasts.

Microtubules and their associated proteins may also provide a long-term marking system of the old end in nitrogen-starved cells. These cells become very short and arrest in $G_1$ phase, but upon refeeding they will remember to regrow from the old end, even after several days of starvation (J. Bähler, unpublished observations). In starved cells, the actin cytoskeleton becomes completely depolarized, whereas the arrangement of microtubules is not visibly affected (190). Interestingly, cells that lack Tea1p, Tea2p, or Tip1p will branch very frequently while recovering from nutrient starvation, but much less so while logarithmically growing (M. Arellano, H. Browning, D. Brunner, and J. Hayles, personal communication).

Pom1p belongs to a novel class of protein kinases (180). It does not seem to be a part of the dynamic complex at microtubule ends. Rather, its localization is more peripheral and proximal to the cell membrane (J. Bähler and P. Nurse, unpublished data). Pom1p requires microtubules as well as Tea1p for its localization to the cell ends, whereas Tea1p can localize independently of Pom1p. Thus, Pom1p may function downstream of Tea1p. Pom1p localization shows a negative correlation with actin localization (Plate 2B). Pom1p is highly concentrated at the new ends during and after cell division; in its absence, cells will grow at either end after division, and they fail to switch to bipolar growth (180). Pom1p therefore seems to be part of the system that enables the cell to distinguish the old end from the new end. Because *pom1* mutants grow in a unipolar way, some daughter cells have two ends that have never grown in the previous cell cycle (two 'new' ends). It is noteworthy that these cells will branch with high efficiency (180). Similar observations have been made with *tea1* mutants (D. Brunner, personal communication), raising the possibility that ends are only efficiently recognized after cell division if they were growing during the previous cell cycle. It is possible that a tag deposited during polarized growth, or components of the growth machinery itself, remain at the ends, thus marking the old ends of both daughter cells after division. Thus, the switch to bipolar growth before cell division might contribute to the cell's ability to recognize the old ends after cell division.

## 3.2 Defining a landmark for cell division

In *S. cerevisiae*, the bud site marks the position of polarity establishment and subsequently also becomes the site of cytokinesis (Fig. 2). In contrast, in *S. pombe* cells, the site of cytokinesis is spatially separated from the sites of polarized growth (Fig. 6A–C) and will become the new end in the next generation (Fig. 6D). During mitosis, an actin-based contractile ring is formed at the cell centre. Later, actin patches become polarized towards this medial ring. Ring contraction is accompanied by polarized secretion of cell-wall material at the cell centre, resulting in the formation of a septum that generates two equal daughter cells upon cleavage (Plate 1D; 191). Below, we will review factors that are involved in the proper positioning of the cell division site in *S. pombe*. For a more general discussion of cytokinesis and its regulation in fission yeast, we refer to recent reviews (191, 192).

The site of the medial ring coincides with the position of the premitotic nucleus. This correlation is also observed in cells that have more than one nucleus or show aberrant nuclear locations due to drugs or mutation (166, 193, 194), suggesting that the nucleus determines the site of ring formation. There is good evidence that the nucleus itself is maintained in the cell centre by a SPB-mediated interaction with the interphase microtubule cytoskeleton (164, 167, 168, 176, 195).

How might the nuclear position generate a landmark for cell division? Mid1/ Dmf1p is a PH domain containing protein with some similarities to *Drosophila* Anillin (196, 197) and may function in signalling the nuclear position to the cell cortex (198). Mid1p is located in the nucleus throughout interphase. During mitosis it leaves the nucleus and forms a ring at the cell centre. In *mid1* mutants, the septum is positioned at aberrant locations and orientations, whereas the nuclei are positioned normally at the cell centre (Fig. 6F; 194, 198). Thus, Mid1p is required for proper localization of cell division, probably by recruiting actin and other medial ring components to the cell centre (see Section 3.5.3). Some temperature-sensitive mutants in *plo1*, encoding the *S. pombe* homologue of the Polo-like kinase (199), show a very similar phenotype to *mid1* mutants (197). Polo-like kinases are widely conserved and have several functions throughout mitosis, such as spindle formation and activation of anaphase onset (200, 201). In *plo1* mutants, Mid1p does not form a ring during mitosis but remains in the nucleus (197). This is intriguing, because Mid1p becomes hyperphosphorylated during ring formation (198), raising the possibility that phosphorylation of Mid1p by the Plo1p kinase is required for nuclear exit and ring formation (Fig. 7). Consistent with this model, Mid1p exits the nucleus prematurely and becomes phosphorylated upon overexpression of Plo1p. Moreover, genetic and two-hybrid analyses suggest that Mid1p and Plo1p act together in the same pathway for septum positioning (197). Given its functions in mitosis and cytokinesis (197, 199), Plo1p appears to be a key protein for the spatial and temporal co-ordination of mitosis with cytokinesis (Fig. 7; see Section 3.6).

At least one protein seems to function in both the growth and cytokinesis landmarks: Pom1p is not only required for the localization of polarized growth, but also for the localization of medial rings and septa (180). The phenotype of *pom1* mutants in medial ring placement and organization is less severe compared to *mid1* and *plo1* mutants (180, 197). Unlike Plo1p, Pom1p is not required for the formation of the Mid1p ring, but for its proper placement. Nuclear localization of Mid1p appears to be impaired in the absence of Pom1p (197), which could explain why the position of

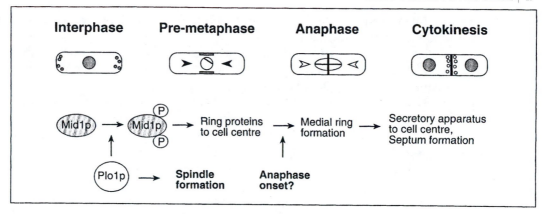

**Fig. 7** Model for the generation of a landmark and the polarization towards the cell centre during cytokinesis in fission yeast. In interphase cells, Mid1p localizes in the nucleus. Upon phosphorylation by the Polo-like kinase, Plo1p, before mitotic metaphase, Mid1p leaves the nucleus and forms a cortical band in the cell centre. This band functions as a landmark to recruit F-actin and other components of the medial ring to the cell centre. Upon anaphase onset, these ring proteins coalesce into a functional contractile ring. This medial ring then functions to recruit F-actin patches, cell-wall-synthesizing enzymes, and other components of the secretory apparatus to the cell centre. After nuclear division, the ring contracts and a septum is formed to divide the cell symmetrically. Plo1p also functions in spindle formation and possibly anaphase onset, and may thus co-ordinate mitosis with cytokinesis, both spatially and temporally. See text for further details.

the medial ring does not correlate well with the nuclear position in *pom1* mutants. Clearly, more work is required to obtain a more complete understanding of the Pom1p roles in directing both polarized growth and cytokinesis. Identification of the proteins phosphorylated by Pom1p is now an important goal. Other mutants (*pos1–pos3*) that also show defects in septum positioning have been described (202), but their relationship with *mid1*, *plo1*, and *pom1* mutants is not known at present.

## 3.3 Defining a landmark in response to an extracellular signal

During sexual differentiation, the intrinsic programme of cell polarity is overidden by extrinsic signals generated by gradients of mating pheromones. When cells of the P and M mating types are starved for nitrogen, they secrete mating-type-specific pheromones. The pheromones are recognized by specific receptors on the surface of cells of the opposite mating type (P-cells contain the M-factor receptor, Map3p, whereas M-cells contain the P-factor receptor, Mam2p). The pheromone receptors belong to the family of serpentine receptors, and they are similar to the pheromone receptors of *S. cerevisiae*. As in *S. cerevisiae* (see Section 2.3 and Chapter 7 of this volume), binding of pheromone to the corresponding receptor triggers dissociation of a heterotrimeric G protein into Gα and Gβγ and activation of a MAPK signal transduction pathway. Interestingly, however, the Gα subunit functions in activating the MAPK pathway in *S. pombe*, whereas in *S. cerevisiae* the Gβγ subunit is responsible for propagating the signal. Activation of the MAPK pathway, which also requires Ras1p and Shk1/Pak1p, brings about the changes in cell behaviour required for mating (for reviews, see refs 203, 204).

Both nitrogen starvation and the presence of mating pheromones lead to $G_1$ arrest in the cell cycle. The distribution of actin first becomes bipolar, but later the localization of all actin is polarized again towards one end only (15). From this end, polarized growth is then initiated towards a cell of opposite mating type ('shmoo' formation), until the mating partners touch each other and fuse at their shmoo tips (203, 204). Shmoo formation is always initiated from a cell end, but growth is no longer in a straight axis, and the growing end becomes thinner than in vegetative cells (Plate 1E). As in *S. cerevisiae*, *S. pombe* cells seem to polarize towards the highest concentration of pheromone (205–207), which thus provides a spatial extracellular signal during mating (for a related discussion, see Chapter 7). The bipolar distribution of actin before shmoo formation may reflect a courtship period during which the cell decides upon a mating partner and then initiates shmoo formation at the end closest to that partner. Accordingly, cells are apparently able to form shmoos at either the old or new end (15, 208).

The landmark that recognizes the pheromone signal and directs shmoo formation is not known. Both actin and microtubules are essential for shmoo formation. Actin is concentrated at the shmoo tip, and microtubules are directed towards this tip (Plate 1E; 15, 187, 209). Upon cell-to-cell contact during mating, microtubules are nucleated at the shmoo tip. Interestingly, microtubules are required for localization of actin to the shmoo tip, whereas the microtubule organization is independent of actin (187). Thus, microtubules may also be involved in directing polarized growth during mating, but it is not known what guides microtubules to the shmoo tip. It seems likely, however, that the mating receptors and/or heterotrimeric G protein play some role in directing shmoo formation analogous to the situation in *S. cerevisiae* and other eukaryotes (210). Microtubule-dependent marker proteins such as Tea1p and Pom1p, which function in the spatial organization of growth in vegetative cells (see Section 3.1.2), show highly reduced levels of localization upon pheromone treatment (178); T. Niccoli, J. Bähler, and P. Nurse, unpublished data). Downregulation of marker proteins for vegetative growth might therefore allow the cell to respond to a more flexible landmark during shmoo formation.

## 3.4 Establishment of cell polarity

Cell polarization culminates in the actin cytoskeleton being organized towards the landmarks for growth, cytokinesis, or mating. It is thought that special polarity establishment proteins recognize the landmarks and signal to the actin cytoskeleton, similar to the situation in *S. cerevisiae* (see Section 2.4). However, knowledge of these polarity establishment proteins, and especially their relationship to the landmarks and actin cytoskeleton, is still limited in *S. pombe*, but they seem to include conserved proteins.

The fission yeast *ras* oncogene homologue, *ras1*, functions in cell mating by contributing to activation of the pheromone response pathway that consists of a MAPK cascade (reviewed in refs 64, 204). However, cells lacking *ras1* also show aberrant shapes, being short and plump when growing exponentially and round upon nitrogen starvation (211, 212). The function of *ras1* in cell polarity has to be independent of its role in the pheromone response pathway, because mutants in genes functioning in the MAPK cascade show normal cell shape (204). Furthermore, cells lacking *ste6*, encoding a guanine nucleotide exchange factor (GEF) that activates the Ras1p-GTPase, are sterile but show normal morphology (213), indicating that there might be a morphology-specific factor to catalyse nucleotide exchange on Ras1p. Thus, Ras1p regulates at least two different pathways, similar to its mammalian homologue (64). Other genes involved in regulation of *ras1* are *ral2* and *gap1* (214, 215). Ral2p is thought to activate Ras1p, but its sequence shows no clues to its function. Gap1 is a GTPase-activating protein (GAP), which negatively regulates Ras1p (Fig. 8); cells lacking Gap1p show hyperpolarization in response to mating factor, as do activated *ras1* mutants.

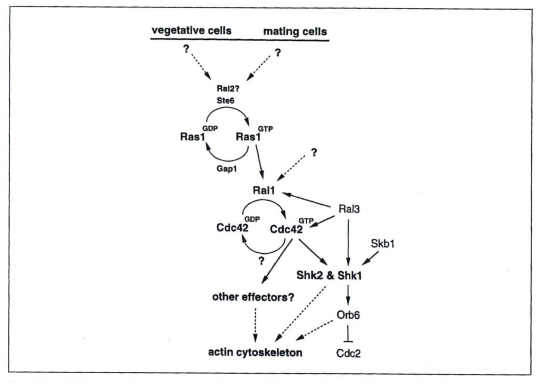

**Fig. 8** Model for events upstream and downstream of the Cdc42p-GTPase module in fission yeast. Ras1p-GTP activates Cdc42p by binding to Ral1p both in vegetative cells and during mating. Events upstream of Ras1p are not known, and Cdc42p may also be activated by Ras1p-independent mechanisms. Cdc42p-GTP then organizes the actin cytoskeleton by binding to different effectors such as the PAK-like kinases Shk1p and Shk2p. Shk1p may also function in a pathway that delays mitosis by downregulating Cdc2p. Ral3p is an adaptor protein that seems to bind and activate several components of the Cdc42p pathway. Note that Ras1p, Shk1p, and Shk2p also function in activating the pheromone response pathway (not shown). Hatched arrows represent interactions that probably require several steps. See text for further details.

Insight into the morphogenetic role of Ras1p in vegetative cells and during mating has been gained through identification of mutants with similar phenotypes as the *ras1* deletion (216, 217). Ral1/Scd1p is a protein homologous to *S. cerevisiae* Cdc24p, a GEF for Cdc42p (see Section 2.4). The Rho-type small GTPase, Cdc42p, is also present in fission yeast (218). *Schizosaccharomyces pombe cdc42* is an essential gene controlling polarized growth: null mutants are very small, round, and dense, whereas over-expression of activated alleles results in large cells of abnormal shape (95, 219), and partial loss of *cdc42* leads to round and sterile cells (T. Toda, personal communication). GTPase-activating enzymes (GAPs) for *S. pombe* Cdc42p have not been identified yet, but there is a protein similar to *S. cerevisiae* Rga1p in the databank (Table1). Activated Ras1p is thought to bind to and activate Ral1p, which then activates Cdc42p, although it has not been shown that Ral1p acts as an exchange factor for Cdc42p (Fig. 8; 216, 217). Thus, Ras1p seems to be upstream of a pathway involving Cdc42p, similar to the Ras-related protein Bud1/Rsr1p in *S. cerevisiae* (compare Fig. 3 with Fig. 8). However, there is no evidence that the Ras1p-GTPase module is directly involved in recognizing polarity landmarks, as is the case in *S. cerevisiae* (see Section 2.2.4), and Ras1p is important for polarity establishment both during vegetative growth and (even more so) during mating. It is likely that Cdc42p can also be activated independently of Ras1p, at least in vegetative cells, because deletion mutants of *ral1* and *cdc42* have more severe morphological defects than deletion mutants of *ras1*. There are some indications that the Ras1p pathway also affects microtubule organization (220). Moreover, Moe1p, which is conserved in Metazoa but not in budding yeast, interacts with Ral1p; in its absence cells show abnormal spindles, and microtubules are longer and more stable than in wild-type cells (221). These data raise the possibility of an interaction between microtubule-dependent landmarks and the Ras1p pathway. Ras1p has also some poorly characterized functions in cytokinesis (222, 223), but it is not known whether it is required to establish cell polarity for cell division. Ral3p/Scd2p is an SH3 domain protein homologous to *S. cerevisiae* Bem1p. Bem1p may be an adaptor protein that interacts with a number of proteins involved in cell polarity, including Cdc42p and Cdc24p (see Section 2.3). Ral3p seems to stabilize the Ral1p–Cdc42p complex; in its absence, the cells show similar shape aberrations as *ras1* and *ral1* deletion mutants (216, 217). Ral3p localizes to the sites of both polarized growth and cytokinesis (165) as well as to the shmoo tip (T. Niccoli and P. Nurse, personal communication), suggesting that the Cdc42p module localizes to all the sites of cell polarity.

Interestingly, Tea1p shows some similarity to components of the actin cytoskeleton, including Kelch and Ezrin (178, 181), and Ezrin association with the plasma membrane may be regulated by Rho-GTPase (224). Moreover, *pom1* mutations show genetic interaction with several mutations that affect both the actin and microtubule cytoskeleton (180). Thus, Tea1p and Pom1p might directly interact with the actin cytoskeleton. These two landmark proteins therefore have the potential to transmit the positional information provided by the microtubule cytoskeleton to the actin cytoskeleton. Finally, it is noteworthy that three mutants (*orb3, orb8,* and *orb9*) have been identified that show a specific defect in polarity establishment (rather than maintenance) after cell division (171). Identification of the corresponding genes should prove informative.

## 3.5 Signalling to the actin cytoskeleton

### 3.5.1 Actin and cell polarity

F-actin is essential for polarized growth, cytokinesis, mating, and sporulation in *S. pombe*, and it is recruited to sites of cell polarity (Plate 1). Actin might stabilize sites of cell growth through osmotic regulation (225, and references cited therein) and is also thought to be required for the polarized delivery of secretory vesicles to the target sites. Secretory vesicles become delocalized when actin distribution is disturbed (226–228). There is a close spatial and temporal relationship between actin localization and cell-wall formation (Plate 1C–E). This relationship is even more evident in reverting protoplasts (188, 189). Actin is polarized to the peripheral region of the round protoplasts where the cell wall starts to regenerate, and the presence of an actin inhibitor prevents cell-wall formation (188). Moreover, it has been shown recently that actin is required to localize α-glucan synthase to the sites of cell-wall formation (229). Mutations in the *S. pombe* actin gene (230–232) as well as over-expression of actin (J. Bähler, unpublished data) lead to a variety of defects in polarized growth and cytokinesis. Furthermore, mutation or overexpression of genes encoding actin-binding proteins such as profilin (*cdc3*; 233), tropomyosin (*cdc8*; 234), components of the Arp2/3 complex (235–237), and fimbrin (J.-Q. Wu, J. Bähler, and J. Pringle, in preparation) lead to aberrations in cell shape. Treatment of cells with the actin inhibitor latrunculin-A (22) results in an immediate arrest of polarized growth (J. Bähler, unpublished data), raising the possibility that actin is not only required for targeting secretion, but for secretion itself.

### 3.5.2 Recruitment of actin for polarized growth

Many proteins that function in recruiting the actin cytoskeleton to sites of cell polarity seem to be conserved between eukaryotes (Table 1). Some of these proteins are probably universally used to organize actin for different polarity events, such as growth, cytokinesis, and mating, but there are also some factors that are specific for a given event (see Section 3.5.3). Because of the central role of Cdc42p in establishing cell polarity, many actin-organizing proteins are presumably effectors of the Cdc42p-GTPase (Fig. 8). Shk1/Pak1p and Shk2/Pak2p are fission yeast homologues of mammalian p21-activated (PAK) kinases and *S. cerevisiae* Ste20p, Cla4p, and Skm1p (see Section 2.5.1); like their counterparts, they are effectors of activated Cdc42p, as shown by a range of genetic and biochemical experiments (238–241). Recent data indicate that binding of Cdc42p disrupts the intramolecular interaction of the regulatory domain with the catalytic domain of Shk1/Pak1p, thereby releasing the kinase from autoinhibition (242). Deletion of *shk1/pak1* results in a similar phenotype as deletion of *cdc42*, whereas Shk2/Pak2p plays a minor role and partially overlaps in function with Shk1/Pak1p. Both kinases are involved in polarized growth during mating and in activating the pheromone response pathway (217, 241, 243). However, cells defective in the pheromone response pathway are of normal shape, indicating that at least Shk1/Pak1p must also have other targets. Ral3p (see Section 3.4) may stabilize the Shk1/Pak1p–Cdc42p complex, and seems to be involved in activating the Shk1/Pak1p kinase *in vivo* (S. Marcus, personal communication). Substrates of the PAK-like kinases involved in actin recruitment have not been identified. As in other organisms, other effectors of Cdc42p certainly remain to be identified in fission yeast. It is noteworthy that a protein homologous to WASP has been identified by the fission yeast genome project (Table1); like *S. cerevisiae* Las17p, it does not contain a CRIB domain (see Section 2.5.1). There are also *S. pombe* homologues of the Cdc42p effectors Msb3p and Msb4p (see Section 2.5.1; 133). Other potential effectors of Cdc42p, such as formins and IQGAP proteins, seem to be required specifically for cytokinesis or mating (see Section 3.5.3).

Besides Cdc42p, there are at least four additional Rho-type GTPases in fission yeast (Rho1p–Rho4p; Table 1). Rho1p is an essential protein that may be required for the establishment of cell polarity; in its absence, actin remains randomly localized (244). Moreover, Rho1p regulates β-glucan synthase activity and cell-wall formation during polarized growth and septation, and it localizes to the growing cell ends and to the cell centre before septum formation (244–246,). Unlike Rho1p, Rho2p is not required for the establishment of cell polarity, but it also regulates cell-wall formation and actin localization (247). Rho2p may be involved in activating α-glucan synthesis (T. Mateo, M. Arellano, and P. Perez, personal communication). Thus, Rho1p and Rho2p have some distinct but overlapping roles in actin organization and activation of cell-wall synthesizing enzymes. No biological studies on Rho3p and Rho4p have been published thus far. The two fission-yeast protein kinase C-related genes, *pck1* and *pck2*, are required for normal cell shape, probably by regulating cell-wall synthesis (229, 248, 249). As with Pkc1p in *S. cerevisiae*, Pck1p and Pck2p seem to be downstream targets of Rho1p (cited in ref. 229). Several more genes have been identified that are required for the maintenance of cell polarity and shape (for reviews, see refs 250, 251), and many of them seem to regulate the protein kinase C pathway and/or cell-wall integrity. Most of these genes have been cloned, and they encode proteins required for cell-wall synthesis, such as Cwg2p, Cps1p, and Ags1/Mok1p (229, 252–254); components of a MAPK pathway required for cell-wall integrity, such as Pmk1/Spm1p and Mkh1p (255–257); other kinases and regulatory proteins, such as Kin1p, Orb5p, and Sts5p (258–260); as well as phosphatases and regulatory subunits such as Ppe1p, Ppb1p, Paa1p, and Pab1p (261–263).

Clearly, actin is important for the localization of the cell-wall synthesizing machinery, which controls cell shape (see Section 3.5.1). There are some indications, however, that cell-wall formation is then itself required to maintain a polarized actin localization. First, actin becomes depolymerized upon cell-wall digestion during protoplast formation (188). Secondly, actin shows no polarized localization in protoplasts of *pck2* mutants that fail to regenerate due to a defect in cell-wall formation (249). Finally, overexpression of the α-glucan synthase, Mok1p, leads to actin remaining stuck at one cell end (229). These findings raise the intriguing possibility that the cell wall is required for fixation of cell polarity in *S. pombe*, similar to the cell wall in *Fucus* zygotes (264) or the extracellular matrix and integrins in Metazoa (265).

### 3.5.3 Recruitment of actin for cytokinesis and mating

Mid1p is required for a cytokinesis landmark; it cycles between the nucleus in interphase cells and a medial ring in mitotic cells (see Section 3.2). How might Mid1p direct the formation of the actin-based contractile ring during mitosis? There is an intermediate stage between Mid1p nuclear localization and ring formation: during SPB separation and spindle formation, Mid1p leaves the nucleus and first forms a diffuse cortical band at the cell centre (197). There are some indications that this cortical Mid1p band then functions to recruit actin and other ring components from the cell ends to the centre. In *mid1* mutants, the formation of the medial ring is frequently initiated near the cell ends, where actin is already concentrated (197). Moreover, in wild-type cells actin ring structures accumulate at the cell centre before mitotic metaphase, coincident with formation of the cortical Mid1p band. In cells with a fully formed mitotic spindle (metaphase and/or anaphase), both Mid1p and actin coalesce into a distinct medial ring (197, 266). This self-focusing process may involve actin bundling proteins such as the IQGAP-related Rng2p (267), α-actinin, and fimbrin (J.-Q. Wu, J. Bähler, and J. R. Pringle, in preparation), and possibly tension provided by myosin (194, 223, 268–270). Mutations in all of the corresponding genes result in aberrant ring structures that are either wide and diffuse or consist of several misplaced strands, similar to the transitional structures seen during early stages of ring formation. Thus, the Mid1p band may provide a platform on which actin and actin-binding proteins assemble and organize into a functional contractile ring (Fig. 7). This mechanism could explain why Mid1p is required for medial ring placement, but is then dependent on several medial ring components to become itself organized into the ring (198).

Another characteristic of ring formation is that it seems to start from a single spot at the cell periphery. Some ring components can be detected in a spot before ring formation (223, 267, 271); J.-Q. Wu, J. Bähler, and J. R. Pringle, in preparation). It is possible that several medial ring proteins form a large complex that is instrumental in ring formation. At least some of these proteins can also localize to the SPB at the nuclear periphery (267, 272). This localization might provide an effective way to bring ring proteins to the cell centre. Cdc12p, a member of the formin family of proteins that interact with profilin and Rho-type GTPases (see Section 2.5.1; 273), is required for medial ring formation, and it has been observed in a motile cytoplasmic spot throughout interphase; this spot can move either along cytoplasmic microtubules or actin cables (271; F. Chang, personal communication). Thus, microtubules, in addition to their role in nuclear localization, might also have a direct role in localizing ring components (see Section 3.2). Notably, microtubules also co-localize with the medial ring after nuclear division (176, 220). The involvement of actin cables in ring formation is supported by data on the tropomyosin, Cdc8p, which is required for formation of both actin cables and the medial ring (194, 234, 266).

After a mature medial ring is formed, actin patches become polarized towards this ring. The function of these patches is presumably to localize secretion for septum synthesis. The Arp2/3 protein complex, which localizes to actin patches but not to the medial ring, is involved in patch relocation (191, 237), possibly by providing patch motility (236). Cdc15p, a PSTPIP-like protein that localizes to the medial ring, is specifically required to recruit actin patches towards the ring, and expression of Cdc15p during interphase leads to actin reorganization to the cell centre (274–276). Two polarity establishment proteins, Rho1p and Ral3p, as well as the potential Rho-GTPase substrates Rng2p and Cdc12p also localize to the medial ring (165, 244, 245, 267, 271). The medial ring also functions to localize both α- and β-glucan synthases, which are required for septum formation (229; J. Liu and M. Balasubramanian, personal communication). Thus, the medial ring may provide a platform to guide the secretory apparatus to the cell centre for septum formation (Fig. 7).

Not much is known about cell polarization during mating, but it presumably involves many of the proteins required for other polarization events. However, it also requires some specific factors. Actin association to the shmoo tip is completely lost when microtubules are depolymerized, whereas in vegetative cells actin localization does not seem to depend on microtubules to the same degree (187). Microtubules may be required continuously to target actin to the shmoo tip. Fus1p is a mating-specific formin in *S. pombe*, which localizes to the shmoo tip (272, 277). Formins may link Rho-GTPases to actin organization via profilin (see above), and, at least in *S. cerevisiae*, the formin Bni1p is also involved in capturing microtubules at the cell cortex (see Section 2.2.3). Fus1p and profilin are not required for the establishment of a shmoo tip, but for the maintenance of actin localization at the shmoo tip and cell fusion (15, 272). In conclusion, common factors (e.g. Rho-type GTPases or profilin) as well as specific factors (e.g. the formins Cdc12p or Fus1p) are involved in organizing the actin cytoskeleton for cell polarization during cytokinesis and mating.

## 3.6 Cell-cycle regulation of cell polarity

Cell polarization events are co-ordinated with other cell-cycle events. In vegetative *S. pombe* cells, there are at least three transitions in cell polarity during the cell cycle (Fig. 6):

(1) initiation of unipolar growth after cell division (Fig. 6D);

(2) initiation of bipolar growth during interphase (Fig. 6B); and

(3) termination of end growth/polarization towards the cell centre during mitosis (Fig. 6C).

Knowledge of how these transitions may be integrated with cell-cycle progression is rather cryptic in fission yeast. Essentially nothing is known about the first transition. Below, we will discuss some emerging mechanisms that might regulate the second and third transitions. During mating, the MAPK of the pheromone response pathway is required for polarization in *S. cerevisiae* (see Section 2.6). It is not clear how the pheromone response pathway or the nutritional signalling pathway is involved in establishing polarity during mating in *S. pombe,* but mutants in genes functioning in the MAPK cascades fail to form a shmoo (204). Certainly, the Ras1p-GTPase also plays an important role in shmoo formation (see Section 3.4).

The switch from unipolar to bipolar growth or NETO ('new end take off') is normally dependent on the completion of S phase and on a minimal cell size (278). Five mutants have been described that fail to switch to bipolar growth: *ssp1* (279), *orb2* (allelic to *shk1/pak1*), *ban2, tea1* (171), and *pom1* (180). Except *ssp1*, all these mutants show other defects in cell polarity, suggesting that NETO requires some of the factors used to direct or establish polarized growth in general. Recently, intriguing data on the function of the Ssp1p kinase have been reported (280). Ssp1p localizes mainly in the cytoplasm, but when cells are stressed by high osmolarity, it relocalizes to the active sites of polarity at the cell ends and centre. Ssp1p may then promote the stress-induced relocalization of actin, because it can induce F-actin depolymerization. Accordingly, Ssp1p may function to trigger NETO by increasing the actin monomer pool. Consistent with this view, the authors showed that treatments of cells that destabilize F-actin will induce NETO even before S phase (280). Thus, the landmark at the new end must be in place before the actual NETO event, which is consistent with Tea1p and Pom1p marking the new end throughout interphase (178, 180). Interestingly, destablization of F-actin did not lead to NETO in cells lacking Tea1p (280). It should be noted that Pom1p prevents growth at the new end after cell division (180), and the NETO signal might therefore also involve a modification of Pom1p at the new end. The signal that triggers NETO is not known. It has been proposed that NETO is regulated by the cyclin-dependent kinase, Cdc2p (the homologue of *S. cerevisiae* Cdc28p), the activity of which starts to increase in early $G_2$ (281). This would be parallel to the situation in *S. cerevisiae*, where the switch from polar to isotropic growth in the bud is positively regulated by the Cdc28p/Clb kinase (282). Interestingly, the *S. cerevisiae* Elm1p kinase, which is similar to Ssp1p, is required for the switch from polar to isotropic bud growth (283), in a pathway that also includes Cdc42p and the Shk1/Pak1p homologue Cla4p (284, 285). Thus, further research might reveal some intriguing parallels in regulation of the switch from polar to isotropic bud growth in *S. cerevisiae* and the switch from unipolar to bipolar growth in *S. pombe.*

Recent data suggest that some proteins involved in cell polarity also function to integrate the duration of polarized growth with the onset of mitosis (Fig. 8). Skb1p is a conserved protein that positively regulates Shk1/Pak1p (286). Overexpression of both Skb1p and Shk1/Pak1p leads to a mitotic delay and elongated cell size, whereas absence of Skb1p leads to short cells (287, 288). Thus, the Skb1p–Shk1/Pak1p complex acts as a mitotic inhibitor, possibly by binding to Cdc2p (287). Alternatively, these proteins might regulate the Cdc2p complex through Orb6p. Orb6p is a kinase related to metazoan proteins, such as Rho kinase and Lats/Warts, that may act downstream of Shk1/Pak1p (171, 288). It is also required for both maintenance of polarized growth and co-ordination of growth with mitosis: decrease of Orb6p levels causes loss of cell polarity and a premature mitosis, whereas an increase in Orb6p levels leads to mitotic delay and elongated cell size. Orb6p seems to regulate cell cycle progression via Cdc2p (288). Interestingly, the metazoan homologues of Orb6p are tumour suppressor genes, and human Lats1 has recently been shown to associate with Cdc2p, thereby inhibiting its kinase activity (reviewed in ref. 289). Further work is necessary to reveal the degree of similarity of these regulatory pathways between metazoan cells and fission yeast. Taken together, the data on *S. cerevisiae* and *S. pombe* suggest that Cdc2p can positively regulate PAK-like kinases (see Section 2.6), and it can also be negatively regulated by a PAK-like kinase-dependent pathway. This scheme might help to co-ordinate the cyclic change between polarized growth and mitosis, and it might involve different cyclin–Cdc2p complexes.

How is polarization towards the cell centre during mitosis and cytokinesis regulated by the cell-cycle machinery? Plo1p seems to be important for the spatial and temporal co-ordination of mitosis with medial ring formation (Fig. 7; see Section 3.2). Formation and placement of the Mid1p and medial ring can occur in the absence of the mitotic spindle (194, 198, 290). However, the formation of the Mid1p and medial ring is normally co-ordinated with spindle formation, and a mature ring only forms at around the time of anaphase onset (Fig. 7; see Section 3.5.3). Ring formation may depend on the execution of the Plo1p function in spindle formation because *plo1* mutants transiently blocked in spindle formation do not make medial rings until they escape from this block (197). Plo1p seems to be a limiting factor for cytokinesis because it can induce septum formation in interphase cells when overexpressed (199). Similarly, a small GTPase of the Ras superfamily, Spg1p, can induce premature septum formation in interphase cells upon increased expression. However, the main function of Spg1p seems to be later during septum formation (291, 292). In *S. cerevisiae*, inactivation of Cdc28p / Clb at the end of mitosis leads to actin polarization towards the bud neck (282). Similarly, inactivation of Cdc2p may be required in *S. pombe* to signal actin patch relocation to the medial ring. Cells arrested in mitosis will form a division septum upon inactivation of Cdc2p (293). Targets of this regulatory pathway remain to be identified. It is possible that activation of Cdc2p before mitosis is required for medial ring formation, and subsequent inactivation for mitotic exit then triggers ring constriction and septum formation. Work to investigate these regulatory mechanisms has only just begun. It would also be of great interest to obtain insight into the cell-cycle regulation of polarity establishment at the old end after cell division.

## 4. Concluding remarks

The knowledge reviewed above varies remarkably between the two yeasts. This might reflect partly the interests and biases of the authors, but probably has three main reasons. First, many more investigators studied cell polarity problems in *S. cerevisiae* compared to *S. pombe*, and the detailed knowledge about mechanisms and signalling pathways is correspondingly more advanced in *S. cerevisiae*. Secondly, research in the two yeasts has partly been done on different aspects of cell polarity, mainly because the two yeasts have their own unique strengths. Last, but not least, there seem to be some real differences in how the two yeasts solve problems of cell polarity. Budding and fission yeasts are only distantly related, and they differ remarkably in their morphology and mode of division ( 'budding' versus 'fission'). Thus, these two yeasts, with all their advantages as model organisms, provide two complementary systems to study aspects of cell polarity. Comparisons of both differences and similarities should help to identify principles that will apply to cell polarity in more complex organisms. Such a comparative approach has proven to be fruitful before in identifying basic concepts of the eukaryotic cell cycle.

Below, we will briefly point out some of the major differences and similarities that have emerged from studies on cell polarity in the two yeasts. Strikingly, most of the landmark proteins used for bud-site selection in *S. cerevisiae* are not found thus far in *S. pombe* or in metazoan cells (Table 1). Other landmark proteins of budding yeast, such as septins, Axl1p, or Rsr1p/Bud1p, are conserved, but there is no evidence that they play a similar role in other cells. Landmark proteins of fission yeast, on the other hand, are either not present in budding yeast (such as Pom1p or Mid1p), or the homologues seem to function differently in budding yeast (such as Tea1p versus Kel1 and Kel2p). Some of these differences in landmark proteins may be a reflection of the different roles microtubules play in the two yeasts. In *S. pombe*, microtubules are involved in directing cell polarity to appropriate sites, probably by delivering marker proteins to these sites. While evidence for microtubule-independent landmark systems in *S. pombe* is strong, nothing is known about their nature. In *S. cerevisiae*, there is no evidence that microtubules play any role in marking sites of polarization. Rather, they use these sites to locate the nucleus and orient the spindle. Notably, the strategies to mark sites of cell polarity vary even within each yeast (i.e. different sets of proteins are required to mark sites for vegetative growth, cell division, and mating). In conclusion, systems that provide positional information for cell polarity seem to vary considerably between, and even within, the two yeasts, and they might therefore be poorly conserved between different eukaryotes. This may be in contrast to the proteins that establish cell polarity and signal to the actin cytoskeleton. Many of these proteins are conserved between yeast and metazoan cells (Table 1), and many similarities between the signalling pathways are emerging. Similarly, different polarity events within one yeast seem to rely to a great extent on the same set of proteins. However, some of the conserved proteins seem to be used in different ways by the two yeasts. Ras1p in fission yeast plays a prominent role in mating cells, where it regulates both a MAPK cascade and a cell-polarity pathway involving Cdc42p, whereas Rsr1p/Bud1p in budding yeast is involved mainly in recognizing the bud site in vegetative cells. As expected, Cdc42p and F-actin are crucial for cell polarity in both yeasts. However, in fission yeast, Cdc42p and actin might additionally be required for secretion, as suggested by *cdc42* deletion and actin inhibitor experiments. Both yeasts have formins that probably behave similarly biochemically (e.g. profilin binding), but they show different defects when deleted. Moreover, in fission yeast specific formins are used for different polarization events. Finally, it is notable that the pseudohyphal way of growth of *S. cerevisiae* seems to be more related to the way *S. pombe* grows in several respects (28, 294):

(1)  cells are elongated, because they have no isotropic growth phase;

(2)  cells divide symmetrically;

(3) the timing of cell-cycle phases (e.g. long $G_2$) is similar; and

(4) *S. cerevisiae* Ras2p is used in a similar way to *S. pombe* Ras1p upstream of Cdc42p and an MAPK pathway.

We expect future work on cell polarity to reveal more similarities, but also other differences between the two yeasts. Both should be insightful, because in the end, it does not matter whether you make a bud or a fission, as long as you get yourself organized in a polarized way.

# Acknowledgements

We thank the many colleagues who provided unpublished results, and M. Mitchell and J. Sgouros for help with homology searches for Table 1. M. Arellano, D. Brunner, A.-C. Butty, M.-P. Gulli, J. Hayles, M. Jaquenoud, A. Neiman, D. McCollum, P. Nurse, K. Sawin, V. Simanis, and T. Toda are acknowledged for suggestions and critical reading of the manuscript. Work in the laboratory of M.P. is supported by the Swiss National Science Foundation, the Swiss Cancer League, and a Helmut Horten Incentive Award. J.B. was supported by the Imperial Cancer Research Fund and the Novartis Stiftung.

# References

1. Lew, D. J. and Reed, S. I. (1993) Morphogenesis in the yeast cell cycle: regulation by Cdc28 and cyclins. *J. Cell Biol.*, **120**, 1305–1320.
2. Stone, E. M., Heun, P., Laroche, T., Pillus, L., and Gasser, S. M. (2000) MAP kinase signaling induces nuclear reorganization in budding yeast. *Curr. Biol.*, in press.
3. Longtine, M. S., Fares, H., and Pringle, J. R. (1998) Role of the yeast Gin4p protein kinase in septin assembly and the relationship between septin assembly and septin function. *J. Cell Biol.*, **143**, 719–736.
4. Carroll, C. W., Altman, R., Schieltz, D., Yates, J. R., and Kellogg, D. (1998) The septins are required for the mitosis-specific activation of the Gin4 kinase. *J. Cell Biol.*, **143**, 709–717.
5. Barral, Y., Parra, M., Bidlingmaier, S., and Snyder, M. (1999) Nim1-related kinases coordinate cell cycle progression with the organization of the peripheral cytoskeleton in yeast. *Genes Dev.*, **13**, 176–187.
6. Robinow, C. F. and Hyams, J. S. (1989) In *Molecular biology of the fission yeast*, (ed. A. Nasim, P. Young, and B. F. Johnson), pp. 273–330. Academic Press, San Diego.
7. Kim, H. B. (1996) Immunofluorescence localisation of *Schizosaccharomyces pombe cdc103+* gene product. *J. Microbiol.*, **34**, 248–254.
8. Longtine, M. S., DeMarini, D. J., Valencik, M. L., Al-Awar, O. S., Fares, H., De Virgilio, C., and Pringle, J. R. (1996) The septins: roles in cytokinesis and other processes. *Curr. Opin. Cell Biol.*, **8**, 106–119.
9. Kupiec, M., Byers, B., Esposito, R. E., and Mitchell, A. P. (1995) Meiosis and sporulation in *S. cerevisiae*. In *The molecular and cellular biology of the yeast* Saccharomyces—*cell cycle and cell biology*, Vol. 3 (E. W. Jones, J. R. Pringle, and J. R. Broach), pp. 889–1036. Cold Spring Harbor Laboratory Press, Cold Spring Harbor, NY.
10. Neiman, A. M. (1998) Prospore membrane formation defines a developmentally regulated branch of the secretory pathway in yeast. *J. Cell Biol.*, **140**, 29–37.

11. Chu, S., DeRisi, J., Eisen, M., Mulholland, J., Botstein, D., Brown, P. O., and Herskowitz, I. (1998) The transcriptional program of sporulation in budding yeast. *Science*, **282**, 699–705.

12. Fares, H., Goetsch, L., and Pringle, J. R. (1996) Identification of a developmentally regulated septin and involvement of the septins in spore formation in *Saccharomyces cerevisiae*. *J. Cell Biol.*, **132**, 399–411.

13. De Virgilio, C., DeMarini, D. J., and Pringle, J. R. (1996) SPR28, a sixth member of the septin gene family in *Saccharomyces cerevisiae* that is expressed specifically in sporulating cells. *Microbiology*, **142**, 2897–2905.

14. Tanaka, K. and Hirata, A. (1982) Ascospore development in the fission yeasts *Schizosaccharomyces pombe* and *S. japonicus*. *J. Cell Sci.*, **56**, 263–279.

15. Petersen, J., Nielsen, O., Egel, R., and Hagan, I. M. (1998) F-actin distribution and function during sexual differentiation in *Schizosaccharomyces pombe*. *J. Cell Sci.*, **111**, 867–876.

16. Botstein, D., Amberg, D., Mulholland, J., Huffaker, T., Adams, A., Drubin, D., and Stearns, T. (1995) The yeast cytoskeleton. In *The molecular and cellular biology of the yeast Saccharomyces—cell cycle and cell biology*, Vol. 3, (ed. E. W. Jones, J. R. Pringle, and J. R. Broach), pp. 1–90. Cold Spring Harbor Laboratory Press, Cold Spring Harbor, NY.

17. Ayscough, K. R. and Drubin, D. G. (1996) Actin: general principles from studies in yeast. *Annu. Rev. Cell Dev. Biol.*, **12**, 129–160.

18. Schmidt, A. and Hall, M. N. (1998) Signaling to the actin cytoskeleton. *Annu. Rev. Cell Dev. Biol.*, **14**, 305–338.

19. Ayscough, K. R. and Drubin, D. G. (1998) A role for the yeast actin cytoskeleton in pheromone receptor clustering and signalling. *Curr. Biol.*, **8**, 927–930.

20. Jacobs, C. W., Adams, A. E., Szaniszlo, P. J., and Pringle, J. R. (1988) Functions of microtubules in the *Saccharomyces cerevisiae* cell cycle. *J. Cell Biol.*, **107**, 1409–1426.

21. Yang, S., Ayscough, K. R., and Drubin, D. G. (1997) A role for the actin cytoskeleton of *Saccharomyces cerevisiae* in bipolar bud-site selection. *J. Cell Biol.*, **136**, 111–123.

22. Ayscough, K. R., Stryker, J., Pokala, N., Sanders, M., Crews, P., and Drubin, D. G. (1997) High rates of actin filament turnover in budding yeast and roles for actin in establishment and maintenance of cell polarity revealed using the actin inhibitor latrunculin-A. *J. Cell Biol.*, **137**, 399–416.

23. Chant, J. (1996) Generation of cell polarity in yeast. *Curr. Opin. Cell Biol.*, **8**, 557–565.

24. Roemer, T., Vallier, L. G., and Snyder, M. (1996) Selection of polarized growth sites in yeast. *Trends Cell Biol.*, **6**, 434–441.

25. Pringle, J. R., Bi, E., Harkins, H. A., Zahner, J. E., De Virgilio, C., Chant, J., Corrado, K., and Fares, H. (1995) Establishment of cell polarity in yeast. *Cold Spring Harbor Symp. Quant. Biol.*, **60**, 729–744.

26. Palmieri, S. J. and Haarer, B. K. (1998) Polarity and division site specification in yeast. *Curr. Opin. Microbiol.*, **1**, 678–686.

27. Nasmyth, K. A. (1982) Molecular genetics of yeast mating type. *Annu. Rev. Genet.*, **16**, 439–500.

28. Kron, S. J. and Gow, N. A. (1995) Budding yeast morphogenesis: signalling, cytoskeleton and cell cycle. *Curr. Opin. Cell Biol.*, **7**, 845–855.

29. Chant, J. and Pringle, J. R. (1995) Patterns of bud-site selection in the yeast *Saccharomyces cerevisiae*. *J. Cell Biol.*, **129**, 751–765.

30. Sanders, S. L. and Field, C. M. (1994) Cell division. Septins in common? *Curr. Biol.*, **4**, 907–910.

31. Chant, J. (1996) Septin scaffolds and cleavage planes in *Saccharomyces*. *Cell*, **84**, 187–190.

32. Frazier, J. A., Wong, M. L., Longtine, M. S., Pringle, J. R., Mann, M., Mitchison, T. J., and

Field, C. (1998) Polymerization of purified yeast septins: evidence that organized filament arrays may not be required for septin function. *J. Cell Biol.*, **143**, 737–749.

33. Chant, J., Mischke, M., Mitchell, E., Herskowitz, I., and Pringle, J. R. (1995) Role of Bud3p in producing the axial budding pattern of yeast. *J. Cell Biol.*, **129**, 767–778.

34. Sanders, S.L. and Herskowitz, I. (1996) The BUD4 protein of yeast, required for axial budding, is localized to the mother/BUD neck in a cell cycle-dependent manner. *J. Cell Biol.*, **134**, 413–427.

35. Roemer, T., Madden, K., Chang, J., and Snyder, M. (1996) Selection of axial growth sites in yeast requires Axl2p, a novel plasma membrane glycoprotein. *Genes Dev.*, **10**, 777–793.

36. Halme, A., Michelitch, M., Mitchell, E. L., and Chant, J. (1996) Bud10p directs axial cell polarization in budding yeast and resembles a transmembrane receptor. *Curr. Biol.*, **6**, 570–579.

37. Sanders, S. L., Gentzsch, M., Tanner, W., and Herskowitz, I. (1999) *O*-glycosylation of Ax12/Bud10p by Pmt4p is required for its stability, localization, and function in daughter cells. *J. Cell Biol.*, **145**, 1177–1188.

38. Adames, N., Blundell, K., Ashby, M. N., and Boone, C. (1995) Role of yeast insulin-degrading enzyme homologs in propheromone processing and bud site selection. *Science*, **270**, 464–467.

39. Fujita, A., Oka, C., Arikawa, Y., Katagai, T., Tonouchi, A., Kuhara, S., and Misumi, Y. (1994) A yeast gene necessary for bud-site selection encodes a protein similar to insulin-degrading enzymes. *Nature*, **372**, 567–570.

40. Valtz, N. and Herskowitz, I. (1996) Pea2 protein of yeast is localized to sites of polarized growth and is required for efficient mating and bipolar budding. *J. Cell Biol.*, **135**, 725–739.

41. Sheu, Y. J., Santos, B., Fortin, N., Costigan, C., and Snyder, M. (1998) Spa2p interacts with cell polarity proteins and signaling components involved in yeast cell morphogenesis. *Mol. Cell Biol.*, **18**, 4053–4069.

42. Arkowitz, R. A. and Lowe, N. (1997) A small conserved domain in the yeast spa2p is necessary and sufficient for its polarized localization. *J. Cell Biol.*, **138**, 17–36.

43. Amberg, D. C., Zahner, J. E., Mulholland, J. W., Pringle, J. R., and Botstein, D. (1997) Aip3p/Bud6p, a yeast actin-interacting protein that is involved in morphogenesis and the selection of bipolar budding sites. *Mol. Biol. Cell*, **8**, 729–753.

44. Finger, F. P. and Novick, P. (1998) Spatial regulation of exocytosis: lessons from yeast. *J. Cell Biol.*, **142**, 609–612.

45. Haarer, B. K., Corbett, A., Kweon, Y., Petzold, A. S., Silver, P., and Brown, S. S. (1996) SEC3 mutations are synthetically lethal with profilin mutations and cause defects in diploid-specific bud-site selection. *Genetics*, **144**, 495–510.

46. Finger, F. P., Hughes, T. E., and Novick, P. (1998) Sec3p is a spatial landmark for polarized secretion in budding yeast. *Cell*, **92**, 559–571.

47. Brizzio, V., Gammie, A. E., and Rose, M. D. (1998) Rvs161p interacts with Fus2p to promote cell fusion in *Saccharomyces cerevisiae*. *J. Cell Biol.*, **141**, 567–584.

48. Sil, A. and Herskowitz, I. (1996) Identification of asymmetrically localized determinant, Ash1p, required for lineage-specific transcription of the yeast HO gene. *Cell*, **84**, 711–722.

49. Jansen, R. P., Dowzer, C., Michaelis, C., Galova, M., and Nasmyth, K. (1996) Mother cell-specific HO expression in budding yeast depends on the unconventional myosin myo4p and other cytoplasmic proteins [see comments]. *Cell*, **84**, 687–697.

50. Long, R. M., Singer, R. H., Meng, X., Gonzalez, I., Nasmyth, K., and Jansen, R. P. (1997) Mating type switching in yeast controlled by asymmetric localization of ASH1 mRNA. *Science* **277**, 383–387.

51. Takizawa, P. A., Sil, A., Swedlow, J. R., Herskowitz, I., and Vale, R. D. (1997) Actin-dependent localization of an RNA encoding a cell-fate determinant in yeast. *Nature*, **389**, 90–93.

52. Nasmyth, K. and Jansen, R. P. (1997) The cytoskeleton in mRNA localization and cell differentiation. *Curr. Opin. Cell Biol.*, **9**, 396–400.

53. Heil-Chapdelaine, R. A., Adames, N. R., and Cooper, J. A. (1999) Formin' the connection between microtubules and the cell cortex. *J. Cell Biol.*, **144**, 809–811.

54. Lee, L., Klee, S. K., Evangelista, M., Boone, C., and Pellman, D. (1999) Control of mitotic spindle position by the *Saccharomyces cerevisiae* formin Bni1p. *J. Cell Biol.*, **144**, 947–961.

55. Marsh, L. and Rose, M. D. (1995) The pathway of nuclear fusion during mating in *S. cerevisiae*. In *The molecular and cellular biology of the yeast* Saccharomyces — *cell cycle and cell biology*, Vol. 3, (ed. E. W. Jones, J. R. Pringle, and J. R. Broach), pp. 827–888. Cold Spring Harbor Laboratory Press, Cold Spring Harbor, NY.

56. Miller, R. K. and Rose, M. D. (1998) Kar9p is a novel cortical protein required for cytoplasmic microtubule orientation in yeast. *J. Cell Biol.*, **140**, 377–390.

57. Miller, R. K., Matheos, D., and Rose, M. D. (1999) The cortical localization of the microtubule orientation protein, Kar9p, is dependent upon actin and proteins required for polarization. *J. Cell Biol.*, **144**, 963–975.

58. Zahner, J. E., Harkins, H. A., and Pringle, J. R. (1996) Genetic analysis of the bipolar pattern of bud site selection in the yeast *Saccharomyces cerevisiae*. *Mol. Cell Biol.*, **16**, 1857–1870.

59. Segall, J. E. (1993) Polarization of yeast cells in spatial gradients of alpha mating factor. *Proc. Natl Acad. Sci., USA*, **90**, 8332–8336.

60. Chenevert, J. (1994) Cell polarization directed by extracellular cues in yeast. [Review]. *Mol. Biol. Cell*, **5**, 1169–1175.

61. Arkowitz, R. A. (1999) Responding to attraction: chemotaxis and chemotropism in *Dictyostelium* and yeast. *Trends Cell Biol.*, **9**, 20–27.

62. Sprague, G. F. and Thorner, J. W. (1992) Pheromone response and signal transduction during the mating process of *Saccharomyces cerevisiae*. In *The molecular and cellular biology of the yeast* Saccharomyces, (ed. E. W. Jones, J. R. Pringle, and J. R. Broach), pp. 657–744. Cold Spring Harbor Laboratory Press, Cold Spring Harbor.

63. Leberer, E., Thomas, D. Y., and Whiteway, M. (1997) Pheromone signalling and polarized morphogenesis in yeast. *Curr. Opin. Genet. Dev.*, **7**, 59–66.

64. Banuett, F. (1998) Signalling in the yeasts: an informational cascade with links to the filamentous fungi. *Microbiol. Mol. Biol. Rev.*, **62**, 249–274.

65. Whiteway, M. S., Wu, C., Leeuw, T., Clark, K., Fourest-Lieuvin, A., Thomas, D. Y., and Leberer, E. (1995) Association of the yeast pheromone response G protein beta gamma subunits with the MAP kinase scaffold Ste5p. *Science*, **269**, 1572–1575.

66. Feng, Y., Song, L. Y., Kincaid, E., Mahanty, S. K., and Elion, E. A. (1998) Functional binding between Gbeta and the LIM domain of Ste5 is required to activate the MEKK Ste11. *Curr. Biol.*, **8**, 267–278.

67. Pryciak, P. M. and Huntress, F. A. (1998) Membrane recruitment of the kinase cascade scaffold protein Ste5 by the G beta gamma complex underlies activation of the yeast pheromone response pathway. *Genes Dev.*, **12**, 2684–2697.

68. Leeuw, T., Wu, C., Schrag, J. D., Whiteway, M., Thomas, D. Y., and Leberer, E. (1998) Interaction of a G-protein beta-subunit with a conserved sequence in Ste20/PAK family protein kinases. *Nature*, **391**, 191–195.

69. Herskowitz, I. (1995) MAP kinase pathways in yeast: for mating and more. *Cell*, **80**, 187–197.

70. Eby, J. J., Holly, S. P., van Drogen, F., Grishin, A. V., Peter, M., Drubin, D. G., and Blumer, K. J. (1998) Actin cytoskeleton organization regulated by the PAK family of protein kinases. *Curr. Biol.*, **8**, 967–970.

71. Elion, E. A. (1995) Ste5p: a meeting place for MAP kinases and their associates. *Trends Cell Biol.*, **5**, 322–327.

72. Cook, J. G., Bardwell, L., Kron, S. J., and Thorner, J. (1996) Two novel targets of the MAP kinase Kss1 are negative regulators of invasive growth in the yeast *Saccharomyces cerevisiae*. *Genes Dev.*, **10**, 2831–2848.

73. Tedford, K., Kim, S., Sa, D., Stevens, K., and Tyers, M. (1997) Regulation of the mating pheromone and invasive growth responses in yeast by two map kinase substrates. *Curr. Biol.*, **7**, 228–238.

74. Fields, S. and Herskowitz, I. (1987) Regulation by the yeast mating-type locus of STE12, a gene required for cell-type-specific expression. *Mol. Cell Biol.*, **7**, 3818–3821.

75. Peter, M. and Herskowitz, I. (1994) Direct inhibition of the yeast cyclin-dependent kinase Cdc28-Cln by Far1. *Science*, **265**, 1228–1231.

76. Jackson, C. L. and Hartwell, L. H. (1990) Courtship in *Saccharomyces cerevisiae*: an early cell–cell interaction during mating. *Mol. Cell Biol.*, **10**, 2202–2213.

77. Hicke, L. (1999) Gettin' down with ubiquitin: turning off cell-surface receptors, transporters and channels. *Trends Cell Biol.*, **9**, 107–112.

78. Schrick, K., Garvik, B., and Hartwell, L. H. (1997) Mating in *Saccharomyces cerevisiae*: the role of the pheromone signal transduction pathway in the chemotropic response to pheromone. *Genetics*, **147**, 19–32.

79. Valtz, N., Peter, M., and Herskowitz, I. (1995) FAR1 is required for oriented polarization of yeast cells in response to mating pheromones. *J. Cell Biol.*, **131**, 863–873.

80. Butty, A. C., Pryciak, P. M., Huang, L. S., Herskowitz, I., and Peter, M. (1998) The role of Far1p in linking the heterotrimeric G protein to polarity establishment proteins during yeast mating. *Science*, **282**, 1511–1516.

81. Nern, A. and Arkowitz, R. A. (1999) A Cdc24p–Far1p–Gβγ protein complex required for yeast orientation during mating. *J. Cell Biol.*, **144**, 1187–1202.

82. Madden, K. and Snyder, M. (1992) Specification of sites for polarized growth in *Saccharomyces cerevisiae* and the influence of external factors on site selection. *Mol. Biol. Cell*, **3**, 1025–1035.

83. Dorer, R., Pryciak, P. M., and Hartwell, L. H. (1995) *Saccharomyces cerevisiae* cells execute a default pathway to select a mate in the absence of pheromone gradients. *J. Cell Biol.*, **131**, 845–861.

84. Nern, A. and Arkowitz, R. A. (1998) A GTP-exchange factor required for cell orientation. *Nature*, **391**, 195–198.

85. Blondel, M., Huang, L. S., Alepuz, P. M., Ammerer, G., and Peter, M. (1999) Nuclear export of Far1p in response to pheromones requires the export receptor Msn5p/Ste21p. *Genes Dev.*, **13**, 2284–2300.

86. Drubin, D. G. and Nelson, W. J. (1996) Origins of cell polarity. *Cell*, **84**, 335–344.

87. Peterson, J., Zheng, Y., Bender, L., Myers, A., Cerione, R., and Bender, A. (1994) Interactions between the bud emergence proteins Bem1p and Bem2p and Rho-type GTPases in yeast. *J. Cell Biol.*, **127**, 1395–1406.

88. Stevenson, B. J., Ferguson, B., De, V. C., Bi, E., Pringle, J. R., Ammerer, G., and Sprague, G. J. (1995) Mutation of RGA1, which encodes a putative GTPase-activating protein for the polarity-establishment protein Cdc42p, activates the pheromone-response pathway in the yeast *Saccharomyces cerevisiae*. *Genes Dev.*, **9**, 2949–2963.

89. Zheng, Y., Hart, M. J., Shinjo, K., Evans, T., Bender, A., and Cerione, R. A. (1993) Biochemical comparisons of the *Saccharomyces cerevisiae* Bem2 and Bem3 proteins. Delineation of a limit Cdc42 GTPase-activating protein domain. *J. Biol. Chem.*, **268**, 24629–24634.

90. Zheng, Y., Cerione, R., and Bender, A. (1994) Control of the yeast bud-site assembly GTPase Cdc42. Catalysis of guanine nucleotide exchange by Cdc24 and stimulation of GTPase activity by Bem3. *J. Biol. Chem.* **269**, 2369–2372.

91. Koch, G., Tanaka, K., Masuda, T., Yamochi, W., Nonaka, H., and Takai, Y. (1997) Association of the Rho family small GTP-binding proteins with Rho GDP dissociation inhibitor (Rho GDI) in *Saccharomyces cerevisiae*. *Oncogene*, **15**, 417–422.

92. Jaquenoud, M., Gulli, M. P., Peter, K., and Peter, M. (1998) The Cdc42p effector Gic2p is targeted for ubiquitin-dependent degradation by the SCFGrr1 complex. *EMBO J.*, **17**, 5360–5373.

93. Ziman, M., Preuss, D., Mulholland, J., O'Brien, J. M., Botstein, D., and Johnson, D. I. (1993) Subcellular localization of Cdc42p, a *Saccharomyces cerevisiae* GTP-binding protein involved in the control of cell polarity. *Mol. Biol. Cell*, **4**, 1307–1316.

94. Stowers, L., Yelon, D., Berg, L. J., and Chant, J. (1995) Regulation of the polarization of T cells toward antigen-presenting cells by Ras-related GTPase CDC42. *Proc. Natl Acad. Sci.*, *USA*, **92**, 5027–5031.

95. Miller, P. J. and Johnson, D. I. (1997) Characterization of the *Saccharomyces cerevisiae* cdc42–1ts allele and new temperature-conditional-lethal cdc42 alleles. *Yeast*, **13**, 561–572.

96. Park, H. O., Bi, E., Pringle, J. R., and Herskowitz, I. (1997) Two active states of the Ras-related Bud1/Rsr1 protein bind to different effectors to determine yeast cell polarity. *Proc. Natl Acad. Sci.*, *USA*, **94**, 4463–4468.

97. Bender, A. and Pringle, J. R. (1989) Multicopy suppression of the cdc24 budding defect in yeast by CDC42 and three newly identified genes including the ras-related gene RSR1. *Proc. Natl Acad. Sci.*, *USA*, **86**, 9976–9980.

98. Herskowitz, I., Park, H. O., Sanders, S., Valtz, N., and Peter, M. (1995) Programming of cell polarity in budding yeast by endogenous and exogenous signals. *Cold Spring Harbor Symp. Quant. Biol.*, **60**, 717–727.

99. Elia, L. and Marsh, L. (1998) A role for a protease in morphogenic responses during yeast cell fusion. *J. Cell Biol.*, **142**, 1473–1485.

100. Henchoz, S., Chi, Y., Catarin, B., Herskowitz, I., Deshaies, R. J., and Peter, M. (1997) Phosphorylation- and ubiquitin-dependent degradation of the cyclin-dependent kinase inhibitor Far1p in budding yeast. *Genes Dev.*, **11**, 3046–3060.

101. Johnson, D. I. (1999) Cdc42: An essential Rho-type GTPase controlling eukaryotic cell polarity. *Microbiol. Mol. Biol. Rev.*, **63**, 54–105.

102. Leberer, E., Dignard, D., Harcus, D., Thomas, D. Y., and Whiteway, M. (1992) The protein kinase homologue Ste20p is required to link the yeast pheromone response G-protein beta gamma subunits to downstream signalling components. *EMBO J.*, **11**, 4815–4824.

103. Cvrckova, F., De Virgilio, C., Manser, E., Pringle, J. R., and Nasmyth, K. (1995) Ste20-like protein kinases are required for normal localization of cell growth and for cytokinesis in budding yeast. *Genes Dev.*, **9**, 1817–1830.

104. Martin, H., Mendoza, A., Rodriguez-Pachon, J. M., Molina, M., and Nombela, C. (1997) Characterization of SKM1, a *Saccharomyces cerevisiae* gene encoding a novel Ste20/PAK-like protein kinase. *Mol. Microbiol.*, **23**, 431–444.

105. Brown, J. L., Jaquenoud, M., Gulli, M. P., Chant, J., and Peter, M. (1997) Novel Cdc42-binding proteins Gic1 and Gic2 control cell polarity in yeast. *Genes Dev.*, **11**, 2972–2982.

106. Chen, G. C., Kim, Y. J., and Chan, C. S. (1997) The Cdc42 GTPase-associated proteins Gic1 and Gic2 are required for polarized cell growth in *Saccharomyces cerevisiae*. *Genes Dev.*, **11**, 2958–2971.

107. Burbelo, P. D., Drechsel, D., and Hall, A. (1995) A conserved binding motif defines numerous candidate target proteins for both Cdc42 and Rac GTPases. *J. Biol. Chem.*, **270**, 29071–29074.

108. Peter, M., Neiman, A. M., Park, H. O., Vanlohuizen, M., and Herskowitz, I. (1996) Functional analysis of the interaction between the small Gtp binding protein Cdc42 and the Ste20 protein kinase in yeast. *EMBO J.*, **15**, 7046–7059.

109. Leberer, E., Wu, C. L., Leeuw, T., Fourestlieuvin, A., Segall, J. E., and Thomas, D. Y. (1997) Functional characterization of the Cdc42p binding domain of yeast Ste20p protein kinase. *EMBO J.*, **16**, 83–97.

110. Wu, C., Leeuw, T., Leberer, E., Thomas, D. Y., and Whiteway, M. (1998) Cell cycle- and Cln2p-Cdc28p-dependent phosphorylation of the yeast Ste20p protein kinase. *J. Biol. Chem.*, **273**, 28107–28115.

111. Li, R., Zheng, Y., and Drubin, D. G. (1995) Regulation of cortical actin cytoskeleton assembly during polarized cell growth in budding yeast. *J. Cell Biol.*, **128**, 599–615.

112. Leeuw, T., Fourest-Lieuvin, A., Wu, C., Chenevert, J., Clark, K., Whiteway, M., Thomas, D. Y., and Leberer, E. (1995) Pheromone response in yeast: association of Bem1p with proteins of the MAP kinase cascade and actin. *Science*, **270**, 1210–1213.

113. Kohno, H., Tanaka, K., Mino, A., Umikawa, M., Imamura, H., Fujiwara, T., Fujita, Y., Hotta, K., Qadota, H., Watanabe, T., Ohya, Y., and Takai, Y. (1996) Bni1p implicated in cytoskeletal control is a putative target of Rho1p small GTP binding protein in *Saccharomyces cerevisiae*. *EMBO J.*, **15**, 6060–6068.

114. Evangelista, M., Blundell, K., Longtine, M. S., Chow, C. J., Adames, N., Pringle, J. R., Peter, M., and Boone, C. (1997) Bni1p, a yeast formin linking cdc42p and the actin cytoskeleton during polarized morphogenesis. *Science*, **276**, 118–122.

115. Imamura, H., Tanaka, K., Hihara, T., Umikawa, M., Kamei, T., Takahashi, K., Sasaki, T., and Takai, Y. (1997) Bni1p and Bnr1p: downstream targets of the Rho family small G-proteins which interact with profilin and regulate actin cytoskeleton in *Saccharomyces cerevisiae*. *EMBO J.*, **16**, 2745–2755.

116. Fujiwara, T., Tanaka, K., Mino, A., Kikyo, M., Takahashi, K., Shimizu, K., and Takai, Y. (1998) Rho1p–Bni1p–Spa2p interactions: Implication in localization of Bni1p at the bud site and regulation of the actin cytoskeleton in *Saccharomyces cerevisiae*. *Mol. Biol. Cell*, **9**, 1221–1233.

117. Ayscough, K. R. (1998) *In vivo* functions of actin-binding proteins. *Curr. Opin. Cell Biol.*, **10**, 102–111.

118. Machesky, L. M. (1998) Cytokinesis: IQGAPs find a function. *Curr. Biol.*, **8**, R202–205.

119. Lippincott, J. and Li, R. (1998) Sequential assembly of myosin II, an IQGAP-like protein, and filamentous actin to a ring structure involved in budding yeast cytokinesis. *J. Cell Biol.*, **140**, 355–366.

120. Epp, J. A. and Chant, J. (1997) An IQGAP-related protein controls actin-ring formation and cytokinesis in yeast. *Curr. Biol.*, **7**, 921–929.

121. Osman, M. A. and Cerione, R. A. (1998) Iqg1p, a yeast homologue of the mammalian IQGAPs, mediates cdc42p effects on the actin cytoskeleton. *J. Cell Biol.*, **142**, 443–455.

122. Hirano, H., Tanaka, K., Ozaki, K., Imamura, H., Kohno, H., Hihara, T., Kameyama, T., Hotta, K., Arisawa, M., Watanabe, T., Qadota, H., Ohya, Y., and Takai, Y. (1996) ROM7/BEM4 encodes a novel protein that interacts with the Rho1p small GTP-binding protein in *Saccharomyces cerevisiae*. *Mol. Cell Biol.*, **16**, 4396–4403.

123. Mack, D., Nishimura, K., Dennehey, B. K., Arbogast, T., Parkinson, J., Toh-e, A., Pringle, J. R., Bender, A., and Matsui, Y. (1996) Identification of the bud emergence gene BEM4 and its interactions with rho-type GTPases in *Saccharomyces cerevisiae*. *Mol. Cell Biol.*, **16**, 4387–4395.

124. Bender, L., Lo, H. S., Lee, H., Kokojan, V., Peterson, V., and Bender, A. (1996) Associations among PH and SH3 domain-containing proteins and Rho-type GTPases in yeast. *J. Cell Biol.*, **133**, 879–894.

125. Rohatgi, R., Ma, L., Miki, H., Lopez, M., Kirchhausen, T., Takenawa, T., and Kirschner, M. W. (1999) The interaction between N-WASP and the Arp2/3 complex links Cdc42-dependent signals to actin assembly. *Cell*, **97**, 221–231.

126. Ramesh, N., Anton, I. M., Martinez-Quiles, N., and Geha, R. S. (1999) Waltzing with WASP. *Trends Cell Biol.*, **9**, 15–19.

127. Symons, M., Derry, J. M. J., Karlak, B., Jiang, S., Lemahieu, V., McCormick, F., Francke, U., and Abo, A. (1996) Wiskott–Aldrich syndrome protein, a novel effector for the GTPase CDC42Hs, is implicated in actin polymerization. *Cell*, **84**, 723–734.

128. Li, R. (1997) Bee1, a yeast protein with homology to Wiscott–Aldrich syndrome protein, is critical for the assembly of cortical actin cytoskeleton. *J. Cell Biol.*, **136**, 649–658.

129. Moreau, V., Madania, A., Martin, R. P., and Winson, B. (1996) The *Saccharomyces cerevisiae* actin-related protein Arp2 is involved in the actin cytoskeleton. *J. Cell Biol.*, **134**, 117–132.

130. Winter, D., Lechler, T., and Li, R. (1999) Activation of the yeast Arp2/3 complex by Bee1p, a WASP-family protein. *Curr. Biol.*, **9**, 501–504.

131. Bi, E. and Zigmond, S. H. (1999) Actin polymerization: Where the WASP stings. *Curr. Biol.*, **9**, R160–R163.

132. Machesky, L. M. and Gould, K. L. (1999) The Arp2/3 complex: a multifunctional actin organizer. *Curr. Opin.Cell Biol.*, **11**, 117–121.

133. Bi, E., Chiavetta, J. B., Chen, H., Chen, G.-C., Chan, C. S. M., and Pringle, J. R. (2000) Identification of novel, evolutionarily conserved Cdc42p-interacting proteins and of redundant pathways linking Cdc24p and Cdc42p to actin polarization in yeast. *Mol. Biol. of the Cell*, **11**, 773–793.

134. Bi, E. and Pringle, J. R. (1996) ZDS1 and ZDS2, genes whose products may regulate Cdc42p in *Saccharomyces cerevisiae*. *Mol. Cell. Biol.*, **16**, 5264–5275.

135. Yu, Y., Jiang, Y. W., Wellinger, R. J., Carlson, K., Roberts, J. M., and Stillman, D. J. (1996) Mutations in the homologous ZDS1 and ZDS2 genes affect cell cycle progression. *Mol. Cell. Biol.*, **16**, 5254–5263.

136. Ma, X.J., Lu, Q., and Grunstein, M. (1996) A search for proteins that interact genetically with histone H3 and H4 amino termini uncovers novel regulators of the Swe1 kinase in *Saccharomyces cerevisiae*. *Genes Dev.*, **10**, 1327–1340.

137. Lew, D. J. and Reed, S. I. (1995) A cell cycle checkpoint monitors cell morphogenesis in budding yeast. *J. Cell Biol.*, **129**, 739–749.

138. McMillan, J. N., Sia, R. A. L., and Lew, D. J. (1998) A morphogenesis checkpoint monitors the actin cytoskeleton in yeast. *J. Cell Biol.*, **142**, 1487–1499.

139. Field, C., Li, R., and Oegema, K. (1999) Cytokinesis in eukaryotes: a mechanistic comparison. *Curr. Opin. Cell Biol.*, **11**, 68–80.

140. Benton, B. K., Tinkelenberg, A., Gonzalez, I., and Cross, F. R. (1997) Cla4p, a *Saccharomyces cerevisiae* Cdc42p-activated kinase involved in cytokinesis, is activated at mitosis. *Mol. Cell Biol.*, **17**, 5067–5076.

141. O'Rourke, S. M. and Herskowitz, I. (1998) The Hog1 MAPK prevents cross talk between

the HOG and pheromone response MAPK pathways in *Saccharomyces cerevisiae*. *Genes Dev.*, **12**, 2874–2886.

142. Simon, M. N., De Virgilio, C., Souza, B., Pringle, J. R., Abo, A., and Reed, S. I. (1995) Role for the Rho-family GTPase Cdc42 in yeast mating-pheromone signal pathway. *Nature*, **376**, 702–705.

143. Zhao, Z.S. , Leung, T., Manser, E., and Lim, L. (1995) Pheromone signalling in *Saccharomyces cerevisiae* requires the small GTP-binding protein Cdc42p and its activator Cdc24p. *Mol. Cell Biol.*, **15**, 5246–5257.

144. Oehlen, L. J. and Cross, F. R. (1998) Potential regulation of Ste20 function by the Cln1-Cdc28 and Cln2-Cdc28 cyclin-dependent protein kinases. *J. Biol. Chem.*, **273**, 25089–25097.

145. Oehlen, L. J. and Cross, F. R. (1998) The role of Cdc42 in signal transduction and mating of the budding yeast *Saccharomyces cerevisiae*. *J. Biol. Chem.*, **273**, 8556–8559.

146. Benton, B. K., Tinkelenberg, A. H., Jean, D., Plump, S. D., and Cross, F. R. (1993) Genetic analysis of Cln/Cdc28 regulation of cell morphogenesis in budding yeast. *EMBO J.*, **12**, 5267–5275.

147. Cvrckova, F. and Nasmyth, K. (1993) Yeast G1 cyclins CLN1 and CLN2 and a GAP-like protein have a role in bud formation. *EMBO J.*, **12**, 5277–5286.

148. Davis, C. R., Richman, T. J., Deliduka, S. B., Blaisdell, J. O., Collins, C. C., and Johnson, D. I. (1998) Analysis of the mechanisms of action of the Saccharomyces cerevisiae dominant lethal cdc42G12V and dominant negative cdc42D118A mutations. *J. Biol. Chem.*, **273**, 849–858.

149. Kaiser, C. A., Gimeno, R. E., and Shaywitz, D. A. (1997) In *The molecular and cellular biology of the yeast* Saccharomyces—*cell cycle and cell biology*, Vol. 3, (ed. E. W. Jones, J. R. Pringle, and J. R. Broach), pp. 91–228. Cold Spring Harbor Laboratory Press, Cold Spring Harbor, NY.

150. Orlean, P. (1997) In *The molecular and cellular biology of the yeast* Saccharomyces—*cell cycle and cell biology*, Vol. 3 (E. W. Jones, J. R. Pringle, and J. R. Broach), pp. 229–362. Cold Spring Harbor Laboratory Press, Cold Spring Harbor, NY.

151. Tanaka, K. and Takai, Y. (1998) Control of reorganization of the actin cytoskeleton by Rho family small GTP-binding proteins in yeast. *Curr. Opin. Cell Biol.*, **10**, 112–116.

152. Chant, J. and Stowers, L. (1995) GTPase cascades choreographing cellular behavior: movement, morphogenesis, and more. *Cell*, **81**, 1–4.

153. Ayscough, K. R., Eby, J. J., Lila, T., Dewar, H., Kozminski, K. G., and Drubin, D. G. (1999) Sla1p is a functionally modular component of the yeast cortical actin cytoskeleton required for correct localization of both Rho1p-GTPase and Sla2p, a protein with talin homology. *Mol. Biol. Cell*, **10**, 1061–1075.

154. Madaule, P., Axel, R., and Myers, A. M. (1987) Characterization of two members of the rho gene family from the yeast *Saccharomyces cerevisiae*. *Proc. Natl Acad. Sci., USA*, **84**, 779–783.

155. Yamochi, W., Tanaka, K., Nonaka, H., Maeda, A., Musha, T., and Takai, Y. (1994) Growth site localization of Rho1 small GTP-binding protein and its involvement in bud formation in *Saccharomyces cerevisiae*. *J. Cell Biol.*, **125**, 1077–1093.

156. Helliwell, S. B., Schmidt, A., Ohya, Y., and Hall, M. N. (1998) The Rho1 effector Pkc1, but not Bni1, mediates signalling from Tor2 to the actin cytoskeleton. *Curr. Biol.*, **8**, 1211–1214.

157. Mazur, P. and Baginsky, W. (1996) *In vitro* activity of 1,3-beta-D-glucan synthase requires the GTP-binding protein Rho1. *J. Biol. Chem.*, **271**, 14604–14609.

158. Qadota, H., Python, C. P., Inoue, S. B., Arisawa, M., Anraku, Y., Zheng, Y., Watanabe, T., Levin, D. E., and Ohya, Y. (1996) Identification of yeast Rho1p GTPase as a regulatory subunit of 1,3-beta-glucan synthase. *Science*, **272**, 279–281.

159. Kamada, Y., Qadota, H., Python, C. P., Anraku, Y., Ohya, Y., and Levin, D. E. (1996) Activation of yeast protein kinase C by Rho1 GTPase. *J. Biol. Chem.*, **271**, 9193–9196.

160. Zhao, C., Jung, U.S., Garrett-Engele, P., Roe, T., Cyert, M. S., and Levin, D. E. (1998) Temperature-induced expression of yeast FKS2 is under the dual control of protein kinase C and calcineurin. *Mol. Cell Biol.*, **18**, 1013–1022.

161. Nishi, K., Shimoda, C., and Hayashibe, M. (1978) Germination and outgrowth of *Schizosaccharomyces pombe* ascospores isolated by Urografin density gradient centrifugation. *Can. J. Microbiol.*, **24**, 893–897.

162. Marks, J. and Hyams, J. S. (1985) Localisation of F-actin through the cell division cycle of *S. pombe*. *Eur. J. Cell Biol.*, **39**, 27–32.

163. Marks, J., Hagan, I. M., and Hyams, J. S. (1986) Growth polarity and cytokinesis in fission yeast: the role of the cytoskeleton. *J. Cell Sci. Suppl.*, **5**, 229–241.

164. Walker, G. M. (1982) Cell cycle specificity of certain antimicrotubular drugs in *Schizosaccharomyces pombe*. *J. Gen. Microbiol.*, **128**, 61–71.

165. Sawin, K. E. and Nurse, P. (1998) Regulation of cell polarity by microtubules in fission yeast. *J. Cell Biol.*, **142**, 457–471.

166. Toda, T., Umesono, K., Hirata, A., and Yanagida, M. (1983) Cold-sensitive nuclear division arrest mutants of the fission yeast *Schizosaccharomyces pombe*. *J. Mol. Biol.*, **168**, 251–270.

167. Umesono, K., Toda, T., Hayashi, S., and Yanagida, M. (1983) Cell division cycle genes nda2 and nda3 of the fission yeast *Schizosaccharomyces pombe* control microtubular organization and sensitivity to anti-mitotic benzimidazole compounds. *J. Mol. Biol.*, **168**, 271–284.

168. Hiraoka, Y., Toda, T., and Yanagida, M. (1984) The NDA3 gene of fission yeast encodes beta-tubulin: a cold-sensitive nda3 mutation reversibly blocks spindle formation and chromosome movement in mitosis. *Cell*, **39**, 349–358.

169. Yaffe, M. P., Harata, D., Verde, F., Eddison, M., Toda, T., and Nurse, P. (1996) Microtubules mediate mitochondrial distribution in fission yeast. *Proc. Natl Acad. Sci., USA*, **93**, 11664–11668.

170. Radcliffe, P., Hirata, D., Childs, D., Vardy, L., and Toda, T. (1998) Identification of novel temperature-sensitive lethal alleles in essential beta-tubulin and nonessential alpha 2-tubulin genes as fission yeast polarity mutants. *Mol. Biol. Cell*, **9**, 1757–1771.

171. Verde, F., Mata, J., and Nurse, P. (1995) Fission yeast cell morphogenesis: identification of new genes and analysis of their role during the cell cycle. *J. Cell Biol.*, **131**, 1529–1538.

172. Hirata, D., Masuda, H., Eddison, M., and Toda, T. (1998) Essential role of tubulin-folding cofactor D in microtubule assembly and its association with microtubules in fission yeast. *EMBO J.*, **17**, 658–666.

173. Beinhauer, J. D., Hagan, I. M., Hegemann, J. H., and Fleig, U. (1997) Mal3, the fission yeast homologue of the human APC-interacting protein EB-1 is required for microtubule integrity and the maintenance of cell form. *J. Cell Biol.*, **139**, 717–728.

174. Nakaseko, Y., Nabeshima, K., Kinoshita, K., and Yanagida, M. (1996) Dissection of fission yeast microtubule associating protein p93Dis1: regions implicated in regulated localization and microtubule interaction. *Genes Cells*, **1**, 633–644.

175. Hagan, I. M. and Hyams, J. S. (1988) The use of cell division cycle mutants to investigate the control of microtubule distribution in the fission yeast *Schizosaccharomyces pombe*. *J. Cell Sci.*, **89**, 343–357.

176. Hagan, I. M. (1998) The fission yeast microtubule cytoskeleton. *J. Cell Sci.*, **111**, 1603–1612.

177. Ding, R., West, R. R., Morphew, D. M., Oakley, B. R., and McIntosh, J. R. (1997) The spindle pole body of *Schizosaccharomyces pombe* enters and leaves the nuclear envelope as the cell cycle proceeds. *Mol. Biol. Cell*, **8**, 1461–1479.

178. Mata, J. and Nurse, P. (1997) tea1 and the microtubular cytoskeleton are important for generating global spatial order within the fission yeast cell. *Cell*, **89**, 939–949.

179. Ding, D. Q., Chikashige, Y., Haraguchi, T., and Hiraoka, Y. (1998) Oscillatory nuclear movement in fission yeast meiotic prophase is driven by astral microtubules, as revealed by continuous observation of chromosomes and microtubules in living cells. *J. Cell Sci.*, **111**, 701–712.

180. Bahler, J. and Pringle, J. R. (1998) Pom1p, a fission yeast protein kinase that provides positional information for both polarized growth and cytokinesis. *Genes Dev.*, **12**, 1356–1370.

181. Vega, L. R. and Solomon, F. (1997) Microtubule function in morphological differentiation: growth zones and growth cones. *Cell*, **89**, 825–828.

182. Gundersen, G. G. and Cook, T. A. (1999) Microtubules and signal transduction. *Curr. Opin. Cell Biol.*, **11**, 81–94.

183. Mata, J. and Nurse, P. (1998) Discovering the poles in yeast. *Trends Cell Biol.*, **8**, 163–167.

184. Sipiczki, M. and Grallert, A. (1997) Polarity, spatial organisation of cytoskeleton, and nuclear division in morphologically altered cells of *Schizosaccharomyces pombe*. *Can. J. Microbiol.*, **43**, 991–998.

185. Sipiczki, M., Miklos, I., and Grallert, A. (1997) In *Proceedings of the international symposium on theoretical biophysics and biomathematics*, (ed. L. Luo, Q. Li, and W. Lee), pp. 129–132. Inner Mongolia University Press, Hohhot, China.

186. Ayscough, K., Hajibagheri, N. M., Watson, R., and Warren, G. (1993) Stacking of Golgi cisternae in *Schizosaccharomyces pombe* requires intact microtubules. *J. Cell Sci.*, **106**, 1227–1237.

187. Petersen, J., Heitz, M. J., and Hagan, I. M. (1998) Conjugation in *S. pombe*: identification of a microtubule-organising centre, a requirement for microtubules and a role for Mad2. *Curr. Biol.*, **8**, 963–966.

188. Kobori, H., Yamada, N., Taki, A., and Osumi, M. (1989) Actin is associated with the formation of the cell wall in reverting protoplasts of the fission yeast *Schizosaccharomyces pombe*. *J. Cell Sci.*, **94**, 635–646.

189. Osumi, M., Yamada, N., Kobori, H., Taki, A., Naito, N., Baba, M., and Nagatani, T. (1989) Cell wall formation in regenerating protoplasts of *Schizosaccharomyces pombe*: study by high resolution, low voltage scanning electron microscopy. *J. Electron. Microsc. (Tokyo)*, **38**, 457–468.

190. Rupes, I., Jochova, J., and Young, P. G. (1997) Markers of cell polarity during and after nitrogen starvation in Schizosaccharomyces pombe. *Biochem. Cell Biol.* **75**, 697–708.

191. Gould, K. L. and Simanis, V. (1997) The control of septum formation in fission yeast. *Genes Dev.*, **11**, 2939–2951.

192. LeGoff, X., Utzig, S., and Simanis, V. (1999) Controlling septation in fission yeast: finding the middle, and timing it right. *Curr. Genet.*, **35**, 571–584.

193. Chang, F. and Nurse, P. (1996) How fission yeast fission in the middle. *Cell*, **84**, 191–194.

194. Chang, F., Woollard, A., and Nurse, P. (1996) Isolation and characterization of fission yeast mutants defective in the assembly and placement of the contractile actin ring. *J. Cell Sci.*, **109**, 131–142.

195. Hagan, I. and Yanagida, M. (1997) Evidence for cell cycle-specific, spindle pole body-mediated, nuclear positioning in the fission yeast *Schizosaccharomyces pombe*. *J. Cell Science*, **110**, 1851–1866.

196. Field, C. M., and Alberts, B. M. (1995) Anillin, a contractile ring protein that cycles from the nucleus to the cell cortex. *J. Cell Biol.* **131,** 165–178.

197. Bahler, J., Steever, A. B., Wheatley, S., Wang, Y., Pringle, J. R., Gould, K. L., and McCollum, D. (1998) Role of polo kinase and Mid1p in determining the site of cell division in fission yeast. *J. Cell Biol.*, **143**, 1603–1616.

198. Sohrmann, M., Fankhauser, C., Brodbeck, C., and Simanis, V. (1996) The dmf1/mid1 gene is essential for correct positioning of the division septum in fission yeast. *Genes Dev.*, **10**, 2707–2719.

199. Ohkura, H., Hagan, I. M., and Glover, D. M. (1995) The conserved *Schizosaccharomyces pombe* kinase plo1, required to form a bipolar spindle, the actin ring, and septum, can drive septum formation in G1 and G2 cells. *Genes Dev.*, **9**, 1059–1073.

200. Glover, D. M., Hagan, I. M., and Tavares, A. A. (1998) Polo-like kinases: a team that plays throughout mitosis. *Genes Dev.*, **12**, 3777–3787.

201. Nigg, E. A. (1998) Polo-like kinases: positive regulators of cell division from start to finish. *Curr. Opin. Cell Biol.*, **10**, 776–783.

202. Edamatsu, M. and Toyoshima, Y. Y. (1996) Isolation and characterization of pos mutants defective in correct positioning of septum in *Schizosaccharomyces pombe*. *Zool. Sci.*, **13**, 235–239.

203. Yamamoto, M., Imai, Y., and Watanabe, Y. (1997) In *The molecular and cellular biology of the yeast* Saccharomyces, (ed. E. W. Jones, J. R. Pringle, and J. R. Broach), Cold Spring Laboratory Press, Cold Spring Harbor, NY.

204. Davey, J. (1998) Fusion of a fission yeast. *Yeast*, **14**, 1529–1566.

205. Fukui, Y., Kaziro, Y., and Yamamoto, M. (1986) Mating pheromone-like diffusible factor released by *Schizosaccharomyces pombe*. *EMBO J.*, **5**, 1991–1993.

206. Leupold, U. (1987) Sex appeal in fission yeast. *Curr. Genet.*, **12**, 543–545.

207. Leupold, U., Sipiczki, M., and Egel, R. (1991) Pheromone production and response in sterile mutants of fission yeast. *Curr. Genet.*, **20**, 79–85.

208. Miyata, H. and Miyata, M. (1981) Mode of conjugation in homothallic cells of *Schizosaccharomyces pombe*. *J. Gen. Appl. Microbiol.*, **27**, 365–371.

209. Svoboda, A., Bahler, J., and Kohli, J. (1995) Microtubule-driven nuclear movements and linear elements as meiosis-specific characteristics of the fission yeasts *Schizosaccharomyces versatilis* and *Schizosaccharomyces pombe*. *Chromosoma*, **104**, 203–214.

210. Parent, C. A. and Devreotes, P. N. (1999) A cell's sense of direction. *Science*, **284**, 765–770.

211. Fukui, Y., Kozasa, T., Kaziro, Y., Takeda, T., and Yamamoto, M. (1986) Role of a ras homolog in the life cycle of *Schizosaccharomyces pombe*. *Cell*, **44**, 329–336.

212. Pichova, A. and Streiblova, E. (1992) Features of the cell periphery in the deformed *ras1⁻* cells of *Schizosaccharomyces pombe*. *Exp. Mycol.*, **16**, 178–187.

213. Hughes, D. A., Fukui, Y., and Yamamoto, M. (1990) Homologous activators of ras in fission and budding yeast. *Nature*, **344**, 355–357.

214. Fukui, Y., Miyake, S., Satoh, M., and Yamamoto, M. (1989) Characterization of the *Schizosaccharomyces pombe* ral2 gene implicated in activation of the ras1 gene product. *Mol. Cell Biol.*, **9**, 5617–5622.

215. Imai, Y., Miyake, S., Hughes, D. A., and Yamamoto, M. (1991) Identification of a GTPase-activating protein homolog in *Schizosaccharomyces pombe*. *Mol. Cell Biol.*, **11**, 3088–3094.

216. Fukui, Y. and Yamamoto, M. (1988) Isolation and characterization of *Schizosaccharomyces pombe* mutants phenotypically similar to ras1. *Mol. Gen.Genet.*, **215,** 26–31.

217. Chang, E. C., Barr, M., Wang, Y., Jung, V., Xu, H. P., and Wigler, M. H. (1994) Cooperative interaction of *S. pombe* proteins required for mating and morphogenesis. *Cell*, **79**, 131–141.

218. Fawell, E., Bowden, S., and Armstrong, J. (1992) A homologue of the ras-related CDC42 gene from *Schizosaccharomyces pombe*. *Gene*, **114**, 153–154.

219. Miller, P. J. and Johnson, D. I. (1994) Cdc42p GTPase is involved in controlling polarized cell growth in *Schizosaccharomyces pombe*. *Mol. Cell Biol.*, **14**, 1075–1083.

220. Pichova, A., Kohlwein, S. D., and Yamamoto, M. (1995) New arrays of cytoplasmic microtubules in the fission yeast *Schizosaccharomyces pombe*. *Protoplasma*, **188**, 252–257.

221. Chen, C. R., Li, Y. C., Chen, J., Hou, M. C., Papadaki, P., and Chang, E. C. (1999) Moe1, a conserved protein in *Schizosaccharomyces pombe*, interacts with a Ras effector, Scd1, to affect proper spindle formation. *Proc. Natl Acad. Sci.*, *USA*, **96**, 517–522.

222. Song, K., Mach, K. E., Chen, C. Y., Reynolds, T., and Albright, C. F. (1996) A novel suppressor of ras1 in fission yeast, byr4, is a dosage-dependent inhibitor of cytokinesis. *J. Cell Biol.*, **133**, 1307–1319.

223. Kitayama, C., Sugimoto, A., and Yamamoto, M. (1997) Type II myosin heavy chain encoded by the myo2 gene composes the contractile ring during cytokinesis in *Schizosaccharomyces pombe*. *J. Cell Biol.*, **137**, 1309–1319.

224. Hirao, M., Sato, N., Kondo, T., Yonemura, S., Monden, M., Sasaki, T., Takai, Y., and Tsukita, S. (1996) Regulation mechanism of ERM (ezrin/radixin/moesin) protein/plasma membrane association: possible involvement of phosphatidylinositol turnover and Rho-dependent signaling pathway. *J. Cell Biol.*, **135**, 37–51.

225. Mulholland, J., Preuss, D., Moon, A., Wong, A., Drubin, D., and Botstein, D. (1994) Ultrastructure of the yeast actin cytoskeleton and its association with the plasma membrane. *J. Cell Biol.*, **125**, 381–391.

226. Kanbe, T., Akashi, T., and Tanaka, K. (1993) Effect of cytochalasin A on actin distribution in the fission yeast *Schizosaccharomyces pombe* studied by fluorescent and electron microscopy. *Protoplasma*, **176**, 24–32.

227. Kanbe, T., Akashi, T., and Tanaka, K. (1994) Changes in the distribution of F-actin in the fission yeast *Schizosaccharomyces pombe* by arresting growth in distilled water: correlative studies with fluorescence and electron microscopy. *J. Electron. Microsc. (Tokyo)*, **43**, 20–24.

228. Kanbe, T., Kobayashi, I., and Tanaka, K. (1989) Dynamics of cytoplasmic organelles in the cell cycle of the fission yeast *Schizosaccharomyces pombe*: three-dimensional reconstruction from serial sections. *J. Cell Sci.*, **94**, 647–656.

229. Katayama, S., Hirata, D., Arellano, M., Perez, P., and Toda, T. (1999) Fission yeast alpha-glucan synthase Mok1 requires the actin cytoskeleton to localize the sites of growth and plays an essential role in cell morphogenesis downstream of protein kinase C function. *J. Cell Biol.*, **144**, 1173–1186.

230. Mertins, P. and Gallwitz, D. (1987) A single intronless action gene in the fission yeast *Schizosaccharomyces pombe*: nucleotide sequence and transcripts formed in homologous and heterologous yeast. *Nucleic Acids Res.*, **15**, 7369–7379.

231. Ishiguro, J. and Kobayashi, W. (1996) An actin point-mutation neighboring the 'hydrophobic plug' causes defects in the maintenance of cell polarity and septum organization in the fission yeast *Schizosaccharomyces pombe*. *FEBS Lett.*, **392**, 237–241.

232. McCollum, D., Balasubramanian, M. K., and Gould, K. L. (1999) Identification of cold-sensitive mutations in the *S. pombe* actin locus. *FEBS Lett.*, **451**, 321–326.

233. Balasubramanian, M. K., Hirani, B. R., Burke, J. D., and Gould, K. L. (1994) The *Schizosaccharomyces pombe* cdc3+ gene encodes a profilin essential for cytokinesis. *J. Cell Biol.*, **125**, 1289–1301.

234. Balasubramanian, M. K., Helfman, D. M., and Hemmingsen, S. M. (1992) A new tropomyosin essential for cytokinesis in the fission yeast *S. pombe*. *Nature*, **360**, 84–87.

235. Balasubramanian, M. K., Feoktistova, A., McCollum, D., and Gould, K. L. (1996) Fission yeast Sop2p: a novel and evolutionarily conserved protein that interacts with Arp3p and modulates profilin function. *EMBO J.*, **15**, 6426–6437.

236. Machesky, L. M. and Gould, K. L. (1999) The Arp2/3 complex: a multifunctional actin organizer. *Curr. Opin. Cell Biol.*, **11**, 117–121.

237. McCollum, D., Feoktistova, A., Morphew, M., Balasubramanian, M., and Gould, K. L. (1996) The *Schizosaccharomyces pombe* actin-related protein, Arp3, is a component of the cortical actin cytoskeleton and interacts with profilin. *EMBO J.*, **15**, 6438–6446.

238. Marcus, S., Polverino, A., Chang, E., Robbins, D., Cobb, M. H., and Wigler, M. H. (1995) Shk1, a homolog of the *Saccharomyces cerevisiae* Ste20 and mammalian p65PAK protein kinases, is a component of a Ras/Cdc42 signaling module in the fission yeast *Schizosaccharomyces pombe*. *Proc. Nat. Acad. Sci., USA*, **92**, 6180–6184.

239. Ottilie, S., Miller, P. J., Johnson, D. I., Creasy, C. L., Sells, M. A., Bagrodia, S., Forsburg, S. L., and Chernoff, J. (1995) Fission yeast pak1+ encodes a protein kinase that interacts with Cdc42p and is involved in the control of cell polarity and mating. *EMBO J.*, **14**, 5908–5919.

240. Sells, M. A., Barratt, J. T., Caviston, J., Ottilie, S., Leberer, E., and Chernoff, J. (1998) Characterization of Pak2p, a pleckstrin homology domain-containing, p21-activated protein kinase from fission yeast. *J. Biol. Chem.*, **273**, 18490–18498.

241. Yang, P., Kansra, S., Pimental, R. A., Gilbreth, M., and Marcus, S. (1998) Cloning and characterization of shk2, a gene encoding a novel p21-activated protein kinase from fission yeast. *J. Biol. Chem.*, **273**, 18481–18489.

242. Tu, H. and Wigler, M. (1999) Genetic evidence for Pak1 autoinhibition and its release by Cdc42. *Mol. Cell. Biol.*, **19**, 602–611.

243. Tu, H., Barr, M., Dong, D. L., and Wigler, M. (1997) Multiple regulatory domains on the Byr2 protein kinase. *Mol. Cell Biol.*, **17**, 5876–5887.

244. Nakano, K., Arai, R., and Mabuchi, I. (1997) The small GTP-binding protein Rho1 is a multifunctional protein that regulates actin localization, cell polarity, and septum formation in the fission yeast *Schizosaccharomyces pombe*. *Genes Cells*, **2**, 679–694.

245. Arellano, M., Duran, A., and Perez, P. (1997) Localisation of the *Schizosaccharomyces pombe* rho1p GTPase and its involvement in the organisation of the actin cytoskeleton. *J. Cell Sci.*, **110**, 2547–2555.

246. Arellano, M., Duran, A., and Perez, P. (1996) Rho 1 GTPase activates the (1–3)beta-D-glucan synthase and is involved in *Schizosaccharomyces pombe* morphogenesis. *EMBO J.*, **15**, 4584–4591.

247. Hirata, D., Nakano, K., Fukui, M., Takenaka, H., Miyakawa, T., and Mabuchi, I. (1998) Genes that cause aberrant cell morphology by overexpression in fission yeast: a role of a small GTP-binding protein Rho2 in cell morphogenesis. *J. Cell Sci.*, **111**, 149–159.

248. Toda, T., Shimanuki, M., and Yanagida, M. (1993) Two novel protein kinase C-related genes of fission yeast are essential for cell viability and implicated in cell shape control. *EMBO J.*, **12**, 1987–1995.

249. Kobori, H., Toda, T., Yaguchi, H., Toya, M., Yanagida, M., and Osumi, M. (1994) Fission yeast protein kinase C gene homologues are required for protoplast regeneration: a functional link between cell wall formation and cell shape control. *J. Cell Sci.*, **107**, 1131–1136.

250. Ishiguro, J. (1998) Genetic control of fission yeast cell wall synthesis: the genes involved in wall biogenesis and their interactions in *Schizosaccharomyces pombe*. *Genes Genet. Syst.*, **73**, 181–191.

251. Snell, V. and Nurse, P. (1993) Investigations into the control of cell form and polarity: the use of morphological mutants in fission yeast. *Dev. Suppl.*, 289–299.

252. Diaz, M., Sanchez, Y., Bennett, T., Sun, C. R., Godoy, C., Tamanoi, F., Duran, A., and Perez, P. (1993) The *Schizosaccharomyces pombe* cwg2+ gene codes for the beta subunit of a geranylgeranyltransferase type I required for beta-glucan synthesis. *EMBO J.*, **12**, 5245–5254.

253. Hochstenbach, F., Klis, F. M., van den Ende, H., van Donselaar, E., Peters, P. J., and Klausner, R. D. (1998) Identification of a putative alpha-glucan synthase essential for cell wall construction and morphogenesis in fission yeast. *Proc. Natl Acad. Sci., USA*, **95**, 9161–9166.

254. Ishiguro, J., Saitou, A., Duran, A., and Ribas, J. C. (1997) cps1+, a *Schizosaccharomyces pombe* gene homolog of *Saccharomyces cerevisiae* FKS genes whose mutation confers hypersensitivity to cyclosporin A and papulacandin B. *J. Bacteriol.*, **179**, 7653–7662.

255. Sengar, A. S., Markley, N. A., Marini, N. J., and Young, D. (1997) Mkh1, a MEK kinase required for cell wall integrity and proper response to osmotic and temperature stress in *Schizosaccharomyces pombe*. *Mol. Cell Biol.*, **17**, 3508–3519.

256. Toda, T., Dhut, S., Superti-Furga, G., Gotoh, Y., Nishida, E., Sugiura, R., and Kuno, T. (1996) The fission yeast pmk1+ gene encodes a novel mitogen-activated protein kinase homolog which regulates cell integrity and functions coordinately with the protein kinase C pathway. *Mol. Cell Biol.*, **16**, 6752–6764.

257. Zaitsevskaya-Carter, T. and Cooper, J. A. (1997) Spm1, a stress-activated MAP kinase that regulates morphogenesis in *S. pombe*. *EMBO J.*, **16**, 1318–1331.

258. Levin, D. E. and Bishop, J. M. (1990) A putative protein kinase gene (kin1+) is important for growth polarity in *Schizosaccharomyces pombe*. *Proc. Natl Acad. Sci., USA*, **87**, 8272–8276.

259. Snell, V. and Nurse, P. (1994) Genetic analysis of cell morphogenesis in fission yeast—a role for casein kinase II in the establishment of polarized growth. *EMBO J.*, **13**, 2066–2074.

260. Toda, T., Niwa, H., Nemoto, T., Dhut, S., Eddison, M., Matsusaka, T., Yanagida, M., and Hirata, D. (1996) The fission yeast sts5+ gene is required for maintenance of growth polarity and functionally interacts with protein kinase C and an osmosensing MAP-kinase pathway. *J. Cell Sci.*, **109**, 2331–2342.

261. Shimanuki, M., Kinoshita, N., Ohkura, H., Yoshida, T., Toda, T., and Yanagida, M. (1993) Isolation and characterization of the fission yeast protein phosphatase gene ppe1+ involved in cell shape control and mitosis. *Mol. Biol. Cell*, **4**, 303–313.

262. Yoshida, T., Toda, T., and Yanagida, M. (1994) A calcineurin-like gene ppb1+ in fission yeast: mutant defects in cytokinesis, cell polarity, mating and spindle pole body positioning. *J. Cell Sci.*, **107**, 1725–1735.

263. Kinoshita, K., Nemoto, T., Nabeshima, K., Kondoh, H., Niwa, H., and Yanagida, M. (1996) The regulatory subunits of fission yeast protein phosphatase 2A (PP2A) affect cell morphogenesis, cell wall synthesis and cytokinesis. *Genes Cells*, **1**, 29–45.

264. Kropf, D. L., Kloareg, B., and Quatrano, R. S. (1988) Cell wall is required for fixation of the embryonic axis in *Fucus* zygotes. *Science*, **239**, 187–190.

265. Howe, A., Aplin, A. E., Alahari, S. K., and Juliano, R. L. (1998) Integrin signaling and cell growth control. *Curr. Opin. Cell Biol.*, **10**, 220–231.

266. Arai, R., Nakano, K., and Mabuchi, I. (1998) Subcellular localization and possible function of actin, tropomyosin and actin-related protein 3 (Arp3) in the fission yeast *Schizosaccharomyces pombe*. *Eur J. Cell Biol.*, **76**, 288–295.

267. Eng, K., Naqvi, N. I., Wong, K. C., and Balasubramanian, M. K. (1998) Rng2p, a protein required for cytokinesis in fission yeast, is a component of the actomyosin ring and the spindle pole body. *Curr. Biol.*, **8**, 611–621.

268. McCollum, D., Balasubramanian, M. K., Pelcher, L. E., Hemmingsen, S. M., and Gould, K. L. (1995) *Schizosaccharomyces pombe* cdc4+ gene encodes a novel EF-hand protein essential for cytokinesis. *J. Cell Biol.*, **130**, 651–660.

269. Motegi, F., Nakano, K., Kitayama, C., Yamamoto, M., and Mabuchi, I. (1997) Identification of Myo3, a second type-II myosin heavy chain in the fission yeast *Schizosaccharomyces pombe*. *FEBS Lett.*, **420**, 161–166.

270. Naqvi, N. I., Eng, K., Gould, K. L., and Balasubramanian, M. K. (1999) Evidence for F-actin-dependent and -independent mechanisms involved in assembly and stability of the medial actomyosin ring in fission yeast. *EMBO J.*, **18**, 854–862.

271. Chang, F., Drubin, D., and Nurse, P. (1997) cdc12p, a protein required for cytokinesis in fission yeast, is a component of the cell division ring and interacts with profilin. *J. Cell Biol.*, **137**, 169–182.

272. Petersen, J., Nielsen, O., Egel, R., and Hagan, I. M. (1998) FH3, a domain found in formins, targets the fission yeast formin Fus1 to the projection tip during conjugation. *J. Cell Biol.*, **141**, 1217–1228.

273. Wasserman, S. (1998) FH proteins as cytoskeletal organizers. *Trends Cell Biol.*, **8**, 111–115.

274. Fankhauser, C., Reymond, A., Cerutti, L., Utzig, S., Hofmann, K., and Simanis, V. (1995) The *S. pombe* cdc15 gene is a key element in the reorganization of F-actin at mitosis. *Cell*, **82**, 435–444 [published erratum appears in *Cell* (1997), **89**, (7), 1185].

275. Balasubramanian, M. K., McCollum, D., Chang, L., Wong, K. C., Naqvi, N. I., He, X., Sazer, S., and Gould, K. L. (1998) Isolation and characterization of new fission yeast cytokinesis mutants. *Genetics*, **149**, 1265–1275.

276. Demeter, J. and Sazer, S. (1998) imp2, a new component of the actin ring in the fission yeast *Schizosaccharomyces pombe*. *J. Cell Biol.*, **143**, 415–427.

277. Petersen, J., Weilguny, D., Egel, R., and Nielsen, O. (1995) Characterization of fus1 of *Schizosaccharomyces pombe*: a developmentally controlled function needed for conjugation. *Mol. Cell Biol.*, **15**, 3697–3707.

278. Mitchison, J. M. and Nurse, P. (1985) Growth in cell length in the fission yeast *Schizosaccharomyces pombe*. *J. Cell Sci.*, **75**, 357–376.

279. Matsusaka, T., Hirata, D., Yanagida, M., and Toda, T. (1995) A novel protein kinase gene ssp1+ is required for alteration of growth polarity and actin localization in fission yeast. *EMBO J.*, **14**, 3325–3338.

280. Rupes, I., Jia, Z., and Young, P. G. (1999) Ssp1 promotes actin depolymerization and is involved in stress response and new end take-off control in fission yeast. *Mol. Biol. Cell*, **10**, 1495–1510.

281. Verde, F. (1998) On growth and form: control of cell morphogenesis in fission yeast. *Curr. Opin. Microbiol.*, **1**, 712–718.

282. Lew, D. J. and Reed, S. I. (1995) Cell cycle control of morphogenesis in budding yeast. *Curr. Opin. Genet. Dev.*, **5**, 17–23.

283. Blacketer, M. J., Koehler, C. M., Coats, S. G., Myers, A. M., and Madaule, P. (1993) Regulation of dimorphism in *Saccharomyces cerevisiae*: involvement of the novel protein kinase homolog Elm1p and protein phosphatase 2A. *Mol. Cell Biol.*, **13**, 5567–5581.

284. Tjandra, H., Compton, J., and Kellogg, D. (1998) Control of mitotic events by the Cdc42 GTPase, the Clb2 cyclin and a member of the PAK kinase family. *Curr. Biol.*, **8**, 991–1000.

285. Sreenivasan, A. and Kellogg, D. (1999) The elm1 kinase functions in a mitotic signaling network in budding yeast. *Mol. Cell Biol.*, **19**, 7983–7994.

286. Gilbreth, M., Yang, P., Wang, D., Frost, J., Polverino, A., Cobb, M. H., and Marcus, S. (1996) The highly conserved skb1 gene encodes a protein that interacts with Shk1, a fission yeast Ste20/PAK homolog. *Proc. Natl Acad. Sci., USA*, **93**, 13802–13807.

287. Gilbreth, M., Yang, P., Bartholomeusz, G., Pimental, R. A., Kansra, S., Gadiraju, R., and Marcus, S. (1998) Negative regulation of mitosis in fission yeast by the shk1 interacting protein skb1 and its human homolog, Skb1Hs. *Proc. Natl Acad. Sci., USA*, **95**, 14781–14786.

288. Verde, F., Wiley, D. J., and Nurse, P. (1998) Fission yeast orb6, a ser/thr protein kinase related to mammalian rho kinase and myotonic dystrophy kinase, is required for maintenance of cell polarity and coordinates cell morphogenesis with the cell cycle. *Proc. Natl Acad. Sci., USA*, **95**, 7526–7531.

289. Kemp, C. J. (1999) You don't need a backbone to carry a tumour suppressor gene. *Nature Genet.*, **21**, 147–148.

290. Hagan, I. and Yanagida, M. (1990) Novel potential mitotic motor protein encoded by the fission yeast cut7+ gene. *Nature*, **347**, 563–566.

291. Schmidt, S., Sohrmann, M., Hofmann, K., Woollard, A., and Simanis, V. (1997) The Spg1p GTPase is an essential, dosage-dependent inducer of septum formation in *Schizosaccharomyces pombe*. *Genes Dev.*, **11**, 1519–1534.

292. Sohrmann, M., Schmidt, S., Hagan, I., and Simanis, V. (1998) Asymmetric segregation on spindle poles of the *Schizosaccharomyces pombe* septum-inducing protein kinase Cdc7p. *Genes Dev.*, **12**, 84–94.

293. He, X., Patterson, T. E., and Sazer, S. (1997) The *Schizosaccharomyces pombe* spindle checkpoint protein mad2p blocks anaphase and genetically interacts with the anaphase-promoting complex. *Proc. Natl Acad. Sci., USA*, **94**, 7965–7970.

294. Mösch, H.-U., Roberts, R. L., and Fink, G. R. (1996) Ras2 signals via the Cdc42/Ste20/mitogen-activated protein kinase module to induce filamentous growth in *Saccharomyces cerevisiae*. *Proc. Natl Acad. Sci.*, **93**, 5352–5356.

295. Snyder, M. (1989) The SPA2 protein of yeast localizes to sites of cell growth. *J. Cell Biol.*, **108**, 1419–1429.

296. Philips, J. and Herskowitz, I. (1998) Identification of Kel1p, a kelch domain-containing protein involved in cell fusion and morphology in *Saccharomyces cerevisiae*. *J. Cell Biol.*, **143**, 375–389.

297. Becker, W., Weber, Y., Wetzel, K., Eirmbter, K., Tejedor, F. J., and Joost, H. G. (1998) Sequence characteristics, subcellular localization, and substrate specificity of DYRK-related kinases, a novel family of dual specificity protein kinases. *J. Biol. Chem.*, **273**, 25893–25902.

298. Nakano, K. and Mabuchi, I. (1995) Isolation and sequencing of two cDNA clones encoding Rho proteins from the fission yeast *Schizosaccharomyces pombe*. *Gene*, **155**, 119–122.

299. Mellor, H. and Parker, P. J. (1998) The extended protein kinase C superfamily. *Biochem. J.*, **332**, 281–292.

300. Michelitch, M. and Chant, J. (1996) A mechanism of Bud1p GTPase action suggested by mutational analysis and immunolocalization. *Curr. Biol.*, **6**, 446–454.

301. Garcia-Ranea, J. A. and Valencia, A. (1998) Distribution and functional diversication of the ras superfamily in *Saccharomyces cerevisiae*. *FEBS Lett.*, **434**, 219–225.

302. Robinson, N. G., Guo, L., Imai, J., Toh-e, A., Matsui, Y., and Tamanoi, F. (1999) Rho3 of *Saccharomyces cerevisiae*, which regulates the actin cytoskeleton and exocytosis, is a GTPase which interacts with Myo2 and Exo70. *Mol. Cell Biol.*, **19**, 3580–3587.

# 3 | Cell polarity in ciliates

JOSEPH FRANKEL

## 1. Two concepts of polarity

As pointed by Wolpert (1), the term 'polarity' is used by biologists in two different senses (Fig. 1). One is the expression of different spatially segregated qualities or attributes. These might characterize the opposite ends of a system, such as the budding and non-budding poles of a yeast cell (2), or else they might partition the entire system into two relatively homogeneous regions, such as the apical and basolateral domains of an epithelial cell (3). I will call this *domain polarity*. The other sense embodies the notion of a direction, or vector, that pervades the entire system. It is this latter sense that Crick had in mind when he stated that 'a system of polarities ... necessarily implies a vector field' (4, p. 436). I will call this *vectorial polarity*.

Vectorial polarity can come about in two ways, as was stated formally by Crick: 'Whereas one can always derive a vector field from a scalar one, the converse is not always true' (4, p. 436). Vectorial polarity can be an expression of a morphogenetic gradient (a scalar field). The polarity of bristles in an insect segment is one possible example; the bristles might point 'downhill' on a segmentally reiterated gradient (5; but see ref. 6). However, as Crick pointed out, one also can have a vector field, hence vectorial polarity, in a homogeneous system without scalar variation; a classic physical example is a magnet.

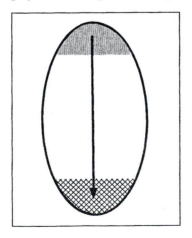

**Fig. 1** Vectorial and domain polarity. A hypothetical elliptical cell (or organism) is shown. The two polar domains are represented by shaded and hatched areas. The arrow indicates the direction of vectorial polarity. For further explanation, see the text.

These two types of polarity are, of course, not mutually exclusive in real biological systems: polar domains can be generated by, or themselves generate, a vectorial polar system. However, the distinction is useful in thinking about polarity in complex unicellular systems, such as ciliates.

# 2. Ciliate organization

## 2.1 Ciliate cortical topology

Ciliates are conspicuously polar in their cell-surface, or cortical, region. They characteristically possess specialized structural domains, which in the simpler ciliates are located near the ends of the cell. In addition, the cell surface of all ciliates manifests vectorial polarity in two orthogonal directions, anteroposterior and circumferential.[1] Since the third orthogonal axis (inside–outside) is fixed, ciliates thereby are asymmetrical objects (Fig. 2). The direction of their asymmetry (or handedness) could, in principle, be reversed by inverting either the anteroposterior or circumferential polar axis (but not the two together).

Ciliates show no trace of bilateral symmetry. Partially for this reason, they typically lack clearly defined dorsal and ventral surfaces, although the oral side of a ciliate is often informally called 'ventral' and the aboral side 'dorsal'. The ciliates' 'right' and 'left' therefore are defined with respect to the internal–external axis shown in Fig. 2, not the more familiar, but less appropriate, dorsal and ventral axis; more

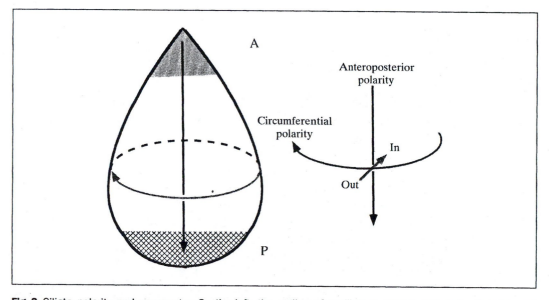

**Fig. 2** Ciliate polarity and asymmetry. On the left, the outline of a ciliate is shown, with polar domains and anteroposterior (A-P) vectorial polarity as in Fig. 1. An orthogonal system of circumferential polarity is superimposed on this. The scheme on the right adds the third cell-external ('out') to cell-internal ('in') axis orthogonal to the other two axes, completing the formal characterization of ciliate asymmetry. For further explanation, see the text.

formally, 'right and left are defined in terms of the observer's right and left, assuming that he stands inside the animal, lines up anteroposteriorly with the animal, and faces the cortical region being examined' (7, p. 233). Ciliatologists violate this rule when describing certain ciliates that are flattened and have prominent oral ('ventral') and aboral ('dorsal') surfaces (8, p. 63), but this need not concern us here.

Ciliate cortical anatomy consists of a repetitive cytoskeletal array on which discrete cortical landmarks are superimposed. The cytoskeletal array typically is made up of ciliary units centred around basal bodies, which are organized into longitudinal ciliary rows. The ciliary units contain accessory structures (fibrillar rootlets and microtubule bands) which are oriented around the basal bodies in a stereotyped manner (Figs 3A and 6). Normally, all of the ciliary units are similarly oriented (Fig. 3A). Hence each ciliary unit, as well as the ciliary array as a whole, manifests a uniform anteroposterior and circumferential polarity.

This array perpetuates itself longitudinally during growth of the ciliate clone by the addition of new ciliary units along the axis of the ciliary row. New basal bodies form at a uniform location anterior to old ones (9) and then follow a uniform path to the cell surface (10), so that the pre-existing polarities of the basal body and its accessory structures are conserved. This conservation extends to the ciliate clone, because cell division is transverse, cutting across the ciliary rows (Figs 3B, 5C, and 5D). Therefore, the polarities of the cytoskeletal array are the same in the two daughter cells derived from the division of a single cell.

The microtubules that make up much of the cortical cytoskeletal array described above are highly stable; the array persists in the cold and is resistant to depolymerizing agents which cause the much more labile internal microtubules to disappear (11–13).

Ciliates also possess distinctive cortical landmarks superimposed upon the more uniform cytoskeletal array. These landmarks are produced anew in every cell cycle, and then are partitioned between daughter cells in a stereotyped manner. Figure 4A shows this abstractly in a hypothetical ciliate with an anterior landmark (X) and a different posterior landmark (O). Before the ciliate divides, it produces new anterior and posterior landmarks for the daughter cells in the mid-region of the parent cell. This equatorial zone is thereby converted into two juxtaposed poles. The cell then divides between these two landmarks, producing two daughter cells that are tandem replicas of the parent cell. The tandem repetition of these specialized landmarks is thereby superimposed upon the uniformity of the ciliary rows.

As first noted by Vance Tartar (14), topologically a ciliate clone is a cylinder (Fig. 4B). Ciliate growth involves the elongation of this 'clonal cylinder', whereas ciliate division is a periodic segmentation of that cylinder.

Given this topology, the vectorial polarity of the cytoskeletal array can potentially be propagated indefinitely. The ciliary rows do not arise de novo in ciliates, with the exception of certain ciliates that form dedifferentiated cysts in which all ciliary structures regress on encystment and reform upon excystment (15). Even in these ciliates there is some question of whether or not templates for these rows are conserved in some form within the apparently dedifferentiated cyst (16).

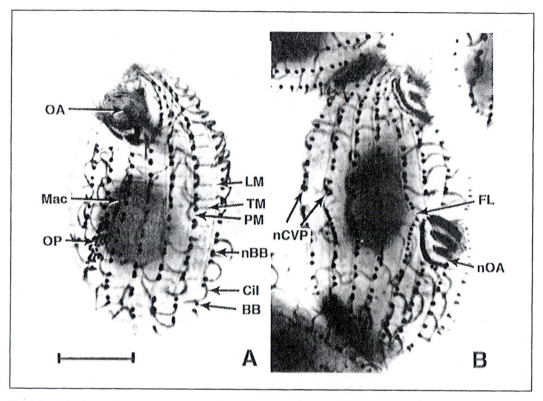

**Fig. 3** Protein-silver preparations of *Tetrahymena thermophila* at an early (A) and a late (B) stage of preparation for cell division. (A) A cell beginning cortical development. The macronucleus (Mac) and the microtubular structures of the cell surface, including cilia (Cil), basal bodies (BB), and longitudinal (LM), transverse (TM), and postciliary (PM) microtubule bands, are stained. Fibrillar rootlets (Fig. 6, FR) are not stained by this method. Unciliated nascent basal bodies (nBB) are visible immediately anterior to ciliated basal bodies. Seven ciliary rows are in focus. Note that all of the ciliary units have uniform polarities. An oral apparatus (OA) is seen (out of focus) near the anterior end of the cell, while the primordium of the new oral apparatus (OP) is visible as a small field of unciliated basal bodies to the left (viewer's right) of the right-postoral ciliary row. (B) A cell that has nearly completed development of a new oral apparatus (nOA) derived from the OP, and formed a fission line (FL). Note the new contractile vacuole pores (nCVP) located next to basal bodies just anterior to the fission line, on the fifth and sixth ciliary rows counting from the right-postoral row as number 1. The scale bar in the left micrograph indicates 10 μM and applies to both photographs.

In contrast, the cortical landmarks symbolized by the Xs and the Os in Fig. 4A do not display any obvious direct continuity. This is a point that we will take up further below, after we describe a real ciliate.

## 2.2 Cortical topography of *Tetrahymena thermophila* and other ciliates

The ciliate that I will describe is *T. thermophila*, the simplest well-studied ciliate (Figs 3, 5). The most prominent anterior cortical landmark is a complex oral apparatus (OA), consisting of multiple aligned rows of closely spaced ciliary units making up

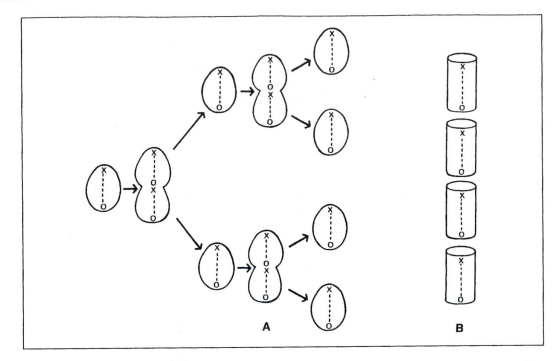

**Fig. 4** A highly schematic representation of ciliate clonal organization. (A) Two successive division cycles. The vertical dashed line indicates a ciliary row, X indicates an anterior cortical landmark, and O indicates a posterior cortical landmark. (B) A four-cell clone of (A) is represented as a clonal cylinder. From Joseph Frankel (text) and Jane A. Otto (illustrations), *Pattern formation: ciliate studies and models*, Fig. 2.2, © Oxford University Press, 1989, with permission.

three 'membranelles' (M1, M2, M3) on the left (viewer's right) and one 'undulating membrane' (UM) on the right (Fig. 5A). Two ciliary rows, the 'right-postoral' and 'left-postoral', subtend the OA (Fig. 5). One of the two posterior landmarks, the cell anus or cytoproct (Cyp), typically is located next to the posterior end of the right-postoral ciliary row. A pair of contractile vacuole pores (CVPs) is located several rows to the right of the right-postoral ciliary row (Fig. 3B); these are the outlet pores of the single osmoregulatory organelle of the cell, the contractile vacuole (not shown).

Cortical development is initiated by the formation of a field of basal bodies to the left of the mid-region of the right-postoral ciliary row (Figs 3A and 5B). This is the oral primordium (OP), which subsequently goes through an elaborate process of development into the membranelles and undulating membrane of the OA destined for the posterior division product. As these structures are being completed, a fission line appears in the form of aligned equatorial gaps in the ciliary rows (Figs. 3B and 5C). New CVPs are formed immediately anterior to these gaps (Fig. 3B). A new contractile vacuole appears internal to the pores at about the same time that the pores appear on the surface. The cell then begins to cleave along the fission line, at which

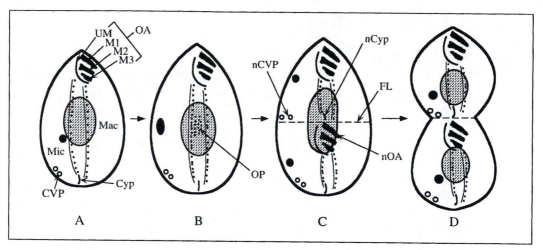

**Fig. 5** Structural features of the oral ('ventral') side of *Tetrahymena* at different stages of the cell-division process. The micronucleus (Mic) and macronucleus (Mac) are shown in each of these diagrams, as well as the two postoral ciliary rows, and three major cortical landmarks, the oral apparatus (OA), cytoproct (Cyp), and contractile vacuole pores (CVP). (A) A non-developing cell. The four major compound structures that make up the oral apparatus, the three membranelles (M) and the undulating membrane (UM), are separately labelled. (B) A cell that has begun micronuclear mitosis and formation of an oral primordium (OP). (C) A cell in which the micronucleus has divided and a fission line (FL) has formed with a new cytoproct (nCyp) and new contractile vacuole pores (nCVP) immediately anterior to the FL. The new oral apparatus (nOA) is almost fully developed. (D) A cell that has completed macronuclear division and is undergoing cell division. For further explanation, see the text.

time a new Cyp is formed anterior to the oral apparatus (Fig. 5C). The cell divides into an anterior daughter with the old OA and new CVPs and Cyp, and a posterior daughter with the new OA and old CVPs and Cyp. Internally, the diploid micronucleus divides mitotically while oral development is being completed, whereas the compound macronucleus divides non-mitotically during the first half of cytokinesis (Fig. 5).

Although patterns of ciliate cortical development are very varied (see ref. 8, Chapter 2 for a brief account, ref. 17 for an extended one), the essential pattern described above, with spatially discontinuous development of cortical landmarks superimposed on continuous extension of ciliary rows, is widespread. Two variations observed in well-studied ciliates deserve mention. In *Paramecium* and its relatives, new oral structures form adjacent to the corresponding old ones and then get to their final positions by differential growth or migration (18). In the hypotrich ciliates, including well-studied genera such as *Euplotes*, *Oxytricha*, and *Paraurostyla*, the ciliature on the oral ('ventral') surface consists not of simple ciliary rows but rather of rows of more complex ciliary structures known as cirri, which are replaced at division by new cirri which may or may not be derived from the old ones (19–22). The account below will centre on *Tetrahymena* and *Paramecium*, in which most cell-biological and genetic studies have been done, but will occasionally bring in relevant information drawn from other ciliates.

## 3.  Anteroposterior polarity

### 3.1  Vectorial polarity of the ciliary rows

Since ciliary rows grow longitudinally by addition of new ciliary units (Fig. 3A) and are bisected transversely (Fig. 3B, 5C), they are potentially immortal. If so, then even genetically identical clones with pre-existing differences in the number of ciliary rows ought to be able to propagate these differences. This has been shown in *Tetrahymena* (23, 24) and in *Euplotes* (25), although the fidelity of this propagation is not perfect (23, 25). Furthermore, ciliary rows sometimes can be gained or lost systematically. In *Tetrahymena*, new ciliary rows can arise from the undulating membrane of the oral apparatus under certain circumstances (26). In *Paramecium*, ciliary rows are newly formed on the right side of the oral apparatus and resorbed on the left side, resulting in a gradual clockwise (as seen from the anterior end) shift of individual ciliary rows in successive cell generations (18 and references cited therein).

The autonomous polarity of the ciliary rows was demonstrated most spectacularly by the propagation of inverted (180°-rotated) ciliary rows for thousands of cell generations in *Paramecium tetraurelia* (27) and for hundreds of generations in *Tetrahymena thermophila* (28) (Fig. 6). The internal organization of the inverted ciliary rows is identical in every way to that of normally oriented rows in *Paramecium* (29) and *Tetrahymena* (30); when normal and inverted ciliary rows are seen side by side, one sometimes can not tell which row is inverted and which is normally oriented. In keeping with this intrinsic identity, new basal bodies are formed at identical positions in both normal and inverted ciliary rows (i.e. to the cell's anterior of old basal bodies in normally oriented ciliary rows (Fig. 3A) and to the cell's posterior of old basal bodies in inverted rows).

Ciliary-row inversions generally arise as a result of an injection of a ciliary row from one cell into a conjoined partner that was bent into an opposite anteroposterior organization (see ref. 8, pp. 75–6 for details). Relevant genic differences between clones that bear ciliary-row inversions and those that do not were excluded in *Paramecium* (27) and are very unlikely in *Tetrahymena* (28).

The influence of the vectorial polarity of ciliary rows extends beyond the ciliary row itself. Ciliary rows exert a determinative influence on the 'fine-positioning' of the CVPs, as these structures are formed to the posterior-left of a basal body in a normally oriented ciliary row of *Tetrahymena* (Fig. 3B) but to the anterior-right of a basal body in an inverted row (31). Ciliary rows also can exert a strong influence on the positioning of cortically located mitochondria in this organism (32, 33). In addition, the ciliary rows indirectly influence the location of the neighbouring longitudinal microtubule bands (LM in Figs 3A and 6B), which themselves are polar and self-propagating (see ref. 8, pp. 77–9, and references cited therein). Finally, the orientation of ciliary rows influences the form of other structures on which they impinge, such as the anterior suture of *Paramecium* (34) and the structurally comparable anterior tip of *Tetrahymena* (30, 35).

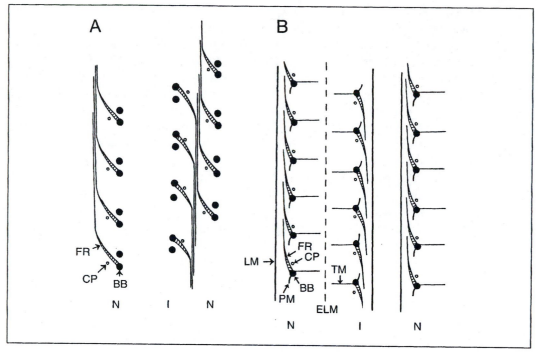

**Fig. 6** Normal (N) and inverted (I) ciliary rows of *Paramecium tetraurelia* (A) and *Tetrahymena thermophila* (B). Basal bodies (BB), fibrillar rootlets (FR), and coated pits (CP) are shown in both diagrams, while the transverse (TM), postciliary (PM), and longitudinal (LM) microtubule bands are drawn only in diagram (B). Cilia are omitted. The vertical dashed line in diagram (B) indicates the location of an extra longitudinal microtubule band (ELM) that is sometimes formed to the cell's right of an inverted ciliary row. Diagram (A) is drawn using information from Sonneborn (36); diagram (B) is redrawn with additions and modifications from Aufderheide, K.J., Frankel, J., and Williams, N.E. (1980) Formation and positioning of surface-related structures in protozoa. *Microbiological Reviews*, **44**, 252-302, Fig. 6, © 1980, American Society of Microbiologists, with permission. Both are reproduced with minor changes from Fig. 4.3 of Joseph Frankel (text) and Jane A. Otto (illustrations) *Pattern formation: ciliate studies and models*, © 1989, Oxford University Press, with permission.

What is it that maintains the self-propagating vectorial polarity of the ciliary rows? Sonneborn (36) argued that this polarity is determined intrinsically by the relevant pre-existing cell structures. Thus, the pre-existing basal body might first specify the location at which the new basal body is formed.[2] Then the new basal body might determine the location of the accessory structures that develop in its neighbourhood. Some of these accessory structures, in particular the overlapping fibrillar rootlets, might maintain the longitudinal integrity of the ciliary row (Fig. 4B).

None the less, some cellular influence must govern the orientation of new ciliary rows, such as those that arise from basal body couplets of the oral undulating membrane in *Tetrahymena* (26) and the new ciliary rows that '. . . must appear along the [oral] anarchic field . . .' (18, p. 165) in *Paramecium*. External factors also can influence the propagation of established ciliary rows. In *Tetrahymena*, in which the perpetuation of ciliary-row inversions is somewhat less stable than in *Paramecium*, basal-body proliferation is less extensive in inverted ciliary rows than in nearby normally oriented rows (37), and inverted ciliary rows may sometimes even be re-inverted spontaneously (38). Additionally, mutations may disrupt the fidelity of inheritance of differences in ciliary row numbers (*bbd* in *Euplotes*; 39), generate localized disorder resulting in ciliary-row inversions (*kin241* (40) and *crochu* (41) in *Paramecium*), or grossly disrupt ciliary rows (*disA* in *Tetrahymena* (35)). Such effects may (*kin241*, *disA*) or may not (*crochu*) be associated with detectable abnormalities in structural components of the ciliary units themselves. *kin241* is particularly interesting because of its high degree of pleiotropy, with effects both on ciliary units as well as on cortical domains. These results indicate that '. . . the geometry of basal body and cortical unit duplication, although constrained within the pre-existing unit organization, also depends on other global physiological parameters' (40).

The molecular basis for the vectorial polarity of ciliary rows is not well understood, although the recent demonstration that inactivation of γ-tubulin inhibits basal-body duplication in *P. tetraurelia* (42) provides an important clue. Also, there are indications that the *crochu* mutation may affect a microtubule-associated protein, perhaps a kinase or phosphatase (41).

## 3.2  Domain polarity: the fission zone

The new structures that develop near the fission line all form next to ciliary rows. We have already seen that the fine-positioning of the CVPs is controlled by the geometry of the basal bodies near which they develop; it is likely that the same holds true for the oral primordium and the cytoproct as well. So where do new CVPs develop when a ciliary row is inverted? Imagine for a moment that the vectorial polarity of the ciliary rows were a reflection of intrinsic differences in the basal bodies that make up each ciliary row. Then, in a normally oriented ciliary row, the unit at the anterior end of the row would be inherently number 1 and the unit at the posterior end would be number 10. If we were to invert that row, unit number 1 would be at the posterior end of the row and number 10 would be at the anterior end. If a new CVP were always to form next to unit number 10, then it would form near the posterior end of a normally oriented ciliary row, and near the anterior end of an inverted row, as shown in Fig. 7B. But this is not what happens. The CVP always forms near the basal body just anterior to the fission line, both in normally oriented and in inverted ciliary rows (Fig. 7C) (7). While the basal body that is assigned to induce a CVP decides where that CVP will form in relation to its internal geometry, something external to the ciliary row decides *which* basal body of a row is to generate a CVP. That external factor is presumably something within the newly generated posterior domain located just anterior to the fission line.

This same supremacy of cellular domain polarity over cytoskeletal vectorial polarity has been demonstrated in *Tetrahymena* for the progression of a wave of active ciliary outgrowth just posterior to the fission line (43). It therefore is not surprising that a major disturbance in the regularity of ciliary rows in the *disorganized* mutant of *T. thermophila* has only a minor effect on the formation of the fission line and on the

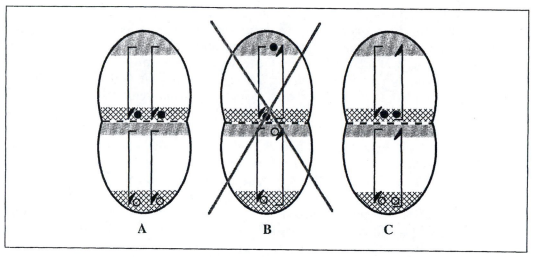

**Fig. 7** A schematic illustration of the positioning of contractile vacuole pores (CVPs) in normally oriented and inverted (180°-rotated) ciliary rows in *Tetrahymena* cells in division. In each diagram, the vertical lines indicate ciliary rows, with their polarities represented by the unilateral arrowheads and horizontal bars. Open circles indicate old CVPs, filled circles indicate new CVPs formed just anterior to the fission line, which is represented by a dashed horizontal line. The anterior end of the cell is oriented upwards on the page. Polar domains are marked as in Figs 1 and 2. (A) Cells with CVPs formed adjacent to normally oriented CVP-rows. (B, C) Cells with CVPs formed next to one inverted (180°-rotated) and one normally oriented ciliary row. (B) The prediction if the position of the CVP along the anteroposterior axis is controlled by the intrinsic polarity of the adjacent ciliary row. The CVP would develop next to the intrinsic posterior end of an inverted ciliary row, which now is at the anterior end of the cell. This is NOT observed. (C) The observed positioning of CVPs. New CVPs develop just anterior to the fission zone, irrespective of the orientation of the ciliary rows next to which they develop.

positioning of cortical landmarks near that line (35). Thus the polarity of the ciliary units in this organism is purely vectorial; whatever differences arise among the basal bodies of ciliary rows are dictated from the outside.

Similar conclusions can also be drawn for other ciliates. In *Paramecium*, the distribution of two kinds of ciliary units (one-basal-body and two-basal-body) '. . . is the same regardless of whether the row is inverted or normally oriented' (36, p. 360). In the large ciliate *Stentor coeruleus*, the oral pouch, which develops at one end of an elongated membranellar band, is formed at the end closest to the tail-pole of the cell, even in bands that are artifically rotated through 180° (44, pp. 174–6). This and other evidence in *Stentor* led to the postulate of an anteroposterior (basal–apical) morphogenetic gradient, which becomes subdivided before cell division (45, 46). This hypothesis was supported by results of microsurgical studies on a highly elongated relative of *Stentor*, *Spirostomum ambiguum* (47; see also ref. 8, pp. 155–8). It is important to make clear that this putative gradient is separate from and superimposed upon the vectorial polarity of ciliary rows, although it could be involved in initiating and stabilizing that polarity.

In both *Tetrahymena* and *Paramecium*, promising beginnings have been made in genetic and molecular analysis of the tandem cellular subdivision. In *Tetrahymena*, the fission line itself is a filamentous structure (48, 49) known to contain actin (50) as well as profilin (51) and a putative fimbrin (52). Temperature-sensitive mutations at several 'cell-division-arrest' (*cda*, formerly *mo*) gene loci prevent formation of the fission line, affect its localization, or interfere with its constriction (53, 54). The spatial and temporal displacement of the fission line caused by the *cdaK1* mutant is enhanced by the kinase inhibitor 6-dimethylaminopurine (6-DMAP) (54).

The *cdaA1* mutation of *T. thermophila*, which prevents the formation of the fission line, brings about an alteration of electrophoretic mobility of a single protein seen in two-dimensional acrylamide gels, called 'p85' because of its apparent molecular weight of 85 kDa (55). Antibodies made against p85 have been localized by immuno-fluorescence to developing basal-body couplets situated just posterior to the fission zone (56). Both the couplets (57) and the p85 localization (56, 58) are absent in *cdaA1* cells kept at the restrictive temperature. Thus the *cdaA* locus may be involved in the maturation of polar domains that are required for fission-line formation. The probable specificity of this action is highlighted by the fact that this mutant does not affect the temporal periodicity of micronuclear divisions and macronuclear DNA replication (59, 60), while it profoundly influences the spatial localization of sub-sequent cortical morphogenetic events (61).

Other immunofluorescence studies indicate that one of the major proteins of the membrane–skeletal layer that underlies the cell membrane of *T. thermophila* partially disappears in the region just posterior to the fission zone in wild-type but not *cdaA1* cells (57). A subepiplasmic 64 kDa protein called 'fenestrin' appears in the same region at the same time (62). These changes are concomitant with cellular subdivision and formation of the fission line; no differences have been seen at earlier stages that might predict the future site of the fission line.

In *Paramecium*, in which localized cell growth and basal-body duplication are conspicuously associated with cell division, processes of cytoskeletal reorganization and duplication spread in a wave-like manner over the surface of dividing cells, initially proceeding in both directions from the fission line. Different cell-surface domains can be defined on the basis of differential spread of these waves of cyto-skeletal change (63). Some of the domain boundaries are altered by mutation of the *kin241* gene (40). Biochemically, a transient wave of phosphorylation of certain cytoskeletal structures is observed during cell division in *P. tetraurelia* (64). Also, changes have been observed in the spatial distribution of acetylated microtubules in dividing paramecia (65, 66). Two distinct theoretical models have been proposed to show how these changes might be consequences of a wave of $Ca^{2+}$ release spreading over the cell surface from the fission line (67–70). Thus far, oscillations of free cytosolic $Ca^{2+}$ have been observed in *P. tetraurelia* (71), but $Ca^{2+}$ waves have eluded detection.

A possibly related area of study concerns the effects of lithium ions on polarity and patterning in ciliates. Lithium has a long history in embryological research, starting with the first demonstration of vegetalization of sea-urchin embryos (72), and pro-ceeding to the large number of recent demonstrations of lithium-induced dorsal-ization and anterior truncation in amphibian embryos (reviewed in ref. 73). The molecular bases of the effects of lithium ions include inhibition of specific components of either (or both) the phosphoinositide (PI) cycle (74) and the Wnt–glycogen synthase kinase (GSK3)–β-catenin pathway (75); the two may be connected (73). The effects of lithium ions on ciliates therefore are of considerable interest because they may provide hints as to the developmental roles of these important pathways, one of which (the PI cycle) is intimately related to $Ca^{2+}$ mobilization (74).

Effects of lithium chloride have been studied in *Stentor*, *Paramecium*, and *Tetrahymena*. In *Stentor*, Tartar (76) reported that LiCl induced a broadening of apical regions and sometimes formation of siamese-twin doublets. Two other investigators claimed that LiCl caused a strengthening of the basal tendency in the putative basal–apical gradient (77, 78), but in my view this claim was insufficiently documented. Two more recent studies, on *Paramecium* and *Tetrahymena* respectively, failed to discover any interference by LiCl with the polarity of ciliary rows, with the proliferation of basal bodies (which was, indeed, enhanced by LiCl in *Tetrahymena*), or with the patterning of cortical landmarks (79, 80). However, in *Paramecium*, which (unlike *Tetrahymena*) grows extensively as it divides, LiCl brought about major disturbances in cytokinesis-associated growth, resulting in misshapen division products (79).

In both of the recent studies, the effects of LiCl were partially rescued by *myo*-inositol (and not, in the *Paramecium* study, by the inactive isomer, *epi*-inositol). However, a detailed study of phosphoinositides in *P. tetraurelia* failed to detect the central player in the classical PI cycle, inositol 1,4,5-trisphosphate (81), and evidence for presence of the Wnt–GSK3–β-catenin pathway in ciliates is non-existent. The roles (if any) of these or other signal-transduction pathways in ciliate polarity and patterning remain to be elucidated.

# 4. Circumferential polarity

## 4.1 The equipotentiality of ciliary rows

Whereas we have shown above that all basal bodies within a ciliary row are intrinsically alike, we have not yet demonstrated that all rows are alike. In view of the potentially indefinite continuity of ciliary rows within clones of some ciliates, we could imagine that the rows might become permanently differentiated. For example, the right-postoral ciliary row of *Tetrahymena*, which is the normal site of formation of both the oral primordium and the cytoproct (Fig. 5), could be uniquely specialized for the generation of these structures. This supposition was disproven by Nanney (82) when he showed that in certain *Tetrahymena* species oral primordia could arise adjacent to ciliary rows other than the right-postoral ciliary row. If there were a systematic direction of this 'cortical slippage' then every row would take its turn as the stomatogenic row. Such a systematic slippage was manifested in some of Nanney's lines, and even more consistently in the *hypoangular* mutant of *T. thermophila* (83).

The issue of the potential differentiation of ciliary rows assumed special poignancy when a serious claim was made that basal bodies in the unicellular alga *Chlamy-*

*domonas* contained DNA (84). However, this claim has been refuted (85) and subsequently withdrawn (86). It is now quite clear that whereas basal bodies might serve as structural templates, they are not informational entities in the conventional sense; it is furthermore clear that, except perhaps for certain structurally modified basal bodies (87, 58), all basal bodies in all ciliary rows of *Tetrahymena* are intrinsically alike. This is almost certainly true for most basal bodies of the ciliary rows of *Paramecium* as well (36, 63).

## 4.2  Contractile-vacuole-pore cytogeometry

We have not yet fully answered the question of which (two) basal bodies in a *Tetrahymena* cell are assigned the task of generating CVPs. We determined above (Section 3.2) that they must be the posteriormost basal bodies of ciliary rows, regardless of whether the rows are normally oriented or inverted. But we did not answer the question of which rows are chosen. We can see in Fig. 3B that the 'CVP rows' are set several rows to the right (viewer's left) of the postoral ciliary rows. How are these rows selected? Most simply, the cell could measure (or count) a fixed distance from a reference ciliary row, such as the right-postoral row.

This hypothesis was tested, and disproven, in a remarkably simple way. The location of the CVPs was observed in *Tetrahymena* cells with varying numbers of ciliary rows (88). In a 'standard' cell, with 18 ciliary rows, if we designate the right-postoral row as number 1 and count to the cell's right (clockwise as viewed from the anterior end of the cell), the two CVPs almost always develop next to rows 5 and 6 (Fig. 8A). However, in a cell with 22 ciliary rows the CVPs are usually generated next to rows 6 and 7 (Fig. 8B).

Examination of many such cells revealed a rough proportionality between the distance of the CVP-rows from the reference ciliary row and the total number of ciliary rows. As Nanney (88) pointed out, this might imply measurement by a 'central angle' of about 80°, with the vertex of this angle along a hypothetical longitudinal axis running through the centre of the cell from its anterior to its posterior end, one side of the angle along an imaginary line from the central axis to ciliary row 1 (the right-postoral row), and the other side from the central axis to the midpoint between the two CVPs (Figs 8A, B).

To test the 'central angle' hypothesis, Nanney took advantage of the fact that ciliates, unlike people, can propagate as back-to-back siamese twins without difficulty. Instead of having one set of cortical landmarks (OA, Cyp, and CVPs), a siamese-twin ciliate (known as a 'homopolar doublet' because the anteroposterior axes of the two components are aligned) has two sets of such landmarks, opposite one another. Asexual propagation is not a problem, as growth is longitudinal, and each 'semicell' can propagate itself as one longitudinal half of the 'clonal cylinder'. A cross-section through such a homopolar-doublet *Tetrahymena* is illustrated schematically in Fig. 8C. This cell has two reference ciliary rows, (1 and 1', bottom and top in the figure) delimiting two 'semicells' with 15 rows apiece (right and left in the figure). The key finding is that the 'central angle' governing the location of the

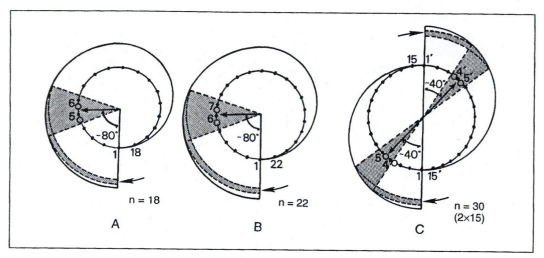

**Fig. 8** Schematic cross-sections representing the geometry of CVP formation around the cell circumference of *Tetrahymena*. The central circles studded with large dots represent cross-sections of *Tetrahymena* cells seen from the anterior, with the dots indicating ciliary rows. (A) and (B) depict singlet cells with 18 and 22 ciliary rows respectively, while (C) symbolizes a homopolar doublet cell with oral areas directly opposite one another and a total of 30 ciliary rows, 15 in each semicell. Ciliary rows are enumerated clockwise starting with the right-postoral ciliary row as number 1. The doublet in (C), with two right-postoral ciliary rows, therefore has two sets of ciliary rows, here numbered as 1 to 15 and 1' to 15', respectively. Nanney's central-angle formulation is shown inside the cells, while a reformulation in terms of thresholds within circumferential gradients is shown outside. The shaded region indicates the width of the area within which CVPs are formed and the corresponding range of values of the postulated gradient that determines formation of CVPs. Redrawn from Fig. 7 of Joseph Frankel, Pattern formation in ciliated protozoa, in *Pattern formation: a primer in developmental biology*, © 1984, Macmillan Publishing Company, with permission of the McGraw-Hill Companies, and reproduced with minor changes from Fig. 6.9 of Joseph Frankel (text) and Jane A. Otto (illustrations), *Pattern formation: ciliate studies and models*, © 1989, Oxford University Press, with permission.

two sets of CVPs is now cut in half, from 80° to 40° (Fig. 8C). This makes it rather unlikely that the cell simply measures a central angle to specify the positions of the CVPs.

How then is the CVP position determined? In comparing Fig. 8C to Figs 8A and 8B, note that what remains the same for positioning of CVPs in both singlets and doublets is the *relative* distance between successive reference (right-postoral) meridians. In singlets, this relative distance is an 80° arc, or 22% of the cell circumference. In homopolar doublets, this distance is an arc of only 40°. While this would be 11% of the cell circumference, it is 22% of the distance between reference meridians. The cell is measuring '. . . a fraction of the distance between one stomatogenic meridian and the next (or possibly between other cortical features) and regulates the field size in relation to that distance' (88, p. 316).

In Fig. 8, I have represented this regulation as being made by reading a narrow range of levels on a uniform circumferential gradient. This implies the existence in ciliates of a circumferential polarity, with poles that happen to be juxtaposed. This is unlike the longitudinal vectorial polarity of the ciliary rows in two fundamental ways: first, the polarity is based on a scalar field, and secondly, the concentration or intensity of a hypothetical substance or condition determines the site of formation of a specific set of cortical landmarks. As the CVPs are internally symmetrical, this example does not inform us of whether a (scalar) gradient can generate a (vectorial) polarity; we will see below (Section 4.4) an instance in which it does.

## 4.3 The locus of stripe contrast and the 'circumferential gradient' in *Stentor*

The hypothesis of a circumferential gradient was first proposed by Uhlig (45) to account for results of microsurgical experiments with the large ciliate, *Stentor coeruleus*. In *Stentor*, there is a structural gradient around the cell circumference, in the form of a progressive gradation in widths of visible pigment stripes that separate adjacent ciliary rows. The necessary geometrical corollary of this gradation is a longitudinal 'locus of stripe contrast' (89), along which the narrowest stripes abut on the broadest. Not only does this locus mark the site at which the oral primordium normally forms, but in numerous grafts it was shown to be necessary, and perhaps sufficient, for formation of an oral primordium (89, 44). Uhlig (but not Tartar) viewed this contrast-zone as a juxtaposed high (and low) point of a circumferential gradient, which Uhlig regarded as orthogonal to the putative basal–apical gradient that was mentioned earlier (for further description and critical discussion, see ref. 8, pp. 120–8).

There is no visible contrast-zone in most other ciliates, and I will present below an alternative formulation of circumferential polarity that is free of discontinuities and (in my view) is more useful in describing a broad range of phenomena.

## 4.4 The reversal of circumferential polarity: 'left-handed' cells

In Section 4.2, I introduced siamese-twin homopolar doublets in ciliates. Such doublets have been well known by ciliatologists for nearly a century (90, 91 and references cited therein), and were analysed extensively in *Paramecium* by Sonneborn (91), who showed that they did not differ genetically from normal singlet cells. In the overwhelming majority of these doublets, the two component 'semicells' have both the same anteroposterior polarity (which makes them 'homopolar') and the same circumferential polarity (Fig. 8C).

The first reported exception was a back-to-back *mirror-image* doublet in a hypotrich ciliate (90). Later on, side-by-side hypotrich mirror-image doublets were repeatedly generated by certain experimental manipulations (92–95). These doublets still have their anteroposterior axes aligned, hence technically still are 'homopolar doublets', *but* one of the two 'semicells' expresses a left–right reversal in the arrangement of cortical landmarks. This reversal of a *single* axis (the circumferential, or left–right axis) gives the two semicells opposite asymmetry or handedness. I refer to the normal semicell as 'right-handed' (RH) and the reversed semicell as 'left-handed' (LH).

As both moieties of a mirror-image hypotrich doublet are nucleated, it is a simple matter to cut them apart into two surviving fragments. The RH component then recovers and divides to produce a perfectly normal singlet clone. The LH component undergoes repeated morphogenetic reorganizations and eventually dies of starvation (96, 97). The reason for this difference is that the internal organization of the ciliary structures in the LH component of a mirror-image doublet is *not* reversed; instead it is either normally oriented or inverted (180°-rotated) (93, 98). The oral membranelles, which generate water currents used in feeding, are inverted (180°-rotated), but the oral pouch forms at the posterior end of the otherwise inverted oral structure (93, 95), perhaps because it is induced there by the unaltered basal–apical gradient. Therefore, in the LH component of a mirror-image doublet, food is swept anteriorwards, *away* from the mouth. The LH component can survive and reproduce itself as long as it is parabiotically fused to its normal RH partner, but can no longer nourish itself as soon as it is cut loose. Its repeated reorganizations are futile, because they always restore the pre-existing non-functional organization; the LH cell is trapped within its organizational straitjacket. Nuclear transplantation studies have demonstrated that the abnormality is not of nuclear origin (96). The inability of a normal nucleus to extricate an LH cell from its topological trap is a dramatic demonstration that DNA is not omnipotent in determining cellular organization.

Whereas LH singlets are unviable in the large and complex hypotrich ciliates, they have proven to be viable in the smaller ciliates, *T. thermophila* (99) and the closely related *Glaucoma scintillans* (100). Such LH singlets were derived microsurgically in *Glaucoma* from a mirror-image doublet (100), whereas in *Tetrahymena* they arose spontaneously within a culture containing regulating doublet cells (see below). There are two reasons why LH singlets can survive despite their inverted oral structures. First, in these ciliates the mouth opening is inverted concordantly with the oral membranelles (100, 101); secondly, (at least in *Tetrahymena*) cells in rich axenic media can grow even when the oral apparatus is non-functional, probably by carrier-mediated uptake through the cell membrane (102).

The difference in organization of LH and RH *Tetrahymena* cells is shown in Fig. 9, here using a schematic polar projection rather than a cross-section as in Fig. 8. Figure 9 summarizes the principal features of the LH organization:

(1) the LH and RH cells are mirror images with respect to the arrangement of cortical structures, for example, the CVPs are located clockwise with respect to the oral meridian in RH cells, counter-clockwise in LH cells;

(2) the LH organization, like the RH organization, is clonally propagated (101) despite the lack of a relevant genic difference between the two forms (99);

(3) the reversed arrangement of the cortical landmarks is superimposed on a normal cytoskeletal array of ciliary rows and associated structures; and

(4) oral structures are inverted (i.e. rotationally permuted), or sometimes disorganized, because of an incompatibility between the reversed global field, which influences the orientation of the oral membranelles, and the unaltered intrinsic organization of basal bodies and associated cytoskeletal structures (103, 104).

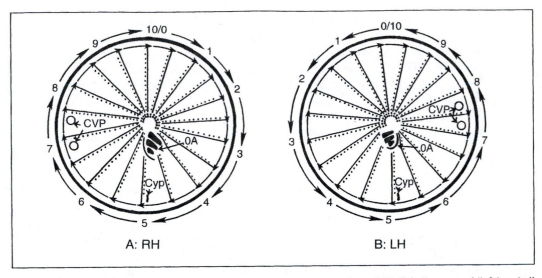

**Fig. 9** A polar representation of (A) normal 'right-handed' (RH) singlet cells and (B) globally reversed 'left-handed' (LH) singlet cells of *T. thermophila*. This diagram is a polar representation of the cell surface, with the anterior apex shown in the centre and the posterior pole at the periphery (much like a polar map of the Earth). The inner ring of small peripheral arrows represents the local circumferential polarity of the cortical cytoskeletal array, whereas the outer ring of larger arrows represents the polarity of the hypothetical global positional values, indicated by a sequence of numbers. CVP, contractile vacuole pore; Cyp, cytoproct; OA, oral apparatus. For further explanation, see the text. From Fig. 10.1 of Joseph Frankel (text) and Jane A. Otto (illustrations) *Pattern formation: ciliate studies and models*, © 1989, Oxford University Press, with permission.

This last abnormality indicates that the global circumferential order affects the polarity as well as the positioning of newly formed cytoskeletal arrays, much as the insect segmental gradient(s) influence polarity as well as position of bristles (5), albeit in complex ways (6).

One way of thinking about the global reversal is to imagine that the circumferential gradient depicted in Fig. 8 has undergone a reversal in slope. This interpretation is consistent with the observation that the relative distance between the oral meridian and the CVP pores is the same in LH as in RH cells, although oriented in the opposite direction (101). However, detailed observations on the genesis of reversed configurations of cortical landmarks in doublets of *Tetrahymena* (105) and of the hypotrich *Stylonychia mytilus* (106) strongly suggest that ciliates, like developing multicellular organisms (107, 108) do not tolerate positional discontinuities. These observations have stimulated us to adapt elements of the 'polar coordinate model' of French, Bryant, and Bryant (107) to the ciliate context, and replace the circumferential gradient with a seamless 'clockface' of positional values, whose direction of winding is reversed in LH cells (Fig. 9) (8, Chapter 9). One can then imagine that such an array of positional values could be propagated longitudinally during the growth of the ciliate 'clonal cylinder', a process that is mechanistically independent from the longitudinal propagation of the ciliary rows. This qualitative model can be applied to explain the otherwise counter-intuitive transitions from normal homopolar doublets to mirror-image triplet (105) and doublet (109) forms. This model and the observations that provoked it were used as the starting point for a much more rigorous 'vector-field' model of ciliate circumferential organization (110, 111) that could explicitly account for some of these phenomena. Unfortunately, the molecular basis of this field is still unknown.

## 4.5 Mutations affecting circumferential polarity

Three classes of mutations are known that affect circumferential polarity in *T. thermophila*. Their principal effects are depicted schematically, in simplified polar projections, in Fig. 10.

Two recessive mutant alleles have been selected at the *broadened cortical domain* (*bcd*) locus. These mutations bring about a broadening of the territories within which specific cortical landmarks can form; this phenotype is more severe at a higher temperature (112, 113). Mutations with similar broadening of cortical domains associated with a pervasive ciliary hypertophy have also been described extensively

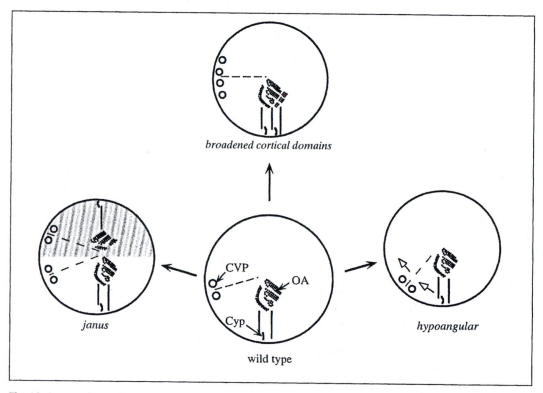

**Fig. 10** Schematic diagrams of wild-type and mutant *T. thermophila*. The pictorial convention here is a polar diagram as in Fig. 9, but with only the oral apparatus (OA), cytoproct (Cyp), contractile vacuole pores (CVP), postoral ciliary rows (solid lines), and CVP midpoints (dashed lines) shown. The shading indicates regions of reversed global handedness. The open arrows indicate the direction of slippage of newly formed structures during predivisional development. For further explanation, see the text.

in the hypotrich ciliate *Paraurostyla weissei* (114–118). An ingenious model postulating a cyclic instability of circumferential positional values has been proposed to account for aspects of this phenotype (115).

The *janus* mutations of *T. thermophila* include five known recessive alleles at three gene loci, all with a similar phenotype (119). Only one (*janB*) is temperature sensitive. In these mutants, a second set of 'ventral' structures, most notably a second oral apparatus and second set of CVPs, is maintained on the 'dorsal' side of the cell (120, 121). The second oral apparatus exhibits abnormalities similar to those that were subsequently analysed in more detail in wild-type LH cells (compare ref. 121 to ref. 103).

It would be easy to think of *janus* cells as back-to-back mirror-image doublets, but they are not. A study of the emergence of the janus phenotype after replacement of the wild-type (*janA*⁺) allele by homozygous *janA* showed that a broadening of the domain of CVP expression (Fig. 11B) precedes a split into two separate CVP domains (Fig. 11C). This is followed by the appearance of a second set of oral structures on the aboral ('dorsal') side (Fig. 11C) (122). During this entire process, the number of ciliary rows remains constant. The essence of the transformation from a wild-type to a janus organization, therefore, is the transformation of the aboral ('dorsal') surface of the cell to a mirror-image oral ('ventral') surface. This can easily be accounted for within the 'clock-face' coordinate system presented above (Fig. 11A). 'Dorsal' positional values that require gene products of the wild-type *janA*⁺ allele (also *janB*⁺, *jan C*⁺) are lost as the homozygous mutant allele comes to full expression. This loss generates a major positional discontinuity (Fig. 11B) that provokes intercalation of the remaining permitted values to form a cell with a mirror-image duplication of 'ventral' organization (Fig. 11C).

This representation generates a testable prediction: any 'dorsal' molecular marker

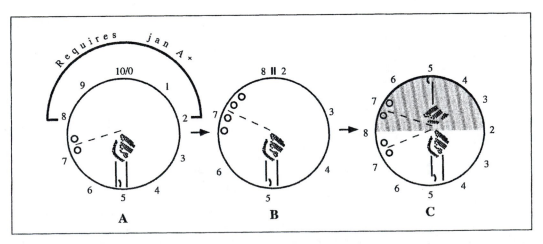

**Fig. 11** An interpretation of the conversion of a normal (RH) singlet *T. thermophila* into a *janus* mirror-image cell following a change of genotype from *janA*⁺/*janA*⁺ to *janA*/*janA*. Diagrammatic conventions are the same as in Fig. 10. The reverse-duplicated region is shaded. For further explanation, see the text.

should be lost in *janus* cells. Unfortunately, searches for a suitable marker, using first monoclonal antibodies and later a GFP-linked cDNA library, have thus far been unsuccessful.

The *hypoangular* phenotype (Fig. 10) has been detected in four separate non-complementing mutations, all of which are more strongly expressed at higher temperatures. The phenotype is complex (83), but its essence is a combination of a clockwise slippage of cortical structures in successive cell generations, and a diminution of the 'central angle' between the oral meridian and the CVPs. None the less, the location of the CVPs is still proportional to the total number of ciliary rows; all that has changed in the mutants is the constant of proportionality. Hence, it is reasonable to conclude that the *hpo* gene may encode a component of the system of intracellular positional information. This makes it especially important to know what this gene is and what it does at a molecular level.

## 5. The evolution of ciliate polarity

Polarity in ciliates is remarkably complex, as it is expressed in at least three levels. The first level is the stable cortical cytoskeletal array, which propagates its organization during the longitudinal growth of a ciliate clone. This array probably is maintained and perpetuated by mechanisms of 'directed assembly' acting over short distances (123), although other cellular factors may contribute. The second level is expressed in the specialized cortical domains found near the fission line, and often elsewhere as well. These domains are generated by the large-scale processes that subdivide the ciliate in every cell cycle. The third level is the heritable circumferential positional system that maintains the cell's cortical landmarks at their correct relative longitudes. While we can assume that the first level of ciliate polarity is associated with the cytoskeleton, we do not know the cellular substrata of the second and third levels; the plasma membrane and the underlying membrane skeleton (reviewed in ref. 124) are plausible candidates.

Considering the pervasiveness and complexity of ciliate polarity, it is not surprising that 30 years ago A. D. Hershey, referring to the early studies on *Paramecium* and on *Stentor*, wrote that 'polarity pervades the cell much as a magnetic field pervades space without the help of the iron filings that bring it to light' (125, p. 700). This is a far cry from polarity as it is now understood in simpler cells such as the yeasts. In these latter cells, polar axes are generated anew in each cell cycle in relation to localized cell-surface markers that are assembled close to pre-existing markers (126). The new markers re-orient the cytoskeleton (reviewed in refs 2, 127, 128). It therefore is not surprising that in yeasts the direction of polarity expressed by the cytoskeleton can change in each cell generation, which would be inconceivable in a ciliate.

Yeasts, however, are evolutionarily even more distant from ciliates than they are from humans (129, 130) and have lost certain primitive eukaryotic organelles, notably ciliary basal bodies. If we were to try to reconstruct a prototypical common ancestor of ciliates, fungi, and animals, it would probably be a small flagellated cell (131). The experimental model organism that most closely approximates the organization of this probable ancestor is the biflagellate unicellular alga, *Chlamydomonas*. Fortunately, cell polarity has been analysed carefully in this organism, notably by Holmes and Dutcher (132), who provided excellent evidence for control of cellular polarity and asymmetry by the intrinsic geometry of the basal-body pair. Furthermore, this system allows easy selection for mutants generating a reversal of cellular handedness, and several such screens failed to reveal any mutations exhibiting such a reversal (132). So far as we know, polarity in *Chlamydomonas* operates entirely through short-range directed assembly, essentially similar to the first level of polarity summarized above for ciliates (132, 133). Similar conclusions had been derived earlier from studies of recovery following disruption of the microtubule cytoskeleton of another small flagellate, *Ochromonas* (134).

Hence, the common ancestor of ciliates, yeasts, and ourselves may have had only the foundation of the first of the three systems that govern intracellular polarity. One can readily imagine that when the single ciliary unit of an ancestral flagellate became repetitively duplicated and arranged into linear assemblies such as ciliary rows, the inherent polarity of that single ciliary unit could become amplified into the propagated polarities of a ciliary row. The other two levels of ciliate polarity (the polar domains associated with equatorial subdivision and the propagated circumferential polarity) have no known precedent in simple flagellates. Thus they might be ciliate inventions evolved to cope with the major novelties of cellular design that arose in the course of ciliate evolution (135–138). Alternatively, there may be more to polarity in the simpler flagellates than has been uncovered in the studies carried out thus far.

This possible novelty, in my opinion, makes the ciliate system more, rather than less, intriguing. Global patterning mechanisms that operate via cell-extrinsic modes of signalling in multicellular organisms appear to have cell-intrinsic analogues (or homologues?) in ciliates (139). How did ciliates come to invent (or adapt) such mechanisms? This is a major challenge for future study. Perhaps the molecular cloning and analysis of expression of mutants that affect the ciliate 'body plan' (140), such as *kin241* in *Paramecium* or *hypoangular* in *Tetrahymena*, will provide some useful clues.

# References

1. Wolpert, L. (1971) Positional information and pattern formation. *Curr. Top. Dev. Biol.*, **6**, 183.
2. Bähler, J. and Peter, M. (2000) Cell polarity in yeast. In *Cell polarity*, (ed. D. Drubin). IRL Press, London.
3. Nelson, W. J., Yeaman, C. and Grindstaff, K. K. (2000) Spatial cues for cellular asymmetry in polarized epithelia. In *Cell polarity*, (ed. D. Drubin). IRL Press, London.
4. Crick, F. H. C. (1971) The scale of pattern formation. In: *Control mechanisms in growth and differentiaton*, 25th Symposium of the Society for Experimental Biology, (ed. D.D. Davies and M. Balls), p. 429. Cambridge University Press, Cambridge.
5. Lawrence, P. A. (1966) Gradients in the insect segment: the orientation of hairs in the milkweed bug. *J. Exp. Biol.*, **44**, 607.

6. Struhl, G., Barbash, D. A. and Lawrence, P. A. (1997) Hedgehog acts by distinct gradient and signal relay mechanisms to organise cell type and cell polarity in the *Drosophila* abdomen. *Development*, **124**, 2155.

7. Ng, S.F. (1977) Analysis of contractile vacuole pore morphogenesis in *Tetrahymena pyriformis* by 180° rotation of ciliary meridians. *J. Cell Sci.*, **25**, 233.

8. Frankel, J. (1989) *Pattern formation: ciliate studies and models.* Oxford University Press, New York.

9. Dippell, R. V. (1968) The development of basal bodies in *Paramecium. Proc. Natl. Acad. Sci. U.S.*, **61**, 461.

10. Allen, R. D. (1969) The morphogenesis of basal bodies and accessory structures of the cortex of the ciliated protozoan *Tetrahymena pyriformis. J. Cell Biol.*, **40**, 716.

11. Tamura, S., Tsuruhara, T., and Watanabe, Y. (1969) Function of nuclear microtubules in macronuclear division of *Tetrahymena pyriformis. Exp. Cell Res.*, **55**, 351.

12. Torres, A. and Delgado, P. (1989) Effects of cold and nocodazole treatments on the microtubular systems of *Paramecium* in interphase. *J. Protozool.*, **36**, 113.

13. Stargell, L. A., Heruth, D. P., Gaertig, J., and Gorovsky, M. A. (1992). Drugs affecting microtubule dynamics increase α-tubulin accumulation via transcription in *Tetrahymena thermophila. Molec. Cell Biol.*, **12**, 1443.

14. Tartar, V. (1962) Morphogenesis in *Stentor. Adv. Morphogen.*, **2**, 1.

15. Grimes, G. W. (1973a) Morphological discontinuity of kinetosomes during the life cycle of *Oxytricha fallax. J. Cell Biol.*, **57**, 229.

16. Hammersmith, R. L. and Grimes, G. W. (1981) Effects of cystment on cells of *Oxytricha fallax* possessing supernumerary dorsal bristle rows. *J. Embryol. Exp. Morph.*, **63**, 17.

17. Foissner, W. (1996) Ontogenesis in ciliated protozoa, with emphasis on stomatogenesis. In *Ciliates. Cells as organisms*, (ed. K. Hausmann and P. C. Bradbury), p. 95. Gustav Fischer Verlag, Stuttgart.

18. Iftode, F., Fleury, A., and Adoutte, A. (1997) Development of the surface pattern during division in *Paramecium*. III. Study of stomatogenesis in the wild type using antitubulin antibodies and confocal microscopy. *Europ. J. Protistol.*, **33**, 145.

19. Grimes, G. W. (1972) Cortical structure in nondividing and cortical morphogenesis in dividing *Oxytricha fallax. J. Protozool.*, **19**, 428.

20. Ruffolo, J. J., Jr (1976) Cortical morphogenesis during the cell division cycle in *Euplotes*: An integrated study using light optical, scanning electron and transmission electron microscopy. *J. Morphol.*, **148**, 489.

21. Jerka-Dziadosz, M. (1980) Ultrastructural study on development of the hypotrich ciliate *Paraurostyla weissei*. I. Formation and morphogenetic movements of ventral ciliary primordia. *Protistologica*, **16**, 571.

22. Fleury, A. (1991) Dynamics of the cytoskeleton during morphogenesis of the ciliate *Euplotes*. I. Basal-bodies related microtubular systems. *Europ. J. Protistol.*, **27**, 99.

23. Nanney, D. L. (1966) Corticotype transmission in *Tetrahymena. Genetics*, **54**, 955.

24. Frankel, J. (1980) Propagation of cortical differences in *Tetrahymena. Genetics*, **94**, 607.

25. Frankel, J. (1973) Dimensions of control of cortical patterns in *Euplotes*: The role of pre-existing structure, the clonal life cycle, and the genotype. *J. Exp. Zool.*, **183**, 71.

26. Nelsen, E. M. and Frankel, J. (1979) Regulation of corticotype through kinety insertion in *Tetrahymena. J. Exp. Zool.*, **210**, 277.

27. Beisson, J. and Sonneborn, T. M. (1965) Cytoplasmic inheritance of the organization of the cell cortex of *Paramecium aurelia. Proc. Natl Acad. Sci., USA*, **53**, 275.

28. Ng, S. F. and Frankel, J. (1977) 180° rotation of ciliary rows and its morphogenetic implications in *Tetrahymena pyriformis*. *Proc. Natl Acad. Sci., USA*, **74**, 1115.

29. Aufderheide, K. J., Rotolo, T. C., and Grimes, G. W. (1999) Analysis of inverted ciliary rows in *Paramecium*. Combined light and electron microscopical observations. *Europ. J. Protistol.*, **35**, 81.

30. Ng. S. F. and Williams, R. J. (1977) An ultrastructural investigation of 180°-rotated ciliary meridians of *Tetrahymena pyriformis*. *J. Protozool.*, **24**, 257.

31. Ng, S. F. (1979) The precise site of origin of the contractile vacuole pore in *Tetrahymena* and its morphogenetic implications. *Acta Protozool.*, **18**, 305.

32. Aufderheide, K. J. (1979) Mitochondrial associations with specific microtubular components of the cortex of *Tetrahymena thermophila*. I. Cortical patterning of mitochondria. *J. Cell Sci.*, **39**, 299.

33. Aufderheide, K. J. (1980) Mitochondrial associations with specific microtubular components of the cortex of *Tetrahymena thermophila*. II. Response of the mitochondrial pattern to changes in the microtubule pattern. *J. Cell Sci.*, **42**, 247.

34. Sonneborn, T. M. (1977) Local differentiation of the cell surface of ciliates: their determination, effects, and genetics. In *The synthesis, assembly, and turnover of cell surface components*, (ed. E G. Poste and G.L. Nicolson), *Cell Surface Reviews*, **4**, 829. Elsevier/North Holland, Amsterdam.

35. Jerka-Dziadosz, M., Jenkins, L. M., Nelsen, E. M., Williams, N. E., Jaeckel-Williams, R. J., and Frankel, J. (1995) Cellular polarity in ciliates: Persistence of global polarity in a *disorganized* mutant of *Tetrahymena thermophila* that disrupts cytoskeletal organization. *Dev. Biol.*, **169**, 644.

36. Sonneborn, T. M. (1970) Gene action in development. *Proc. Roy. Soc. Lond. B*, **176**, 347.

37. Lau, K. M. and Ng, S. F. (1992) Structural deficiency of 180°-rotated ciliary rows in *Tetrahymena*: implications on the inheritance and development of a non-genetically acquired cortical trait during vegetative propagation. *Devel. Genet.*, **13**, 187.

38. Ng, S. F. (1979) Origin and inheritance of an extra band of microtubules in *Tetrahymena* cortex. *Protistologica*, **15**, 5.

39. Frankel, J. (1973) A genically determined abnormality in the number and arrangement of basal bodies in a ciliate. *Dev. Biol.*, **30**, 336.

40. Jerka-Dziadosz, M., Garreau de Loubresse, N., and Beisson, J. (1992) Development of surface pattern during division in *Paramecium*. II. Defective spatial control in the mutant *kin241*. *Development*, **115**, 319.

41. Jerka-Dziadosz, M., Ruiz, F., and Beisson, J. (1998) Uncoupling of basal body duplication and cell division in *crochu*, a mutant of *Paramecium* hypersensitive to nocodazole. *Development*, **125**, 1305.

42. Ruiz, F., Beisson, J., Rossier, J., and Depuis-Williams, P. (1999) Basal body duplication in *Paramecium* requires gamma tubulin. *Curr. Biol.*, **9**, 43.

43. Frankel, J., Nelsen, E. M., and Martel, E. (1981) Development of the ciliature of *Tetrahymena thermophila*. II. Spatial subdivision prior to cytokinesis. *Dev. Biol.*, **88**, 39.

44. Tartar, V. (1961) *The biology of* Stentor. Pergamon Press, Oxford.

45. Uhlig, G. (1960) Entwicklungsphysiologische Untersuchungen zur Morphogenese von *Stentor coeruleus* Ehrbg. *Arch. Protistenk.*, **105**, 1.

46. Tartar, V. (1964) Morphogenesis in homopolar tandem grafted *Stentor coeruleus*. *J. Exp. Zool.*, **156**, 243.

47. Eberhardt, R. (1962) Untersuchungen zur morphogenese von *Blepharisma* und *Spirostomum*. *Arch. Protistenk.*, **106**, 241.

48. Yasuda, T., Numata, O., Ohnishi, K., and Watanabe, Y. (1980) A contractile ring and cortical changes found in the dividing *Tetrahymena pyriformis*. *Exp. Cell Res.*, **128**, 407.

49. Jerka-Dziadosz, M. (1981) Cytoskeleton-related structures in *Tetrahymena thermophila*: Microfilaments at the apical and division-furrow rings. *J. Cell Sci.*, **51**, 241.

50. Hirono, M., Nakamura, M., Tsunemoto, M., Yasuda, T., Ohba, H., Numata, O., and Watanabe, Y. (1987) *Tetrahymena* actin: Localization and possible biological roles of actin in *Tetrahymena* cells. *J. Biochem. (Tokyo)*, **102**, 537.

51. Edamatsu. M., Hirono, M., and Watanabe, Y. (1992) *Tetrahymena* profilin is localized in the division furrow. *J. Biochem.*, **112**, 637.

52. Watanabe, A., Kurasawa, Y., Watanabe, Y., and Numata, O. (1998) A new *Tetrahymena* actin-binding protein is localized in the division furrow. *J. Biochem. (Tokyo)*, **123**, 607.

53. Frankel, J., Nelsen, E. M., and Jenkins, L. M. (1977) Mutations affecting cell division in *Tetrahymena pyriformis*, syngen 1. II Phenotypes of single and double homozygotes. *Dev. Biol.*, **58**, 255.

54. Krzywicka, A., Kiersnowska, M., Włoga, D., and Kaczanowska, J. (1999) Morphogenesis of the fission line of the ciliate *Tetrahymena*: analysis of *cdaK* phenotypes and rearrangements of the membrane-skeleton during cell division. *Europ. J. Protistol.*, **35**, 342.

55. Watanabe, Y., Ohba, H., Hirono, M., and Yasuda, T. (1990) Analysis of furrow formation and furrowing during cell division in *Tetrahymena* using cell-division-arrest mutants. *Ann. N. Y. Acad. Sci.*, **582**, 166.

56. Ohba, H., Ohmori, I., Numata, O., and Watanabe, Y. (1986) Purification and immuno-fluorescence localization of the mutant gene product of a *Tetrahymena cdaA1* mutant affecting cell division. *J. Biochem. (Tokyo)*, **100**, 797.

57. Kaczanowska, J., Buzanska, L., and Ostrowski, M. (1993) Relationship between spatial pattern of basal bodies and membrane skeleton (epiplasm) during the cell cycle of *Tetrahymena*: *cdaA* mutant and anti-membrane immunostaining. *J. Euk. Microbiol.*, **40**, 747.

58. Numata, O., Suzuki, H., Ohba, H., and Watanabe, Y. (1995) The mutant gene product of a *Tetrahymena* cell-division-arrest mutant *cdaA* is localized in the accessory structure of specialized basal body close the division furrow. *Zool. Sci. (Tokyo)*, **12**, 133.

59. Frankel, J., Jenkins, L. M., and DeBault, L. E. (1976) Causal relations among cell cycle processes in *Tetrahymena pyriformis*: An analysis employing temperature-sensitive mutants. *J. Cell Biol.*, **71**, 242.

60. Cleffmann, G. and Frankel, J. (1978) The DNA replication schedule is not affected in a division blocked mutant of *Tetrahymena thermophila*. *Exp. Cell Res.*, **117**, 191.

61. Kaczanowska, J., Buzanska, L., and Frontczak, M. (1992) The influence of fission line expression on the number and positioning of oral primordia in the *cdaA1* mutant of *Tetrahymena thermophila*. *Devel. Genet.*, **13**, 216.

62. Nelsen, E. M., Williams, N. E., Yi, H., Knaak, J., and Frankel, J. (1994) 'Fenestrin' and conjugation in *Tetrahymena*. *J. Euk. Microbiol.*, **41**, 483.

63. Iftode, F., Cohen, J., Ruiz, F., Torres-Rueda, A., Chen-Shan, L., Adoutte, A., and Beisson, J. (1989) Development of surface pattern during division in *Paramecium*. I. mapping of duplication and reorganization of cortical cytoskeletal structures in the wild-type. *Development*, **105**, 191.

64. Sperling, L., Keryer, G., Ruiz, F., and Beisson, J. (1991) Cortical morphogenesis in *Paramecium*: A transcellular wave of protein phosphorylation involved in ciliary rootlet disassembly. *Dev. Biol.*, **148**, 205.

65. Fleury, A. and Laurent, M. (1995) Microtubule dynamics in *Paramecium*. I. Deployment

and dynamic properties of acetylated microtubules in relation to the invariance of the morphogenetic field. *Europ. J. Protistol.*, **31**, 190.

66. Fleury, A., Callen, A.M., Bre, M.H., Iftode, F., Jeanmarie-Wolf, R., Levelliers, N., and Clerot, J.C. (1995) Where and when is microtubular diversity generated in *Paramecium*? Immunological properties of microtubular networks in interphase and dividing cells. *Protoplasma*, **189**, 37.

67. Le Guyader, H. and Hyver, C. (1991) Duplication of cortical units in the cortex of *Paramecium*: a model involving a Ca2+ wave. *J. Theor. Biol.*, **150**, 261.

68. Le Guyader, H. and Hyver, C. (1994) A single model can account for the main asymmetries of the cortex of *Paramecium*: application to homopolar doublets. *J. Theor Biol.*, **170**, 317.

69. Laurent, M., and Fleury, A. (1995) A model with excitability and relay properties for the generation and propagation of the $Ca^{2+}$ morphogenetic wave in *Paramecium*. *J. Theoret. Biol.*, **174**, 227.

70. Laurent, M. and Fleury, A. (1996) Microtubule dynamics in *Paramecium*: II. Modelling of the conversion of a transient molecular signal into a morphogenetic process. Microtubule dynamics in *Paramecium. Europ. J. Protistol.*, **32**, 134.

71. Prajer, M., Fleury, A., and Laurent, M. (1997). Dynamics of calcium regulation in *Paramecium* and possible morphogenetic implications. *J. Cell Sci.*, **110**, 529.

72. Herbst, C. (1892). Experimentelle Untersuchungen über den Einfluss der veränderten chemischen Zusammensetzung des ungebenden Mediums ouf die Entwicklung der Tiere. I. Versuche an Seeigeleiern. *Z. Wiss. Zool.*, **55**, 446.

73. Kao, K. R. and Elinson, R. P. (1998) The legacy of lithium effects on development. *Biol. Cell*, **90**, 585.

74. Berridge, M. J., Downes, P., and Hanly, M. R. (1989). Neural and developmental actions of lithium: a unifying hypothesis. *Cell*, **59**, 411.

75. Klein, P. S. and Melton, D. A. (1996) A molecular mechanism for the effect of lithium on development. *Proc. Natl Acad. Sci., USA*, **93**, 8455.

76. Tartar, V. (1957) Reactions of *Stentor coeruleus* to certain substances added to the medium. *Exp. Cell Res.*, **13**, 317.

77. Schweikhardt, F. (1966) Zytochemisch-Entwicklungsphysiologische Untersuchungen an *Stentor coeruleus* Ehrbg. *Roux' Arch. f. Entwicklungsmech.*, **157**, 21.

78. König. K. (1967) Wirkung von Lithium- und Rhodanid-Ionen auf die polare Differenzierung und die Morphogenese von *Stentor coeruleus* Ehrbg. *Arch. Protistenkunde*, **110**, 179.

79. Beisson, J. and Ruiz, F. (1992) Lithium-induced respecification of pattern in *Paramecium. Devel. Genet.*, **13**, 194.

80. Jerka-Dziadosz, M. and Frankel, J. (1995) The effects of lithium chloride on pattern formation in *Tetrahymena thermophila. Dev. Biol.*, **171**, 497.

81. Freund, W. D., Mayr, G. W., Tietz, J., and Schultz, J. E. (1992) Metabolism of inositol phosphates in the protozoan *Paramecium*. Characterization of a novel inositol-hexaphosphate-dephosphorylating enzyme. *Eur. J. Biochem.*, **207**, 359.

82. Nanney, D. L. (1967) Cortical slippage in *Tetrahymena. J. Exp. Zool.*, **166**, 163.

83. Frankel, J., Jenkins, L. M., Nelsen, E. M., and Stoltzman, C. A. (1993) *hypoangular*: a gene potentially involved in specifying positional information in a ciliate, *Tetrahymena thermophila. Dev. Biol.*, **160**, 333.

84. Hall J. L., Ramanis Z, and Luck D.J. L. (1989) Basal body/centriolar DNA: molecular genetic studies in *Chlamydomonas. Cell*, **59**, 121.

85. Johnson, K. A. and Rosenbaum, J. L. (1990) The basal bodies of *Chlamydomonas reinhardtii* do not contain immunologically detectable DNA. *Cell*, **62**, 615.

86. Hall, J. L. and Luck, D. J. (1995) Basal-body associated DNA: in situ studies in *Chlamydomonas reinhardtii*. *Proc. Natl Acad. Sci., USA*, **92**, 5129.

87. Jerka-Dziadosz, M. (1981) Patterning of ciliary structures in *janus* mutant of *Tetrahymena* with mirror-image cortical duplications. An ultrastructural study. *Acta Protozool.*, **20**, 337.

88. Nanney, D. L. (1966) Cortical integration in *Tetrahymena*: An exercise in cytogeometry. *J. Exp. Zool.*, **161**, 307.

89. Tartar, V. (1956) Pattern and substance in *Stentor*. In *Cellular mechanisms of differentiation and growth*, Society for Developmental Biology Symposium, Vol. 14, (ed. D. Rudnick), p. 73. Princeton University Press, Princeton, N.J.

90. Fauré-Fremiet, E. (1945) Symétrie et polarité chez les Ciliés bi- ou multicomposites. *Bull. Biol. France Belg.*, **79**, 106.

91. Sonneborn, T. M. (1963) Does preformed cell structure play an essential role in cell heredity? In *The nature of biological diversity*, (ed. J.M. Allen), p. 165. McGraw-Hill, New York.

92. Tchang, T.-r., Shi, X.-b., and Pang, Y.-b. (1964) An induced monster ciliate transmitted through three hundred and more generations. *Scientia Sinica*, **13**, 850.

93. Grimes, G. W., McKenna, M. E., Goldsmith-Spoegler, C. M., and Knaupp, E. A. (1980) Patterning and assembly of ciliature are independent processes in hypotrich ciliates. *Science*, **209**, 281.

94. Jerka-Dziadosz, M. (1983) The origin of mirror-image symmetry doublet cells in the hypotrich ciliate *Paraurostyla weissei*. *Roux's Arch. Dev. Biol.*, **192**, 179.

95. Shi, X.-b., and Frankel, J. (1990) Morphology and development of mirror-image doublets of *Stylonychia mytilus*. *J. Protozool.*, **37**, 1.

96. Tchang, T.-r. and Pang, Y.-b. (1977) The cytoplasmic differentiation of jumelle *Stylonychia*. *Scientia Sinica*, **20**, 234.

97. Shi, X.-b. Lu, L., Qiu, Z., and Frankel, J. (1990) Morphology and development of left-handed singlets derived from mirror-image doublets of *Stylonychia mytilus*. *J. Protozool.*, **37**, 14.

98. Jerka-Dziadosz, M. (1985) Mirror-image configuration in the cortical pattern causes modifications in propagation of microtubular structures in the hypotrich ciliate *Paraurostyla weissei*. *Roux's Arch. Dev. Biol.*, **194**, 311.

99. Nelsen, E. M., Frankel, J., and Jenkins, L. (1989) Non-genic inheritance of cellular handedness. *Development*, **105**, 447.

100. Suhama, M. (1985) Reproducing singlets with an inverted oral apparatus in *Glaucoma scintillans* (Ciliophora, Hymenostomatida). *J. Protozool.*, **32**, 454.

101. Nelsen, E. M. and Frankel, J. (1989) Maintenance and regulation of cellular handedness in *Tetrahymena*. *Development*, **105**, 457.

102. Orias, E. and Rasmussen, L. (1976) Dual capacity for nutrient uptake in *Tetrahymena*. IV. Growth without food vacuoles and its implications. *Exp. Cell Res.*, **102**, 127.

103. Nelsen, E. M., Frankel, J., and Williams, N. E. (1989) Oral assembly in left-handed *Tetrahymena thermophila*. *J. Protozool.*, **36**, 582.

104. Frankel, J. (1991) The patterning of ciliates. *J. Protozool.*, **38**, 519.

105. Nelsen, E. M. and Frankel, J. (1986) Intracellular pattern reversal in *Tetrahymena thermophila*. I. Evidence for reverse intercalation in unbalanced doublets. *Dev. Biol.*, **114**, 53.

106. Shi, X.-b., Lu, L., Qiu, Z., He, W., and Frankel, J. (1991) Microsurgically generated

discontinuities provoke heritable changes in cellular handedness in a ciliate, *Stylonychia mytilus. Development*, **111**, 337.

107. French, V., Bryant, P.J., and Bryant, S.V. (1976) Pattern regulation in epimorphic fields. *Science*, **193**, 969.

108. Mittenthal J. E. (1981) The rule of normal neighbors: a hypothesis for morphogenetic pattern regulation. *Dev. Biol.*, **88,** 15.

109. Frankel, J. and Nelsen, E.M. (1986). Intracellular pattern reversal in *Tetrahymena thermophila*. II. Transient expression of a *janus* phenocopy in balanced doublets. *Dev. Biol.*, **114**, 72.

110. Brandts, W. A. M. and Trainor, L. E. H. (1990) A non-linear field model of pattern formation: intercalation and morphallactic regulation. *J. Theor. Biol.*, **146**, 37.

111. Brandts, W. A. M. and Trainor, L. E. H. (1990) A non-linear field model of pattern formation: application to intracellular pattern reversal in *Tetrahymena. J. Theor. Biol.*, **146**, 57.

112. Cole, E. S, Frankel, J., and Jenkins, L. M. (1987) *bcd*: A mutation affecting the width of organelle domains in the cortex of *Tetrahymena thermophila. Roux's Arch. Dev. Biol.*, **196**, 421.

113. Cole, E. S., Frankel, J., and Jenkins, L. M. (1988) Interactions between the *janus* and *bcd* cortical pattern mutants in *Tetrahymena thermophila*: An investigation into global intra-cellular patterning mechanisms using double-mutant analysis. *Roux's Arch. Dev. Biol.*, **197**, 476.

114. Dubielecka, B. and Jerka-Dziadosz, M. (1989) Defective spatial control in patterning of microtubular structures in mutants of the ciliate *Paraurostyla*. I. Morphogenesis in multi-left-marginal mutant. *Eur. J. Protistol.*, **24**, 308.

115. Jerka-Dziadosz, M. (1989) Defective spatial control in patterning of microtubular structures in mutants of the ciliate *Paraurostyla*. II. Spatial coordinates in a double-recessive mutant. *Eur. J. Protistol.*, **24**, 323.

116. Jerka-Dziadosz, M. and Wiernicka, L. (1992) Ultrastructural studies on the development of cortical structures in the ciliary pattern mutants of the hypotrich ciliate *Paraurostyla weissei. Europ. J. Protistol.*, **28**, 258.

117. Jerka-Dziadosz, M. and Czupryn, A. (1995) Development of ventral primordia in per-vasive ciliary hypertrophy mutants (PCH) of the hypotrich ciliate *Paraurosyla weissei. Acta Protozool.*, **34**, 249.

118. Jerka-Dziadosz, M. and Czupryn, A. (1997) The filaments supporting ciliary primordia and the fission furrow in the hypotrich ciliate *Paraurostyla weissei*, revealed by the monoclonal antibody 12G9: studies on wild-type and ciliary hypertrophy mutant. *Protoplasma*, **197**, 241.

119. Frankel, J., Nelsen, E. M., and Jenkins, L. M. (1987) Intracellular pattern reversal in *Tetrahymena thermophila*: *janus* mutants and their geometrical phenocopies. In *Genetic regulation of development*, Society for Developmental Biology Symposium, Vol. 45, (ed. W.F. Loomis), p. 219. Alan R. Liss, New York.

120. Jerka-Dziadosz, M. and Frankel, J. (1979) A mutant of *Tetrahymena thermophila* with a partial mirror-image duplication of cell surface pattern. I. Analysis of the phenotype. *J. Embryol. Exp. Morph.*, **49**, 167.

121. Frankel, J., Jenkins, L. M., and Bakowska, J. (1984) Selective mirror-image reversal of ciliary patterns in *Tetrahymena thermophila* homozygous for a *janus* mutation. *Roux's Arch. Dev. Biol.*, **194**, 107.

122. Frankel, J. and Nelsen, E. M. (1986) How the mirror-image pattern specified by a *janus* mutation of *Tetrahymena* comes to expression. *Dev. Genet.*, **6**, 213.

123. Grimes, G. W. and Aufderheide, K. J. (1991) *Cellular aspects of pattern formation. the problem of assembly*. Karger Verlag, Basel.

124. Frankel, J. (1999) The cell biology of *Tetrahymena thermophila*. In *Methods in cell biology: Tetrahymena thermophila*, (ed. D. J. Asai and J. D. Forney), **62**, p. 27. Academic Press, Orlando, FL.

125. Hershey, A. D. (1970) Genes and hereditary characters. *Nature*, **226**, 697.

126. Chant, J. and Pringle, J. F. (1995) Patterns of bud site selection in the yeast *Saccharomyces cerevisiae*. *J. Cell Biol.*, **129**, 751.

127. Madden, K. and Snyder, M. (1998) Cell polarity and morphogenesis in budding yeast. *Ann. Rev. Microbiol.*, **52**, 687.

128. Mata, J. and Nurse, P. (1998) Discovering the poles of yeast. *Trends Cell Biol.*, **8**, 163.

129. Baldauf, S. L. and Palmer, J. D. (1993) Animals and fungi are each other's closest relatives: congruent evidence from multiple proteins. *Proc. Natl Acad. Sci., USA*, **90**, 11558.

130. Wainright, P.O., Hinkle, G., Sogin, M.L., and Stickel, S. K. (1993) Monophyletic origins of the metazoa: an evolutionary link with the fungi. *Science*, **260**, 340.

131. Schlegel, M. and Eisler, K. (1996) Evolution of ciliates. In *Ciliates: cells as organisms*, (ed. K. Hausmann and P. C. Bradbury), p. 73. Gustav Fischer Verlag, Stuttgart.

132. Holmes, J. A. and Dutcher, S. K. (1989) Cellular asymmetry in *Chlamydomonas reinhardtii*. *J. Cell Sci.*, **94**, 273.

133. Ehler, L.L., Holmes, J.A., and Dutcher, S.K. (1995) Loss of spatial control of the mitotic spindle apparatus in a *Chlamydomonas reinhardtii* mutant strain lacking basal bodies. *Genetics*, **141**, 945.

134. Brown, D.L. and Bouck, G.B. (1973) Microtubule biogenesis and cell shape in *Ochromonas*. II. The role of nucleating sites in shape development. *J. Cell Biol.*, **56**, 360.

135. Orias, E. E. (1991) On the evolution of the karyorelict ciliate cell cycle: heterophasic ciliates and the origin of ciliate binary fission. *BioSystems*, **25**, 67.

136. Frankel, J. (1992) Positional information in cells and organisms. *Trends Cell Biol.*, **2**, 256.

137. Small, E. B. (1984) An essay on the evolution of ciliophoran oral cytoarchitecture based on descent from within a karyorelictean ancestry. *Origins of Life*, **13**, 217.

138. Eisler, K. (1992) Somatic kineties or paroral membrane: which came first in ciliate evolution? *Biosystems*, **26**, 239.

139. Frankel, J. (1997) Is spatial pattern formation homologous in unicellular and multicellular organisms? In: *Physical theory in biology: foundations and explorations*, (ed. C.J. Lumsden, W.A. Brandts, and L.E.H. Trainor), pp. 245. World Scientific, Singapore.

140. Jerka-Dziadosz, M. and Beisson, J. (1990). Genetic approaches to ciliate pattern formation: from self-assembly to morphogenesis. *Trends Genet.*, **6**, 41.

# Notes

1. The circumferential polarity of cortical structures in ciliates has previously been referred to as their 'asymmetry' by this author and by others. In this review, asymmetry is treated as an attribute of the whole system and not of individual axes.

2. The concept of an intrinsic polarity of basal bodies is developed in detail in the following recent review: Beisson, J. and Jerka-Dziadosz, M. (1999). Polarities of the centriolar structure: morphogenetic consequences. *Biol. Cell.* **91**, 367–378.

# 4 | Spatial cues for cellular asymmetry in polarized epithelia

W. JAMES NELSON, CHARLES YEAMAN, and KENT K. GRINDSTAFF

## 1. Introduction

Transporting epithelia form permeability barriers between two compartments in the body, and vectorially transport ions and solutes between those compartments in order to maintain ionic homeostasis. In the kidney and gastrointestinal tract, the epithelium forms a closed monolayer of cells that surrounds a lumen, which topologically defines the 'outside' compartment of the organism; on the opposite side, the epithelium contacts the interstitium and blood supply, which topologically define the 'inside' compartment. To maintain the structural integrity of the epithelium, and to regulate the diffusion of ions and solutes from one compartment to the other through the paracellular space, the epithelium must form strong and specialized contacts between cells. Access to the paracellular space is regulated by the tight junction (*zonula occludens*), which forms a gasket-like, semipermeable seal around each cell (Fig. 1). For vectorial ionic and solute transport across the epithelium, ion channels, exchangers, and transporters must be segregated into two structurally and functionally different domains of the plasma membrane, termed apical and basolateral, that face these different compartments. Typically, the apical membrane faces the luminal compartment, and the basolateral membrane faces the interstitium (Fig. 1).

An example of the function of a transporting epithelium is the reabsorption of ions and solutes from the ultrafiltrate in the kidney nephron (1). In the kidney, ions, water, and solutes are filtered from the serum by the glomerulus, enter the nephron and are reabsorbed by the epithelium back into the blood supply, resulting in the concentration of waste products and the formation of urine. In the lumen of the proximal tubule, the concentration of $Na^+$ is high compared to that inside the surrounding epithelial cells. Thus, $Na^+$ can enter cells by diffusion down its electrochemical gradient through ion channels, exchangers, and transporters localized in the apical membrane domain facing the nephron lumen. The intracellular concentration of $Na^+$ is strictly regulated, and $Na^+$ is actively pumped out of the cell by the $Na^+,K^+$-ATPase

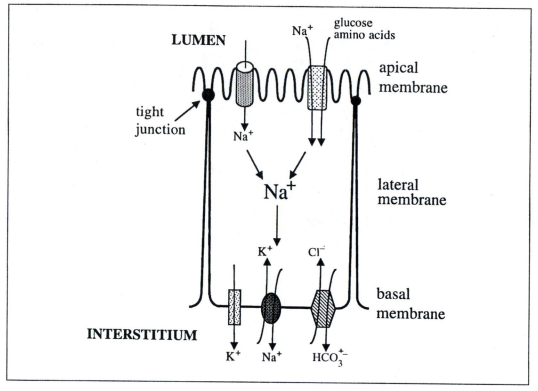

**Fig. 1** Schematic representation of the organization of ion channels, transporters, and exchangers in a polarized, transporting epithelial cell.

that is localized in the basolateral membrane domain facing the interstitium and blood supply. Thereby, the epithelium sets up a steep $Na^+$ gradient across the cells down which other ions and solutes are (co-)transported from the lumen to the interstitium (see Fig. 1).

In this chapter, we discuss current models for how epithelial cell polarity is established and maintained. Early ideas focused on the sorting of membrane proteins in the Golgi complex and their delivery in transport vesicles to either the apical or basolateral membrane (2, 3). One of the major assumptions in the field has been that the mechanisms involved in protein sorting and delivery are unique to polarized cells (4). As we discuss below, there have been important advances in our understanding of these mechanisms, indicating that they are not unique to polarized epithelial cells and are likely to be constitutive processes in many other cell types (5, 6). Thus, the question remains as to the nature of the mechanisms that initiate the development of cell polarity.

In contrast to similarities in protein/membrane sorting mechanisms between polarized and non-polarized cell types, cell–cell adhesion is a characteristic of the multicellular organization of epithelia. Therefore, we have suggested that cell–cell adhesion might act as a spatial cue on the cell surface that generates molecular and structural asymmetry in the plasma membrane (7, 8). We speculate that cell–cell adhesion initiates the reorganization of the actin membrane–cytoskeleton and the recruitment of vesicle targeting patches to different membrane domains. These targeting patches subsequently direct delivery of membrane proteins to specific domains of the plasma membrane, thereby reinforcing and maintaining changes in protein distributions initiated by cell adhesion. Thus, constitutive protein and membrane sorting processes are harnessed to organize structurally and functionally distinct membrane domains through the induction of specialized cell–cell adhesion.

## 2. Sorting of apical and basolateral membrane proteins in the TGN of simple epithelial cells

Generally, the biogenesis of epithelial cell polarity has been viewed from the standpoint of sorting apical and basolateral membrane proteins in the *trans*-Golgi network (TGN) (9, 10). This sorting event has been considered a specialized process in polarized cells and the basis for generating cell surface polarity from the 'inside-out' (Fig. 2). The hypothesis is that the spatial distribution of proteins in the apical and basolateral membrane domain is a direct consequence of their sorting in the TGN and targeted delivery in specialized transport vesicles. Below, we discuss the basis for this view, the underlying mechanisms involved in protein sorting and targeting, and evidence that these processes are constitutive to both polarized and non-polarized cells.

In simple epithelia, such as kidney, sorting of newly synthesized apical and basolateral membrane proteins occurs in the TGN (11, 12). There have been advances in defining sorting signals that are intrinsic to apical and basolateral membrane proteins, but considerably less is known about the sorting machinery that recognizes those signals.

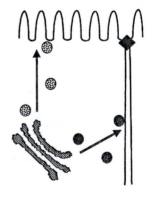

'Inside-out'

1. Membrane proteins sorted into separate 'apical' and 'basolateral' transport vesicles in the TGN

2. Cytoskeleton mediates targeted delivery of vesicles to membrane domains

3. Vesicles dock/fuse with specific membrane domains

**Fig. 2** Early ideas of how epithelial cells generated and maintained cell-surface polarity focused on mechanisms involved in sorting and targeting proteins from the Golgi to the cell surface ('inside-out' mechanism).

Basolateral sorting signals frequently contain a critical tyrosine residue within the amino acid sequence NPXY or YXXØ (where X is any amino acid and Ø is an amino acid with a bulky hydrophobic group) (13). In several cases, these motifs are predicted to form a 'tight β-turn' (14, 15). Tyrosine-based motifs mediate basolateral sorting of several proteins, including receptors for LDL (16, 17) and asialoglyco-proteins (18), lysosomal acid phosphatase (19), envelope glycoproteins of vesicular stomatitis virus (VSV G) (20, 21) and human immunodeficiency virus (HIV) (22), and mutated forms of p75$^{NTR}$ (23, 24) and influenza haemagglutinin (HA) (25).

In some cases, including Fc receptor (26, 27) and MHC class II invariant chain (28, 29), a di-hydrophobic motif serves to target the protein basolaterally. However, many proteins contain basolateral sorting signals that bear little or no sequence similarity to these motifs, including the poly-IgA receptor (pIgA-R) (30), neural cell-adhesion molecule (N-CAM; 31), and transferrin receptor (32). It is noteworthy that the sorting signal in pIgA-R comprises a structure of a tight β-turn (30) that resembles that in the tyrosine-based signals (see above), but it is unknown whether this structure is present in the context of the native protein.

Protein sorting to the apical plasma membrane is dependent on motifs present in the luminal domain or membrane-anchor, rather than the cytoplasmic domain (33). Initial clues into potential sorting mechanisms came from studies of a class of apical membrane proteins that is anchored to the lipid bilayer by a glycosylphosphoinositol (GPI) group (34–36). The identity of signals responsible for apical sorting of membrane-spanning proteins is much less clear, but they appear to be located in the large, highly folded extracellular domain. This conclusion is based on findings that viral protein chimeras (37, 38), 'tail-minus' mutants of basolateral membrane proteins (16, 19, 39), and soluble extracellular domains of apical and basolateral membrane proteins (31, 34, 35, 40–44) are all targeted to the apical membrane of polarized MDCK epithelial cells. Early studies examined a potential role of N-glycans in the extracellular domain in apical sorting (45, 46). There is some experimental evidence for the involvement of N-glycans in apical sorting of certain proteins (47–50), but apical sorting of many proteins occurs independently of N-glycosylation (44, 51–55). Thus, if used, this is not a universal mechanism. For at least the neurotrophin receptor, p75$^{NTR}$, apical sorting information has been localized to an O-glycosylated 'stalk' that is adjacent to the transmembrane domain (44). When this signal is removed, p75$^{NTR}$ is very efficiently targeted to the basolateral membrane, indicating the presence of multiple sorting motifs in this protein. The mechanisms of carbohydrate-mediated apical sorting are not known. One potential mechanism involves recognition of carbohydrate moieties by sorting lectins, such as VIP36 (56, 57), which is hypo-thesized to partition into microdomains of the TGN along with other apical membrane proteins and lipids (see below).

Sorting of different classes of membrane proteins that involve these different apical and basolateral motifs is poorly understood, but the basic principle may involve segregation of different classes by localized clustering in the plane of the lipid bilayer. The cytoplasmic orientation of basolateral sorting signals indicates that the sorting machinery associated with the TGN is cytosolic and that it may involve protein clustering mediated by a cytoplasmic protein lattice. Both YXXØ- and di-hydrophobic-based sorting signals selectively bind clathrin adaptor protein complexes AP-1 and AP-2 (58–62) via the medium ($\mu$) chains of these complexes (58, 60, 61). Although interactions between tyrosine- and di-hydrophobic-based signals and AP-1/AP-2 complexes are responsible for cargo-selective sorting into TGN- and plasma membrane-derived clathrin-coated vesicles, the mechanism responsible for basolateral sorting in polarized epithelial cells remains unknown. A recently described adaptor complex, AP-3, also interacts with tyrosine- and di-hydrophobic-based signals. AP-3 (63–66) appears to be involved in the formation of a novel class of clathrin-coated vesicles in the TGN and endosomes (63, 65, 67). Although the AP-3 complex is required for protein sorting to the yeast vacuole (68) and pigment granule formation in *Drosophila* (69), a function in basolateral protein sorting has not yet been found.

Apical sorting has been proposed to involve formation of glycolipid- and cholesterol-containing membrane domains, or 'rafts', in the exoplasmic leaflet of the Golgi (10, 33, 70). Clustering of some apical-membrane proteins into glycosphingolipid rafts during transport through the Golgi may be mediated by transmembrane domains (71, 72) or GPI anchors (73–75). Following insertion into the apical plasma membrane, GPI-anchored proteins are initially clustered but gradually disperse in the plane of the bilayer (76), whereas HA apparently remains clustered (71). The formation of rafts can be disrupted by inhibiting sphingolipid synthesis with fumonisin $B_1$ (77) or by depleting cells of cholesterol after treatment with methyl-$\beta$-cyclodextrin (71), which results in mis-sorting of GPI-anchored proteins (77) and influenza virus HA (78). However, apical transport of many other proteins is not affected by these inhibitors (79–81) and likely occurs independently of glycolipid rafts. Therefore, apical sorting likely involves more than one mechanism, but the details of these mechanisms are obscure. It is noteworthy that certain basolateral membrane proteins (e.g. $Na^+,K^+$-ATPase) may be sorted into the basolateral pathway by exclusion from glycosphingolipid rafts (77). For this type of protein, efficient basolateral sorting may be achieved in the absence of an active basolateral sorting signal (8).

## 3. Delivery of membrane proteins to specific membrane domains in polarized epithelial cells

In simple epithelia, post-TGN transport vesicles containing apical and basolateral proteins are delivered directly to the appropriate membrane domain (Fig. 2). The cytoskeleton has been the focus of studies to identify mechanisms involved in vesicle trafficking from the TGN to specific membrane domains.

The exact role of actin in vesicle trafficking is unknown. However, vesicle fusion with the plasma membrane may be preceded by the localized depolymerization of actin at the membrane (82–85). Thus, actin may act as a fence to prevent access of vesicles to the plasma membrane prior to membrane docking. In addition, disruption of the actin cytoskeleton appears to inhibit apical endocytosis and transcytosis while basolateral endocytosis and vesicle trafficking between the TGN and apical or basolateral membrane domain remain unaffected (86–91). Furthermore, the presence of myosin I on post-TGN transport vesicles indicates that the actin cytoskeleton is involved in some aspect of vesicle trafficking (92).

Although microtubules have been shown to be involved in basolateral to apical transcytosis and apical recycling, their role in TGN to plasma membrane delivery is somewhat controversial (93–95). It has been suggested that microtubule reorganization and the coincident relocalization of the Golgi complex to apical region of the cell is specialized for efficient delivery of proteins from the TGN to the apical membrane (96). Previous studies have shown that vesicle transport between the TGN and apical membrane is slowed following disruption of microtubules with colchicine or nocodazole, while transport to the basolateral plasma membrane remains relatively unaffected (95–100). Nevertheless, basolateral transport vesicles bind to microtubules *in vitro* (101), and delivery of VSV G protein to the basolateral membrane appears to require microtubule motor proteins (102).

## 4. Specification of vesicle docking and fusion with either the apical or basolateral membrane

Little is known about molecular interactions between TGN-derived transport vesicles and the plasma membrane of polarized epithelial cells (Fig. 2). However, the SNARE hypothesis has provided a conceptual framework to begin characterizing the mechanisms involved (103). This model postulates that correct pairing of addressing proteins on the transport vesicle (termed v-SNAREs) with cognate receptors on the target membrane (termed t-SNAREs) determines the specificity of vesicle docking and fusion (104, 105). Vesicle–target membrane fusion also requires the activity of an ATPase, N-ethylmaleimide-sensitive fusion protein (NSF) (106, 107), and soluble NSF attachment proteins (SNAPs) (108, 109). SNAREs were identified as membrane-anchored receptors for SNAPs (105, 110–112).

Vesicle-associated membrane proteins synonymously referred to as VAMP-1 (113) and VAMP-2 (114), or synaptobrevin 1 (115) and synaptobrevin 2 (116), are the founding members of a growing family of related v-SNAREs (117–122). Syntaxins, SNAP-25, and related proteins comprise an extended family of t-SNAREs defined by a conserved 60-amino-acid 't-SNARE' homology domain (123). Syntaxins occupy distinct cellular compartments (124, 125), and are an ever-growing family of proteins (117, 118, 126–130). SNAP-25 and related proteins are cytoplasmically oriented proteins (131–133) that are anchored to the membrane by palmitoyl groups attached to central cysteine residues (134).

Endogenous t-SNAREs have polarized distributions in several epithelial cell types (135–142). Syntaxin-3 has been found primarily on the apical plasma membrane in MDCK cells (143), enterocytes (144), and hepatocytes (145). In contrast, syntaxin-4 is expressed predominantly on the basolateral membrane domain of MDCK cells (143) and hepatocytes (145). Interestingly, a reversed polarity of these two syntaxin isoforms has been reported in renal collecting duct epithelia (135, 136). SNAP-23, a ubiquitously expressed homologue of SNAP-25, is non-polarized in MDCK and Caco-2 cells (146, 147).

Because t- and v-SNARE components have distinct polarized distributions in several epithelial cell types, it has been suggested that these proteins serve to specify the correct delivery of transport vesicles to apical and basolateral membrane domains (143, 148). However, this conclusion is almost certainly an oversimplification. First, although syntaxin isoforms are differentially distributed between distinct plasma membrane domains, their localization is not restricted to a single membrane domain and their expression patterns clearly overlap (145). Secondly, it has been suggested that exocytosis at the apical membrane domain is independent of SNAREs (149). Treatment of permeabilized MDCK cells with either tetanus toxin or botulinum neurotoxin serotype F, which cleave VAMP-1, -2, and cellubrevin (150, 151), significantly reduced the efficiency of vesicular stomatitis virus G protein (VSV G) transport to the basolateral membrane (149). However, these toxins did not inhibit delivery of influenza virus haemagglutinin (HA) to the apical membrane (149). This result suggested that v-SNAREs are not required for apical secretion. It was also proposed that a novel mechanism, employing annexin XIIIB, is required for transport vesicle fusion with the apical plasma membrane (152). However, the possibility remains that a VAMP homologue that is insensitive to toxin cleavage is required for transport to the apical plasma membrane. This possibility is bolstered by the recent identification of a toxin-insensitive VAMP (Ti-VAMP) in Caco-2 cells, which forms a complex with the apical t-SNARE syntaxin-3 and SNAP-23 (146), and appears to be involved in vesicle docking/fusion with the apical membrane in MDCK cells (153).

## 5. Can sorting and delivery of membrane proteins from the TGN explain the establishment of cell-surface polarity?

Although there has been considerable focus on how protein sorting in the TGN and delivery of post-TGN transport vesicles to the apical and basolateral membranes lead to epithelial polarity, there is now evidence that these processes are not unique to polarized epithelial cells. Two groups have shown that apical and basolateral membrane proteins are sorted at the level of the TGN into separate transport vesicles in fibroblasts (5, 6). However, these proteins had a non-polarized distribution in the plasma membrane, demonstrating that sorting in the TGN is not sufficient to generate cell-surface polarity. Thus, sorting of apical and basolateral proteins in the TGN may be a constitutive process that occurs in non-polarized cells in which the plasma membrane is not specifically segregated into apical and basolateral domains.

The role of microtubules in establishing direct delivery pathways during development of epithelial polarity has been examined in two different clones of MDCK cells (154). These clones differ in their organization of microtubules when the cells have attained full polarity. Efficient and specific direct delivery pathways for vesicles between the TGN and apical or basolateral plasma membrane were established rapidly following the induction of cell–cell and cell–substratum contacts. This occurred prior to the global reorganization of the microtubule/Golgi complex characteristic of polarized epithelial cells. In addition, complete depolymerization of microtubules with nocodazole and low temperature did not affect the fidelity of apical or basolateral delivery of proteins, but decreased the efficiency of their delivery by 25–50% (154). Thus, microtubules appear to facilitate but not specify the delivery of vesicles to either the apical or basolateral plasma membrane in MDCK cells with a microtubule/Golgi distribution characteristic of either non-polarized or polarized epithelial cells (8).

In summary, it is clear that apical and basolateral membrane are sorted in the TGN into separate classes of transport vesicles, and that microtubules regulate the efficiency of vesicle delivery to the plasma membrane domains. However, at the same time, it is now recognized that these processes are in themselves not sufficient to generate distinct cell-surface distributions of apical and basolateral membrane proteins. Thus, other cellular processes must be involved.

## 6. The requirement for spatial cues at the cell surface for the biogenesis of epithelial cell polarity

Cadherin-mediated cell–cell adhesion is a specific characteristic of multicellular epithelial sheets and is conspicuously absent from non-polarized, single cells. We propose that a combination of extrinsic spatial cues mediated by cell–cell and cell–substratum adhesion physically and molecularly define contacting and non-contacting cell surfaces, which are the precursors of basolateral and apical membrane domains, respectively. We suggest that these spatial cues initiate localized assembly of specialized cytoskeletal and signalling protein complexes at the contacting membrane, which position other cytoskeletal complexes and protein-sorting compartments (TGN) in the cytoplasm relative to the spatial cues. Subsequently, protein sorting from these compartments to the cell surface reinforces and maintains the structural and functional specialization of these membrane domains (7, 8). This is the 'outside-in' hypothesis, by which spatial cues on the outside harness constitutive sorting processes in the inside of the cell to drive the assembly of structurally and functionally distinct membrane domains (Fig. 3).

A clear role for cell–cell adhesion in the development of cell-surface polarity is revealed during the early formation of MDCK cell clusters in suspension culture, in which there is little or no extracellular matrix (155–157). The cell surface of single epithelial cells in suspension culture exhibits a random distribution of apical and basolateral membrane proteins. E-cadherin-mediated cell–cell adhesion, in the absence of extracellular matrix (ECM), results in the segregation of apical membrane proteins to the non-contacting (free) surface and basolateral membrane proteins to the contacting surfaces. Cell–ECM adhesion can induce formation of a free (apical)

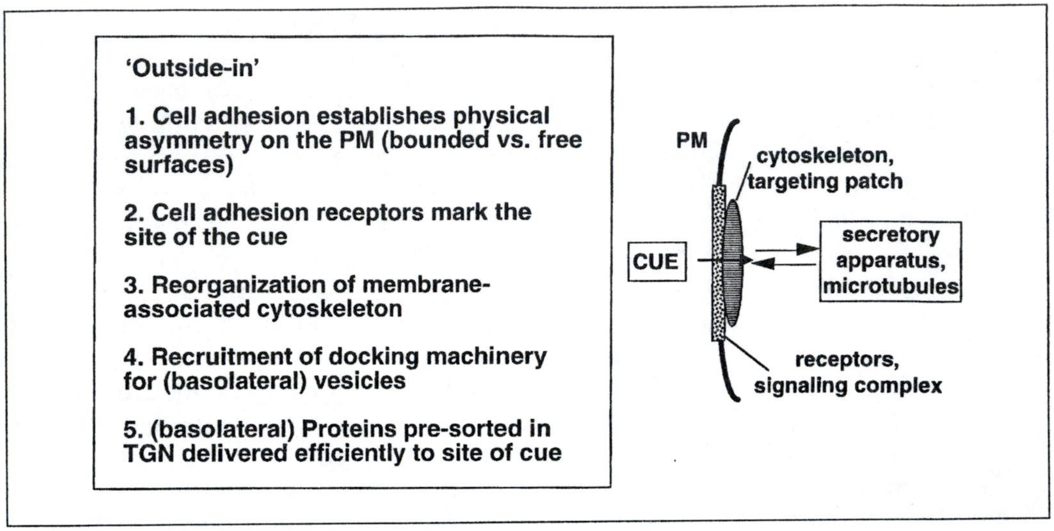

**'Outside-in'**

**1. Cell adhesion establishes physical asymmetry on the PM (bounded vs. free surfaces)**

**2. Cell adhesion receptors mark the site of the cue**

**3. Reorganization of membrane-associated cytoskeleton**

**4. Recruitment of docking machinery for (basolateral) vesicles**

**5. (basolateral) Proteins pre-sorted in TGN delivered efficiently to site of cue**

PM

cytoskeleton, targeting patch

CUE

secretory apparatus, microtubules

receptors, signaling complex

**Fig. 3** The 'outside-in' mechanism for the generation and maintenance of cell-surface polarity by epithelial cells focuses on the roles of extrinsic events at the cell surface in establishing asymmetry; intracellular sorting and delivery are secondary events.

surface, but the distributions of basolateral membrane proteins remains random. However, cell–ECM adhesion, in combination with cell–cell adhesion, is required to establish the apicobasal axis of polarity and position of the tight junction.

# 7. Mechanisms involved in establishing cadherin-mediated cell–cell adhesion

Cadherin-mediated cell–cell adhesion leads to the formation of specific contacts between cells, and initiates changes in the organization of the actin cytoskeleton (158). Here, we discuss recent insights into the cytomechanics of cell–cell adhesion and the consequential changes in actin organization.

Cadherins are single-transmembrane spanning proteins. The N-terminal extra-cellular domain is composed of five repeats that have similar structures and contain $Ca^{2+}$-binding motifs (159). While homotypic binding between the extracellular domains of cadherins on adjacent cells is clearly important for cell–cell recognition (160), the affinity of binding ($\sim1$ μM; 159) may not be sufficient to promote the strong cell–cell adhesion necessary to maintain tissue integrity. X-ray diffraction studies of crystals of the amino-terminal repeat domains of E-cadherin and N-cadherin revealed the presence of dimers and higher-ordered complexes (159, 161). Formation of higher-order complexes between extracellular domains of parallel-oriented cadherins, and clustering of extracellular E-cadherin between cells, might co-operatively increase the strength of adhesion (162, 163).

The cytoplasmic domain of cadherin is also required for cell–cell adhesion. The amino-acid sequences of the cytoplasmic domain of different cadherins are very similar, and contain a highly conserved binding site for a family of sequence-related cytosolic proteins: β-catenin, plakoglobin, and p120[CAS] (164–168). The cadherin/β-catenin complex binds via β-catenin to another cytosolic protein, α-catenin (167, 169, 170), that interacts with actin filaments either directly (171) or through other actin-associated proteins, such as α-actinin (172). Binding of this protein complex to the actin cytoskeleton is consistent with the appearance of a pool of cadherins and catenins at cell–cell contacts that is resistant to extraction by the non-ionic detergent Triton X-100 (166, 173).

During the formation of cadherin-mediated cell–cell contacts, the organization of the actin cytoskeleton changes dramatically. In individual motile cells, actin filaments are continually polymerizing at the free cell-edge of lamellae, and depolymerizing in a transition zone between the cell body and lamellae; this transition zone contains a circumferential ring of actin cables (174). In polarized epithelial cells, actin filaments are organized into a thinner, much more peripherally disposed circumferential ring at the apical surface in association with cadherin cell–cell contacts (175). While circumferential actin structures are thus characteristic of both motile and tissue forms of epithelial cells, a dramatic structural transformation in actin organization occurs during the formation and stabilization of cell–cell contacts.

The dynamics of E-cadherin redistribution during the processes of initial cell–cell contact through development of a polarized monolayer have been poorly described. However, recent studies of the distribution of a fully functional E-cadherin tagged with green fluorescent proteins (EcadGFP) has provided a new dynamic view of sequential stages of cell–cell adhesion that involve specific changes in E-cadherin and the actin cytoskeleton organization (Fig. 4; 176).

Upon cell–cell adhesion, E-cadherin spontaneously clusters into puncta at initial sites of developing cell–cell contacts (176, 177). The formation of E-cadherin puncta results in decreased E-cadherin mobility (176). In new areas of cell–cell contact (less than 15 minutes old), EcadGFP has a high mobile fraction (greater than 90%) and a high diffusion coefficient ($3.6 \pm 1.5 \times 10^{-10}$ cm$^2$ s$^{-1}$). However, where EcadGFP clusters into puncta and associates with the cytoskeleton a smaller fraction is mobile (less than 50%). The puncta formed by EcadGFP are very similar in organization and distribution to structures formed by endogenous E-cadherin and catenins that were previously characterized by retrospective immunocytochemistry (176, 177).

Clustering of E-cadherin into puncta in the membrane may be initiated by weak interactions between extracellular and juxta-membrane domains of cadherins (178). However, interactions between E-cadherin and the actin cytoskeleton are also initiated quickly upon cell–cell contact, and these interactions affect the organization of the adhesion complex (176). As E-cadherin puncta begin to form during this first stage, they always appear to be associated with the ends of thin actin cables which are oriented toward the contact (Fig. 4). These actin filaments branch from circumferential actin cables that are organized parallel to the forming contact and circumscribe the perimeter of migratory cells. Binding of actin filaments to E-cadherin/catenin

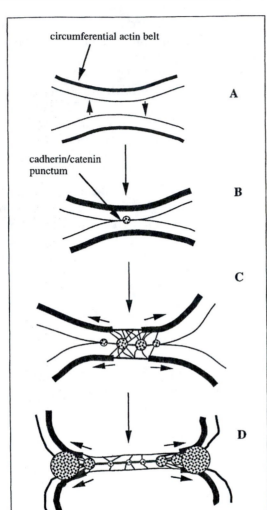

**Fig. 4** Stages in the development of cadherin-mediated cell-cell adhesion. (A) Transition: the cells move towards each other; transient contacts are formed between cells. (B) Initiation: cell-cell contacts are established; cadherin/catenin complexes coalesce into puncta at sites of cell-cell adhesion. (C) Strengthening: as the contact lengthens, additional cadherin/catenin puncta coalesce along the contact (~1 punctum/μm); actin filaments, which have split off from the circumferential actin belt, associate with each punctum; the circumferential actin belt dissociates beneath the contact. (D) Compaction: ends of the circumferential actin belt swing out to the perimeter of the contact; cadherin/catenin puncta sweep to the edge of the contact, where they coalesce further into plaques.

complexes may cause further clustering and stabilization of puncta (176). This type of cadherin/actin organization has been shown to provide a mechanical linkage between fibroblasts (179). Quantitative measurements showed that this initial stage of adhesion coincides with an exponential increase in the strength of adhesion (180). Significantly, this strengthening stage is completely inhibited by treatment of cells with CD (180). Analysis of living cells showed that during this initial stage, CD selectively disassembled contacts and caused formation of aggregates that include cell-surface EcadGFP (176) and probably the barbed ends of actin filaments (181). It is also interesting to note that myosin is involved in the CD-induced aggregation of the barbed ends of actin filaments (181), and that treadmilling of actin ceases in areas of developing cell–cell contacts (182). Thus, E-cadherin puncta may gradually sequester the barbed ends of actin filaments and directly or indirectly anchor them to the membrane at cell–cell contacts which contributes to the gradual strengthening of cell–cell adhesion. These changes in actin organization may also set up cyto-architectural cues for later stages of cellular reorganization (176).

Based on our analysis of changes in the distribution of E-cadherin tagged with green fluorescent protein upon cell–cell adhesion, we proposed a model for how contacts between cells are initiated, strengthened, compacted, and condensed (Fig. 4; 176). Cell–cell adhesion is initiated by weak binding between extracellular domains of E-cadherin, which is present in a highly mobile pool at the plasma membrane. At or near the same time, E-cadherin/catenin complexes attach to actin filaments that branch from actin cables that circumscribe the perimeter of migratory cells. These two processes act synergistically to assemble puncta, which, as a group, are sufficiently adhesive to hold the nascent cell–cell contact together. Subsequently, there is a change in actin dynamics as actin treadmilling ceases in areas of cell–cell contact, perhaps due to sequestration of the barbed ends of actin filaments into E-cadherin puncta. We hypothesize that reduced actin treadmilling causes the dissolution of the circumferential actin cables immediately adjacent to the developing contact. It is also possible that a 'signalling' event at the cell surface induced by cell–cell adhesion causes a change in the organization or polymerized state of the circumferential actin cables adjacent to the contact site. We suggest that stabilization of actin via the clustered cadherin/catenin complex engages the myosin II clutch, thereby inducing a translocation of circumferential actin cables and the rest of the cell body to the cell–cell contact interface and the rapid movement of associated E-cadherin puncta into large plaques. This co-ordinated reorganization of E-cadherin and the actin cytoskeleton results in the establishment of strong, compacted cell–cell contacts and the generation of an actin cable which circumscribes the free edges of the newly contacting cells and is embedded into either side of the E-cadherin plaque at the margins of the contact.

The next stage of adhesion is initiated once another cell joins a two-cell colony. Additional cells join larger cell colonies using the same mechanisms outlined above. When three cells contact, two cell–cell contacts and two free edges flank a central cell. Two perimeter actin cables are localized to the free edges of the central cell, and are further linked, at E-cadherin plaques, to the circumferential actin cables from the two flanking cells. This organization is unstable, and results in further reorganization of E-cadherin puncta and the circumferential actin cytoskeleton. This reorganization is initiated by the lateral translocation of the E-cadherin plaques on one side of the colony towards each other until they coalesce. This triangular organization of E-cadherin undergoes a final rearrangement as the cells condense and maximize contacts between each other. We suggest that one of the perimeter actin cables of the central cell dominates, exerts tension on the E-cadherin plaques, and slowly pulls the plaques from the outside cells together. The colony continues to reorganize into a stable configuration of a circle, with each cell connected together on one side sharing a common vertex in the middle of the colony.

## 8.  Converting cadherin-mediated cell–cell contacts into changes in cell surface polarity

A consequence of cell–cell adhesion is the formation of three membrane-associated structures that reinforce and maintain membrane asymmetry initiated by cell adhesion (8). These structures are:

(1)  membrane–cytoskeleton;

(2)  targeting patches; and

(3)  tight junctions.

## 8.1  Membrane–cytoskeleton

Cell–cell adhesion initiates localized assembly of the spectrin-based membrane–cytoskeleton at sites of contact (183–186). Spectrin is a long, rod-shaped protein that assembles into a protein skeleton with a complex of actin, protein 4.1, adducin, and other proteins (187). The membrane–cytoskeleton may bind directly or indirectly to the cadherin/catenin complex (188). In addition, spectrin binds ankyrin, which, in turn, binds with high affinity to integral membrane proteins, including $Na^+,K^+$-ATPase (189). Thereby, the membrane–cytoskeleton may act as a protein-sorting machine by preferentially sequestering bound proteins (e.g. $Na^+,K^+$-ATPase) which then accumulate at sites of cell–cell contact (190).

Experimental evidence has been obtained for a link between cadherin-mediated cell–cell adhesion and directed recruitment of the membrane–skeleton and $Na^+,K^+$-ATPase to cell–cell contacts (Fig. 5; 191, 192). Fibroblasts express catenins, ankyrin, spectrin and $Na^+,K^+$-ATPase but not cadherin. As a consequence, fibroblasts do not form $Ca^{2+}$-dependent cell–cell contacts and $Na^+,K^+$-ATPase is distributed diffusely over the entire plasma membrane, even when the cells are physically in contact with each other. When E-cadherin is expressed in these cells, it forms a complex with the endogenous catenins and initiates $Ca^{2+}$-dependent cell–cell adhesion (see above). Significantly, in these cells the membrane–skeleton and $Na^+,K^+$-ATPase co-localize with the cadherin/catenin complex at cell–cell contacts in a distribution remarkably similar to the distributions of these protein complexes in polarized MDCK epithelial cells (191). The introduction of a mutant form of E-cadherin that is unable to bind to catenins does not induce this reorganization of the membrane–skeleton and $Na^+,K^+$-ATPase. Thus, a molecular linkage between the cadherin/catenin complex and the $Na^+,K^+$-ATPase/membrane–skeleton complex, possibly through the actin cyto-skeleton, is required for the reorganization of $Na^+,K^+$-ATPase (Fig. 5).

## 8.2  Targeting patches

In order to re-enforce plasma membrane asymmetry that was initiated by cell adhesion and assembly of the actin-based membrane–cytoskeleton, newly synthesized membrane proteins must be delivered from the TGN to the correct nascent plasma

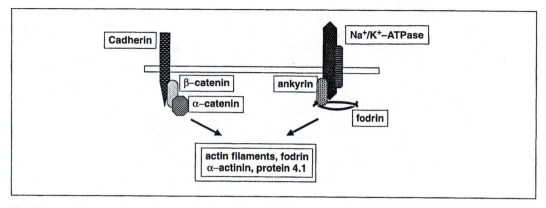

**Fig. 5** Schematic representation of membrane proteins/cytoplasmic adaptor complexes between cadherin/catenins and Na⁺,K⁺-ATPase/ankyrin-fodrin, and how these complexes may be linked through the actin cytoskeleton.

membrane domain. Little is known about molecular interactions between TGN-derived transport vesicles and the plasma membrane of polarized epithelial cells. As discussed above, the SNARE hypothesis postulates that correct pairing of addressing proteins on the transport vesicle, termed v-SNAREs (VAMP-1, VAMP-2) with cognate receptors on the target membrane, termed t-SNAREs (syntaxin, SNAP-25) specifies vesicle docking and fusion (104, 105). In polarized MDCK cells, exogenously expressed syntaxin-3 localizes to the apical plasma membrane, whereas syntaxin-4 is localized predominantly to the basolateral membrane (148). SNAP-23, an ubiquitously expressed homologue of SNAP-25, is non-polarized in MDCK cells. How syntaxins become localized to different membrane domains is unknown.

Although evidence that SNARE complexes determine vesicle–target membrane fusion is strong, it is unclear whether interactions between cognate v- and t-SNARES are sufficient to specify the site of vesicle docking with the target membrane. Much of the work on SNARE interactions has focused on events at the synapse, where synaptic vesicles are generated from a local endosome and are pre-docked at the synaptic membrane (132, 193). In most other cell types, post-TGN transport vesicles must move relatively long distances to their target membrane. Such is the case for vesicle delivery during bud formation in *Saccharomyces cerevisiae*. During the cell cycle, the mother cell forms a new daughter cell bud through asymmetrical growth of the plasma membrane and cell wall (194). Bud formation requires directed delivery of post-TGN transport vesicles from the mother to the daughter cell. Significantly, docking/fusion of these transport vesicles is restricted to a small region of the plasma membrane at the tip of the growing bud (194) (see also Chapters 2 and 9 of this volume). If t-SNARE(s) alone specify transport vesicle docking with the bud tip, they should have a distribution that is restricted to that site. However, the plasma membrane t-SNAREs, Sso1p, Sso2p, and Sec9p, are distributed uniformly over the mother cell and bud plasma membranes (131). Similarly, in neurons, syntaxin-1 and SNAP-25 have uniform distributions along the length of the axon, although the sites of synaptic transmission are restricted (195, 196). These results suggest that v- and t-SNAREs define a set of membranes that have the capacity to fuse, but that other proteins are required to define the site at which fusion occurs.

How are the sites of vesicle docking and fusion with plasma membrane selected? To facilitate interactions with v-SNAREs, t-SNAREs must be activated, perhaps through association with, or dissociating from, other proteins (197). These accessory proteins could be restricted to sites of fusion. Alternatively, t-SNAREs may be competent for fusion everywhere on the plasma membrane, but other structures such as microtubules, actin, or septin filaments direct vesicles only to sites where fusion occurs (198). Apart from the question of how vesicle docking and fusion is controlled is the question of where targeting patches are located. Insight into each of these questions is gained by considering other proteins that play a role in vesicle docking and fusion. Proteins that may control SNARE function include homologues of Sec1, Rop, and Unc-18 proteins, which are ubiquitously expressed (199–203). One isoform appears to be enriched at the apical plasma membrane of polarized epithelial cells (204). Mammalian Unc-18 homologues (Munc-18) lack membrane-spanning domains, but may be recruited from the cytosol to the plasma membrane by association with syntaxins (202, 204). Munc-18a (also known as nsec-1/rb-sec1) and Munc-18b bind syntaxins 1a, 2 and 3 (201, 205), while Munc-18c binds syntaxin-2 and syntaxin-4 (206). These *in vitro* binding studies reveal affinities in the nanomolar range, although the complexes appear to be unstable *in vivo* (205, 206). Interest in Munc-18 isoforms stems from the finding that Munc-18 binding to syntaxin inhibits syntaxin binding to SNAP-25 and VAMP (205). Therefore, Munc-18 proteins could regulate vesicle fusion by controlling the ability of syntaxins to interact with other components of the SNARE complex. A further level of control likely involves regulation of Munc-18–syntaxin binding by phosphorylation (207, 208), or Munc-18 binding proteins such as Mints (209) and DOC2 proteins (210). Regulatory proteins, such as Mints, may be restricted to sites of vesicle docking and fusion through interactions with specific plasma membrane proteins via PDZ domains (209).

A second set of proteins required for transport vesicle docking and fusion with the plasma membrane consists of a family of small GTP-binding proteins related to the yeast Ypt1/Sec4 proteins (211). Sec4p is present on post-Golgi transport vesicles and plasma membrane (212), and these vesicles accumulate in the daughter-cell bud adjacent to the bud tip in *sec4* mutants (213). A closely related mammalian homologue, rab8, is present on TGN-derived basolateral transport vesicles and on the basolateral membrane domain of polarized MDCK cells (214). Perturbation of rab8 function in permeabilized MDCK cells caused a significant reduction in the efficiency of vesicle transport to the basolateral, but not the apical plasma membrane domain (214). Therefore, Sec4, rab8, and closely related proteins such as rab10 (215) and rab13 (216) are likely to be involved in vesicle docking and fusion with the plasma membrane. Although the precise role of these proteins is unclear, they have been proposed to potentiate the binding of v- and t-SNAREs (197, 217). Therefore, these proteins could be key regulators of either the timing or localization of vesicle fusion. Interestingly, both rab8 and rab13 are enriched on the plasma membrane at the apical junctional complex, which includes the tight junction and adherens junctions (214, 216). Furthermore, rab3B, an isoform of a brain-specific protein involved in synaptic vesicle fusion, also has a very restricted distribution at the tight junction (218). Disruption of cell–cell contacts results in redistribution of both rab3B and rab13 from the plasma membrane to the cytoplasm, indicating that the restricted distribution of these proteins is dependent on cell–cell contact (216, 218).

Although rab proteins are likely to be involved in regulating vesicle docking and fusion, they are vesicle-associated proteins and therefore cannot, by themselves, mark a site on the plasma membrane for vesicle docking. Certain rab proteins, such as rab3B, rab8, and rab13, are enriched on the plasma membrane at the tight junction (see above). However, it is not known whether these proteins are recruited to targeting patches prior to transport vesicles or if they are carried to exocytic sites on the vesicles themselves. We propose that rab3B, rab8, and rab13 are deposited at the apical junctional complex following vesicle docking and fusion at this site (see below). Steady state localization of rab proteins at the apical junctional complex may result from slow removal from the plasma membrane relative to the rate of vesicle fusion.

A spatial landmark for vesicle docking and fusion must have the following properties: it is restricted to sites of exocytosis on the plasma membrane, and is localized there prior to the arrival of transport vesicles. A further requirement for a spatial landmark involved in establishing plasma membrane domains during development of epithelial cell polarity is that the proteins that constitute the landmark should become spatially restricted following initiation of cell–cell or cell–ECM contact. A multiprotein complex, consisting of Sec3, Sec5, Sec6, Sec8, Sec10, Sec15, and Exo70 gene products, meets these criteria (219). In yeast, this protein complex (termed the 'exocyst'; 220, 221) is present on the plasma membrane and is restricted to sites of active exocytosis, including the tips of growing buds and the mother-bud neck of large-budded cells (220, 222, 223). Recently, mammalian homologues of Sec5, Sec6, Sec8, Sec10, Sec15, and Exo70 have been identified in neurons (224–227). Similar to yeast, these proteins, referred to here as the Sec6/8 complex, are present in a large complex in neurons and MDCK cells (219, 228).

The organization, distribution, and function of the Sec6/8 complex have the characteristics of a candidate targeting patch. In single MDCK cells in contact with the substratum, the Sec6/8 complex is distributed diffusely in the cytosol, but is rapidly recruited ($t_{1/2} \sim$ 3–6 h) to the plasma membrane at sites of cell–cell contact after the induction of calcium-dependent cell adhesion (219). In early contacting cells, the Sec6/8 complex is co-distributed with E-cadherin and the tight-junction-associated protein ZO-1 along the length of each cell–cell contact, but does not extend beyond the boundary of these contacts (219). As the monolayer becomes polarized, the distribution of Sec6/8 becomes restricted to the apex of the lateral membrane and no longer extends along the length of each cell–cell contact (Fig. 6). Although this distribution is distinct from that of E-cadherin in polarized cells, localization of the Sec6/8 complex at the apical junctional complex is still dependent upon calcium-dependent cell–cell adhesion (219). EGTA treatment of cells disrupts E-cadherin-

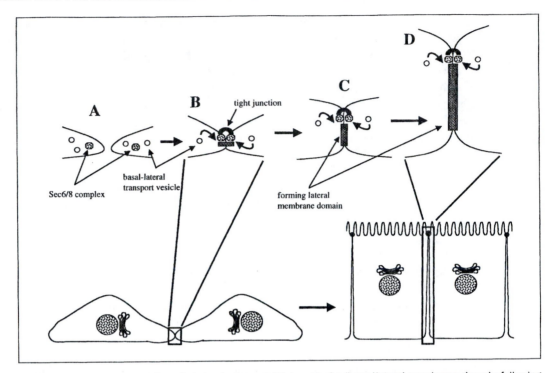

**Fig. 6** Schematic representation of stages in the establishment of a (baso-)lateral membrane domain following induction of cell-cell contact. (A) Prior to cell-cell adhesion, the Sec6/8 complex is cytosolic and basolateral vesicles do not dock efficiently with the plasma membrane. (B) Following cell-cell adhesion, the Sec6/8 complex is recruited to sites of cell-cell adhesion and now causes the efficient and specific recruitment of basolateral transport vesicles to that region of the membrane. (C) and (D) As more and more basolateral transport vesicles fuse with membrane, the new membrane domain grows in area, resulting in the Sec6/8 complex and apical junctional complex (tight junction) becoming restricted to the apex of the lateral membrane.

mediated cell-cell contacts and results in dissociation of Sec6/8 from the plasma membrane. Following re-addition of calcium to the culture medium, the Sec6/8 complex is re-recruited to the plasma membrane and relocalizes to the apical junctional complex with kinetics similar to those observed for ZO-1.

In streptolysin-O (SLO)-permeabilized MDCK cells, antibodies to Sec8 significantly reduce delivery of LDL receptor-containing vesicles to the basolateral membrane, while targeted delivery of p75[NTR] to the apical plasma membrane is unaffected (219). Since the Sec6/8 complex is restricted to the apex of the lateral membrane, this result implies that vesicle docking and fusion occurs near the apical junctional complex. This observation is consistent with previously reported localization of multiple rab proteins at this site (214, 216, 218). Furthermore, the observed redistribution of rab3B and rab13 into cytosolic vesicular structures following extracellular calcium depletion (216, 218) could reflect the accumulation of transport vesicles that are unable to dock efficiently with the plasma membrane in the absence of the Sec6/8 complex.

If the Sec6/8 complex specifies a subdomain of the plasma membrane for efficient vesicle docking, we anticipate that basolateral proteins would be delivered less efficiently to the plasma membrane in contact-naive MDCK cells because the Sec6/8 complex is not membrane bound. Similarly, when the Sec6/8 complex is recruited to the plasma membrane to sites of cell–cell contact, we would anticipate an increased efficiency of delivery of basolateral proteins (8, 219). These predictions are supported by previously published results. Detailed studies of the trafficking of newly synthesized desmoglein I, a basolateral membrane protein, showed than less than 10% was transported to the plasma membrane of contact-naive MDCK cells during a 1 hour chase (229). However, upon initiation of cell–cell contacts, desmoglein I was rapidly ($t_{1/2} \sim 30$ min) and efficiently (>90%) delivered to the plasma membrane (229).

Mechanisms regulating the assembly of targeting patches for transport vesicles on different membrane domains are unknown. Detailed studies of the kinetics and fidelity of membrane protein trafficking between the TGN and the basolateral membrane indicate that direct delivery occurs within 6 hours of induction of cell–cell adhesion (154). That this process is so rapid indicates that cellular components involved in sorting and targeting of membrane proteins to the basolateral membrane domain are present in non-polarized cells prior to cell–cell adhesion. Furthermore, because direct targeting requires cadherin-mediated cell–cell adhesion, it is likely that components of the targeting patch, at least for vesicles containing basolateral proteins, are somehow associated with the cadherin adhesion complex. Specialized cytoskeletal networks assembled at different membrane domains in response to spatial cues may organize vesicle-targeting patches. In turn, these targeting patches specify protein delivery from sorting compartments, thereby reinforcing and maintaining differences in cell-surface protein distributions that were initiated by the spatial cue(s).

If cell–cell adhesion promotes the localized assembly of a basolateral-specific vesicle-targeting patch on the contacting membrane, how does the apical membrane form? Developing epithelial tissues *in vivo* arise from cell aggregates, which lack a free membrane domain (3, 157). Recruitment of the Sec6/8 complex to sites of cell–cell adhesion likely establishes a targeting patch for basolateral transport vesicles, but several lines of evidence indicate that apical transport vesicles fuse very poorly with these contacting plasma membranes. First, contact between a cell and the ECM is sufficient to restrict apical membrane proteins to the free, non-contacting surface (155, 230). Secondly, epithelial cell aggregates grown in suspension culture restrict apical membrane proteins to the free surface (155). Transfer of these aggregates into a collagen gel deprives the cells of their free surface and apical membrane proteins are rapidly removed from the surface and degraded (156, 157). Finally, treatment of epithelial cells with nocodazole significantly reduces the efficiency of vesicle transport to the apical plasma membrane (96, 154). However, apical vesicles do not fuse with the basolateral membrane, but rather accumulate intracellularly and may, in time, fuse with each other to form large vacuole-like structures (231). Therefore, for apical transport vesicles to fuse with the plasma membrane, a cell contacted on all sides by other cells and the ECM must first establish a non-contacting membrane. The SNARE hypothesis may also account for the *de novo* formation of such a membrane domain.

We propose the following scenario for how an apical surface forms *de novo*. TGN-derived transport vesicles containing previously sorted apical protein cargo would fuse very inefficiently with the plasma membrane along cell–cell contacts. If all surfaces of the cell were in contact with other cells or with the ECM, then TGN-derived apical transport vesicles would accumulate in the cytoplasm. Under such conditions, these vesicles may tend to fuse with one another by homotypic vesicle fusion. Homotypic vesicle fusion has been most thoroughly described for yeast vacuoles, where it is clear that v- and t-SNARE proteins are associated in a complex on both 'donor' and 'target' membranes (232–235). This protein complex is dissociated by the action of NSF and α-SNAP prior to docking, which serves to activate the SNARE proteins for subsequent membrane fusion. Because v- and t-SNAREs have been described together on a number of types of vesicles, it is conceivable that post-Golgi apical transport vesicles contain both v- and t-SNAREs (127, 139, 140, 236–238). Under appropriate conditions, these could mediate homotypic fusion between apical transport vesicles, leading to the generation of a large intracellular lumenal compartment. Such compartments have been described (157, 231, 239). Delivery of this lumenal compartment to the plasma membrane may be mediated by the low levels of apical t-SNAREs present on the basolateral plasma membrane. Insertion of apical lumenal compartments into the lateral membrane has been visualized (231, 240, 241). Once inserted in the plasma membrane, the pre-formed apical membrane may be prevented from mixing with the basolateral membrane domain by the rapid movement and assembly of tight junction components to a site between the two membrane domains (see below).

## 8.3   Tight junctions

The tight junction (TJ) assembles at the boundary of the apical and basolateral membranes, and acts as a selective permeability gate to the paracellular space, and an intramembranous fence to prevent free mixing of membrane-domain-specific proteins and lipids (242–245). TJ contains at least two tetra-spanning membrane proteins, occludin and claudin (246). Several cytoplasmic proteins interact either directly or indirectly with occludin, including ZO-1, ZO-2, and ZO-3 (247–252). These proteins may act as a protein scaffold to organize occludin/claudin at the TJ. In addition, these cytoplasmic ZO proteins contain PDZ domains and SH3 domains (248, 250–254) that may recruit the actin cytoskeleton and signalling molecules, including tyrosine kinases. RhoA or Rac1 regulate both TJ function and spatial organization of TJ proteins (255). The TJ may play an even broader role in the organization of membrane domains. Components of the vesicle docking/fusion machinery, including Sec6/8 (219) and rab8 (214) and rab13 (216), are localized to the apical junctional complex, and there is indirect evidence that membrane proteins are delivered to this region of the lateral membrane (256).

# 9. Generating a new membrane domain

Based on the preceding discussion, we propose a model for how cells generate a new membrane domain (see Figs 6 and 7). The simplest scenario is the formation of a new lateral membrane domain following cell–cell adhesion. In the absence of cell–cell contacts, there is not a spatial cue to the cell surface that distinguishes one region of the membrane as being different from another. Under this condition, the Sec6/8 complex is cytosolic and delivery of post-TGN basolateral transport vesicles to the membrane is inefficient, resulting in their accumulation in the cytosol. In addition, cadherin cell-adhesion molecules are distributed diffusely over the cell surface, and the actin cytoskeleton is configured to a motile cell organization. Under these conditions, the free surface is predominantly of an 'apical' composition.

Cell–cell adhesion is mediated by homotypic, weak-affinity interactions between cadherins on adjacent cells, but rapidly results in the reorganization of the actin cytoskeleton in the area defined by the contacting cell membranes. As a result, extensive cell–cell contacts are established rapidly through the coalescence of cadherin into puncta and their association with components of the actin cytoskeleton. It is possible that these cadherin/actin puncta act as nucleation sites for signalling complexes that initiate the recruitment and assembly of the spectrin-based membrane skeleton and targeting patches for basolateral transport vesicles. Although mechanisms involved in recruiting targeting patches to cell–cell contacts is unknown, cadherin-mediated adhesion induces the reorganization of the Sec6/8 complex from the cytoplasm to the contacting membranes. As a consequence, we propose that basolateral transport vesicles are rapidly and specifically targeted to sites on the plasma membrane that are enriched in the Sec6/8 complex, where they dock and

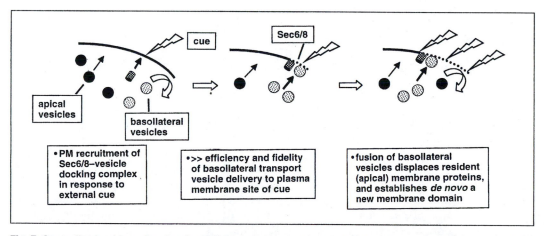

**Fig. 7** Generalized scheme for the formation of a new membrane domain following induction (cue) of the recruitment of the Sec6/8 complex to the membrane. Basolateral vesicles (stippled) fuse with the region of the membrane containing the Sec6/8 complex, a new membrane domain appears (dotted line) that has a structural and functional identity different from that of the original membrane (solid line). A consequence of this new identity is that apical vesicles cannot dock and fuse, thereby maintaining the differences between the two membrane domains.

fuse. This process concentrates both newly delivered membrane in the form of transport vesicles, and specific basolateral membrane proteins to the new plasma membrane domain initiated by cell–cell adhesion. In fact, direct measurements of the surface area of MDCK cells following cell–cell adhesion revealed little or no increase in either the apical or basal surface areas, but a sixfold increase in surface area of the lateral membrane domain. As a consequence, the surface area of the lateral membrane domain increases dramatically. During this period of rapid membrane growth, the distribution of the Sec6/8 complex becomes restricted to the apical junctional complex. Although the mechanism involved in this redistribution is unknown, it is possible that the Sec6/8 complex is either segregated with the tight junction complex from the cadherin/catenin complex, or is linked to the cortical actin bundle that eventually associates with the apical junctional complex. Subsequently, the protein identity of the basolateral membrane is maintained by the direct delivery of basolateral transport vesicles to the Sec6/8 complex and the associated docking/fusion machinery. Finally, as the basolateral identity of this membrane domain is established, the machinery for specifying apical vesicle targeting, docking, and fusion at this new membrane domain is reduced to the point at which apical vesicles are unable to dock with the basolateral membrane domain.

## Acknowledgements

Work from the Nelson laboratory was supported by a grant from the NIH. K. K. Grindstaff is supported additionally by a senior fellowship from the American Cancer Society, California Affiliate, and C. Yeaman is supported by a Walter Berry Fellowship.

## References

1. Berridge, M. J. and Oschman, J. L. (1972) *Transporting epithelia*, p. 91. Academic Press, New York.
2. Simons, K. (1987) Membrane traffic in an epithelial cell line derived from the dog kidney. *Kidney Int. Suppl.*, S201.
3. Rodriguez-Boulan, E. and Nelson, W. J. (1989) Morphogenesis of the polarized epithelial cell phenotype. *Science*, **245**, 718.
4. Simons, K. and Wandinger-Ness, A. (1990) Polarized sorting in epithelia. *Cell*, **62**, 207.
5. Yoshimori, T., Keller, P., Roth, M. G., and Simons, K. (1996) Different biosynthetic transport routes to the plasma membrane in BHK and CHO cells. *J. Cell Biol.*, **133**, 247.
6  Müsch, A., Xu, H., Shields, D., and Rodriguez-Boulan, E. (1996) Transport of vesicular stomatitis virus G protein to the cell surface is signal mediated in polarized and nonpolarized cells. *J. Cell Biol.*, **133**, 543.
7. Drubin, D. G. and Nelson, W. J. (1996) Origins of cell polarity. *Cell*, **84**, 335.
8. Yeaman, C., Grindstaff, K. K., and Nelson, W. J. (1999) New perspectives on mechanisms involved in generating epithelial cell polarity. *Physiol. Rev.*, **79**, 73.
9. Matter, K. and Mellman, I. (1994) Mechanisms of cell polarity: sorting and transport in epithelial cells. *Curr. Opin. Cell Biol.*, **6**, 545.

10. Rodriguez-Boulan, E. and Powell, S. K. (1992) Polarity of epithelial and neuronal cells. *Annu. Rev. Cell Biol.*, **8**, 395.

11. Rodriguez-Boulan, E. and Pendergast, M. (1980) Polarized distribution of viral envelope proteins in the plasma membrane of infected epithelial cells. *Cell*, **20**, 45.

12. Rodriguez-Boulan, E. and Sabatini, D. D. (1978) Asymmetric budding of viruses in epithelial monlayers: a model system for study of epithelial polarity. *Proc. Natl Acad. Sci., USA*, **75**, 5071.

13. Trowbridge, I. S., Collawn, J. F., and Hopkins, C. R. (1993) Signal-dependent membrane protein trafficking in the endocytic pathway. *Annu. Rev. Cell Biol.*, **9**, 129.

14. Collawn, J. F., Stangel, M., Kuhn, L. A., Esekogwu, V., Jing, S. Q., Trowbridge, I. S., and Tainer, J. A. (1990) Transferrin receptor internalization sequence YXRF implicates a tight turn as the structural recognition motif for endocytosis. *Cell*, **63**, 1061.

15. Eberle, W., Sander, C., Klaus, W., Schmidt, B., von Figura, K., and Peters, C. (1991) The essential tyrosine of the internalization signal in lysosomal acid phosphatase is part of a β turn. *Cell*, **67**, 1203.

16. Hunziker, W., Harter, C., Matter, K., and Mellman, I. (1991) Basolateral sorting in MDCK cells requires a distinct cytoplasmic domain determinant. *Cell*, **66**, 907.

17. Matter, K., Hunziker, W., and Mellman, I. (1992) Basolateral sorting of LDL receptor in MDCK cells: the cytoplasmic domain contains two tyrosine-dependent targeting determinants. *Cell*, **71**, 741.

18. Geffen, I., Fuhrer, C., Leitinger, B., Weiss, M., Huggel, K., Griffiths, G., and Spiess, M. (1993) Related signals for endocytosis and basolateral sorting of the asialoglycoprotein receptor. *J. Biol. Chem.*, **268**, 20772.

19. Prill, V., Lehmann, L., von, F. K., and Peters, C. (1993) The cytoplasmic tail of lysosomal acid phosphatase contains overlapping but distinct signals for basolateral sorting and rapid internalization in polarized MDCK cells. *EMBO J.*, **12**, 2181.

20. Thomas, D. C., Brewer, C. B., and Roth, M. G. (1993) Vesicular stomatitis virus glycoprotein contains a dominant cytoplasmic basolateral sorting signal critically dependent upon a tyrosine. *J. Biol. Chem.*, **268**, 3313.

21. Thomas, D. C. and Roth, M. G. (1994) The basolateral targeting signal in the cytoplasmic domain of glycoprotein G from vesicular stomatitis virus resembles a variety of intracellular targeting motifs related by primary sequence but having diverse targeting activities. *J. Biol. Chem.*, **269**, 15732.

22. Lodge, R., Lalonde, J. P., Lemay, G., and Cohen, E. A. (1997) The membrane-proximal intracytoplasmic tyrosine residue of HIV-1 envelope glycoprotein is critical for basolateral targeting of viral budding in MDCK cells. *EMBO J.*, **16**, 695.

23. Le Bivic, A., Sambuy, Y., Patzak, A., Patil, N., Chao, M., and Rodriguez-Boulan, E. (1991) An internal deletion in the cytoplasmic tail reverses the apical localization of human NGF receptor in transfected MDCK cells. *J. Cell Biol.*, **115**, 607.

24. Monlauzeur, L., Rajasekaran, A., Chao, M., Rodriguez-Boulan, E., and Le Bivic, A. (1995) A cytoplasmic tyrosine is essential for the basolateral localization of mutants of the human nerve growth factor receptor in Madin–Darby canine kidney cells. *J. Biol. Chem.*, **270**, 12219.

25. Brewer, C. B. and Roth, M. G. (1991) A single amino acid change in the cytoplasmic domain alters the polarized delivery of influenza virus hemagglutinin. *J. Cell Biol.*, **114**, 413.

26. Hunziker, W. and Fumey, C. (1994) A di-leucine motif mediates endocytosis and basolateral sorting of macrophage IgG Fc receptors in MDCK cells. *EMBO J.*, **13**, 2963.

27. Matter, K., Yamamoto, E. M., and Mellman, I. (1994) Structural requirements and

sequence motifs for polarized sorting and endocytosis of LDL and Fc receptors in MDCK cells. *J. Cell Biol.*, **126**, 991.

28. Odorizzi, G. and Trowbridge, I. S. (1997) Structural requirements for major histo-compatibility complex class II invariant chain trafficking in polarized Madin–Darby canine kidney cells. *J. Biol. Chem.*, **272**, 11757.

29. Simonsen, A., Stang, E., Bremnes, B., Roe, M., Prydz, K., and Bakke, O. (1997) Sorting of MHC class II molecules and the associated invariant chain (Ii) in polarized MDCK cells. *J. Cell Sci.*, **110**, 597.

30. Aroeti, B., Kosen, P. A., Kuntz, I. D., Cohen, F. E., and Mostov, K. E. (1993) Mutational and secondary structural analysis of the basolateral sorting signal of the polymeric immuno-globulin receptor. *J. Cell Biol.*, **123**, 1149.

31. Le Gall, A. H., Powell, S. K., Yeaman, C. A., and Rodriguez-Boulan, E. (1997) The neural cell adhesion molecule expresses a tyrosine-independent basolateral sorting signal. *J. Biol. Chem.*, **272**, 4559.

32. Odorizzi, G. and Trowbridge, I. S. (1997) Structural requirements for basolateral sorting of the human transferrin receptor in the biosynthetic and endocytic pathways of Madin–Darby canine kidney cells. *J. Cell Biol.*, **137**, 1255.

33. Keller, P. and Simons, K. (1997) Post-Golgi biosynthetic trafficking. *J. Cell Sci.*, **110**, 3001.

34. Brown, D. A., Crise, B., and Rose, J. K. (1989) Mechanism of membrane anchoring affects polarized expression of two proteins in MDCK cells. *Science*, **245**, 1499.

35. Lisanti, M. P., Caras, I. W., Davitz, M. A., and Rodriguez-Boulan, E. (1989) A glycophos-pholipid membrane anchor acts as an apical targeting signal in polarized epithelial cells. *J. Cell Biol.*, **109**, 2145.

36. Wilson, J. M., Fasel, N., and Kraehenbuhl, J. P. (1990) Polarity of endogenous and exo-genous glycosyl-phosphatidylinositol-anchored membrane proteins in Madin–Darby canine kidney cells. *J. Cell Sci.*, **96**, 143.

37. McQueen, N., Nayak, D. P., Stephens, E. B., and Compans, R. W. (1986) Polarized ex-pression of a chimeric protein in which the transmembrane and cytoplasmic domains of the influenza virus hemagglutinin have been replaced by those of the vesicular stomatitis virus G protein. *Proc. Natl Acad. Sci., USA*, **83**, 9318.

38. Roth, M. G., Gundersen, D., Patil, N., and Rodriguez-Boulan, E. (1987) The large external domain is sufficient for the correct sorting of secreted or chimeric influenza virus hemagglutinins in polarized monkey kidney cells. *J. Cell Biol.*, **104**, 769.

39. Mostov, K. E., de Bruyn Kops, A., and Deitcher, D. L. (1986) Deletion of the cytoplasmic domain of the polymeric immunoglobulin receptor prevents basolateral localization and endocytosis. *Cell*, **47**, 359.

40. Mostov, K. E., Breitfeld, P., and Harris, J. M. (1987) An anchor-minus form of the polymeric immunoglobulin receptor is secreted predominantly apically in Madin–Darby canine kidney cells. *J. Cell Biol.*, **105**, 2031.

41. Powell, S. K., Lisanti, M. P., and Rodriguez-Boulan, E. J. (1991) Thy-1 expresses two signals for apical localization in epithelial cells. *Am. J. Physiol.*, **260**, C715.

42. Vogel, L. K., Spiess, M., Sjostrom, H., and Noren, O. (1992) Evidence for an apical sorting signal on the ectodomain of human aminopeptidase N. *J. Biol. Chem.*, **267**, 2794.

43. Weisz, O. A., Machamer, C. E., and Hubbard, A. L. (1992) Rat liver dipeptidylpeptidase IV contains competing apical and basolateral targeting information. *J. Biol. Chem.*, **267**, 22282.

44. Yeaman, C., Le Gall, A. H., Baldwin, A. N., Monlauzeur, L., Le Bivic, A., and Rodriguez-Boulan, E. (1997) The O-glycosylated stalk domain is required for apical sorting of neurotrophin receptors in polarized MDCK cells. *J. Cell Biol.*, **139**, 929.

45. Green, R. F., Meiss, H. K., and Rodriguez-Boulan, E. (1981) Glycosylation does not determine segregation of viral envelope proteins in the plasma membrane of epithelial cells. *J. Cell Biol.*, **89**, 230.

46. Roth, M. G., Fitzpatrick, J. P., and Compans, R. W. (1979) Polarity of influenza and vesicular stomatitis virus maturation in MDCK cells: lack of a requirement for glycosylation of viral glycoproteins. *Proc. Natl Acad. Sci., USA*, **76**, 6430.

47. Gut, A., Kappeler, F., Hyka, N., Balda, M. S., Hauri, H.-P., and Matter, K. (1998) Carbohydrate-mediated Golgi to cell surface transport and apical targeting of membrane proteins. *EMBO J.*, **17**, 1919.

48. Kitagawa, Y., Sano, Y., Ueda, M., Higashio, K., Narita, H., Okano, M., Matsumoto, S., and Sasaki, R. (1994) N-glycosylation of erythropoietin is critical for apical secretion by Madin–Darby canine kidney cells. *Exp. Cell Res.*, **213**, 449.

49. Scheiffele, P., Peranen, J., and Simons, K. (1995) N-glycans as apical sorting signals in epithelial cells. *Nature*, **378**, 96.

50. Urban, J., Parczyk, K., Leutz, A., Kayne, M., and Kondor-Koch, C. (1987) Constitutive apical secretion of an 80-kD sulfated glycoprotein complex in the polarized epithelial Madin–Darby canine kidney cell line. *J. Cell Biol.*, **105**, 2735.

51. Gonzalez, A., Nicovani, S., and Juica, F. (1993) Apical secretion of hepatitis B surface antigen from transfected Madin–Darby canine kidney cells. *J. Biol Chem.*, **268**, 6662.

52. Hughey, P. G., Compans, R. W., Zebedee, S. L., and Lamb, R. A. (1992) Expression of the influenza A virus M2 protein is restricted to apical surfaces of polarized epithelial cells. *J. Virol.*, **66**, 5542.

53. Marzolo, M. P., Bull, P., and Gonzalez, A. (1997) Apical sorting of hepatitis B surface antigen (HBsAg) is independent of N-glycosylation and glycosylphosphatidylinositol-anchored protein segregation. *Proc. Natl Acad. Sci., USA*, **94**, 1834.

54. Ragno, P., Estreicher, A., Gos, A., Wohlwend, A., Belin, D., and Vassalli, J. D. (1992) Polarized secretion of urokinase-type plasminogen activator by epithelial cells. *Exp. Cell Res.*, **203**, 236.

55. Ullrich, O., Mann, K., Haase, W., and Koch, B. C. (1991) Biosynthesis and secretion of an osteopontin-related 20-kDa polypeptide in the Madin–Darby canine kidney cell line. *J. Biol. Chem.*, **266**, 3518.

56. Fiedler, K., Parton, R. G., Kellner, R., Etzold, T., and Simons, K. (1994) VIP36, a novel component of glycolipid rafts and exocytic carrier vesicles in epithelial cells. *EMBO J.*, **13**, 1729.

57. Fiedler, K. and Simons, K. (1996) Characterization of VIP36, an animal lectin homologous to leguminous lectins. *J. Cell Sci.*, **109**, 271.

58. Boll, W., Ohno, H., Songyang, Z., Rapoport, I., Cantley, L. C., Bonifacino, J. S., and Kirchhausen, T. (1996) Sequence requirements for the recognition of tyrosine-based endocytic signals by clathrin AP-2 complexes. *EMBO J.*, **15**, 5789.

59. Honig, S., Griffith, J., Geuze, H. J., and Hunziker, W. (1996) The tyrosine-based lysosomal targeting signal in lamp-1 mediates sorting into Golgi-derived clathrin-coated vesicles. *EMBO J.*, **15**, 5230.

60. Ohno, H., Fournier, M. C., Poy, G., and Bonifacino, J. S. (1996) Structural determinants of interaction of tyrosine-based sorting signals with the adaptor medium chains. *J. Biol. Chem.*, **271**, 29009.

61. Ohno, H., Stewart, J., Fournier, M. C., Bosshart, H., Rhee, I., Miyatake, S., Saito, T., Gallusser, A., Kirchhausen, T., and Bonifacino, J. S. (1995) Interaction of tyrosine-based sorting signals with clathrin-associated proteins. *Science*, **269**, 1872.

62. Rodionov, D. G. and Bakke, O. (1998) Medium chains of adaptor complexes AP-1 and AP-2 recognize leucine-based sorting signals from the invariant chain. *J. Biol. Chem.*, **273**, 6005.

63. Dell'Angelica, E. C., Ohno, H., Ooi, C. E., Rabinovich, E., Roche, K. W., and Bonifacino, J. S. (1997) AP-3: an adaptor-like protein complex with ubiquitous expression. *EMBO J.*, **16**, 917.

64. Dell'Angelica, E. C., Ooi, C. E., and Bonifacino, J. S. (1997) Beta3A-adaptin, a subunit of the adaptor-like complex AP-3. *J. Biol. Chem.*, **272**, 15078.

65. Simpson, F., Bright, N. A., West, M. A., Newman, L. S., Darnell, R. B., and Robinson, M. S. (1996) A novel adaptor-related protein complex. *J. Cell Biol.*, **133**, 749.

66. Simpson, F., Peden, A. A., Christopoulou, L., and Robinson, M. S. (1997) Characterization of the adaptor-related protein complex, AP-3. *J. Cell Biol.*, **137**, 835.

67. Honig, S., Sandoval, I. V., and Von Figura, K. (1998) A di-leucine-based motif inthe cytoplmasic tail of LIMP-II and tyrosinase mediates selective binding of AP-3. *EMBO J.*, **17**, 1304.

68. Cowles, C. R., Odorizzi, G., Payne, G. S., and Emr, S. D. (1997) The AP-3 adaptor complex is essential for cargo-selective transport to the yeast vacuole. *Cell*, **91**, 109.

69. Ooi, C. E., Moreira, J. E., Dell'Angelica, E. C., Poy, G., Wassarman, D. A., and Bonifacino, J. S. (1997) Altered expression of a novel adaptin leads to defective pigment granule biogenesis in the *Drosophila* eye color mutant garnet. *EMBO J.*, **16**, 4508.

70. Simons, K. and van Meer, G. (1988) Lipid sorting in epithelial cells. *Biochemistry*, **27**, 6197.

71. Scheiffele, P., Roth, M. G., and Simons, K. (1997) Interaction of influenza virus haemagglutinin with sphingolipid-cholesterol membrane domains via its transmembrane domain. *EMBO J.*, **16**, 5501.

72. Kundu, A., Avalos, R. T., Sanderson, C. M., and Nayak, D. P. (1996) Transmembrane domain of influenza virus neuraminidase, a type II protein, possesses an apical sorting signal in polarized MDCK cells. *J. Virol.*, **70**, 6508.

73. Brown, D. A. and Rose, J. K. (1992) Sorting of GPI-anchored proteins to glycolipid-enriched membrane subdomains during transport to the apical cell surface. *Cell*, **68**, 533.

74. Garcia, M., Mirre, C., Quaroni, A., Reggio, H., and Le Bivic, A. (1993) GPI-anchored proteins associate to form microdomains during their intracellular transport in Caco-2 cells. *J. Cell Sci.*, **104**, 1281.

75. Zurzolo, C., van't Hof, W., van Meer, G., and Rodriguez-Boulan, E. (1994) VIP21/caveolin, glycosphingolipid clusters and the sorting of glycosylphosphatidylinositol-anchored proteins in epithelial cells. *EMBO J.*, **13**, 42.

76. Hannan, L. A., Lisanti, M. P., Rodriguez-Boulan, E., and Edidin, M. (1993) Correctly sorted molecules of a GPI-anchored protein are clustered and immobile when they arrive at the apical surface of MDCK cells. *J. Cell Biol.*, **120**, 353.

77. Mays, R. W., Siemers, K. A., Fritz, B. A., Lowe, A. W., van Meer, G., and Nelson, W. J. (1995) Hierarchy of mechanisms involved in generating Na/K-ATPase polarity in MDCK epithelial cells. *J. Cell Biol.*, **130**, 1105.

78. Keller, P. and Simons, K. (1998) Cholesterol is required for surface transport of influenza virus hemagglutinin. *J. Cell Biol.*, **140**, 1357.

79. Arreaza, G. and Brown, D. A. (1995) Sorting and intracellular trafficking of a glycosyl-phosphatidylinositol-anchored protein and two hybrid transmembrane proteins with the same ectodomain in Madin–Darby canine kidney epithelial cells. *J. Biol. Chem.*, **270**, 23641.

80. Danielsen, E. M. (1995) Involvement of detergent-insoluble complexes in the intracellular transport of intestinal brush border enzymes. *Biochemistry*, **34**, 1596.

81. Wollner, D. A., Krzeminski, K. A., and Nelson, W. J. (1992) Remodeling the cell surface distribution of membrane proteins during the development of epithelial cell polarity. *J. Cell Biol.*, **116**, 889.

82. Hays, R. M., Condeelis, J., Gao, Y., Simon, H., Ding, G., and Franki, N. (1993) The effect of vasopressin on the cytoskeleton of the epithelial cell. *Pediatr. Nephrol.*, **7**, 672.

83. Hays, R. M., Franki, N., Simon, H., and Gao, Y. (1994) Antidiuretic hormone and exocytosis: lessons from neurosecretion. *Am. J. Physiol.*, **267**, C1507.

84. Mays, R. W. and Nelson, W. J. (1992) Mechanisms for regulating the cell surface distribution of Na/K-ATPase in polarized epithelial cells. *Chest*, **101**, 50S.

85. Mays, R. W., Beck, K. A., and Nelson, W. J. (1994) Organization and function of the cytoskeleton in polarized epithelial cells: a component of the protein sorting machinery. *Curr. Opin. Cell Biol.*, **6**, 16.

86. Gottlieb, T. A., Ivanov, I. E., Adesnik, M., and Sabatini, D. D. (1993) Actin microfilaments play a critical role in endocytosis at the apical but not the basolateral surface of polarized epithelial cells. *J. Cell Biol.*, **120**, 695.

87. Naim, H. Y., Dodds, D. T., Brewer, C. B., and Roth, M. G. (1995) Apical and basolateral coated pits of MDCK cells differ in their rates of maturation into coated vesicles, but not in the ability to distinguish between mutant hemagglutinin proteins with different internalization signals. *J. Cell Biol.*, **129**, 1241.

88. Maples, C. J., Ruiz, W. G., and Apodaca, G. (1997) Both microtubules and actin filaments are required for efficient postendocytotic traffic of the polymeric immunoglobulin receptor in polarized Madin–Darby canine kidney cells. *J. Biol. Chem.*, **272**, 6741.

89. Parczyk, K., Haase, W., and Kondor, K. C. (1989) Microtubules are involved in the secretion of proteins at the apical cell surface of the polarized epithelial cell, Madin–Darby canine kidney. *J. Biol. Chem.*, **264**, 16837.

90. Ojakian, G. K. and Schwimmer, R. (1994) Regulation of epithelial cell surface polarity reversal by beta 1 integrins. *J. Cell Sci.*, **107**, 561.

91. Salas, P. J., Misek, D. E., Vega-Salas, D. E., Gundersen, D., Cereijido, M., and Rodriguez-Boulan, E. (1986) Microtubules and actin filaments are not critically involved in the biogenesis of epithelial cell surface polarity. *J. Cell Biol.*, **102**, 1853.

92. Fath, K. R. and Burgess, D. R. (1993) Golgi-derived vesicles from developing epithelial cells bind actin filaments and possess myosin-I as a cytoplasmically oriented peripheral membrane protein. *J. Cell Biol.*, **120**, 117.

93. Breitfeld, P. P., McKinnon, W. C., and Mostov, K. E. (1990) Effect of nocodazole on vesicular traffic to the apical and basolateral surfaces of polarized MDCK cells. *J. Cell Biol.*, **111**, 2365.

94. Hunziker, W., Male, P., and Mellman, I. (1990) Differential microtubule requirements for transcytosis in MDCK cells. *EMBO J.*, **9**, 3515.

95. Matter, K., Bucher, K., and Hauri, H. P. (1990) Microtubule perturbation retards both the direct and the indirect apical pathway but does not affect sorting of plasma membrane proteins in intestinal epithelial cells (Caco-2). *EMBO J.*, **9**, 3163.

96. van Zeijl, M. J. and Matlin, K. S. (1990) Microtubule perturbation inhibits intracellular transport of an apical membrane glycoprotein in a substrate-dependent manner in polarized Madin–Darby canine kidney epithelial cells. *Cell Regul.*, **1**, 921.

97. Achler, C., Filmer, D., Merte, C., and Drenckhahn, D. (1989) Role of microtubules in polarized delivery of apical membrane proteins to the brush border of the intestinal epithelium. *J. Cell Biol.*, **109**, 179.

98. Eilers, U., Klumperman, J., and Hauri, H. P. (1989) Nocodazole, a microtubule-active drug, interferes with apical protein delivery in cultured intestinal epithelial cells (Caco-2). *J. Cell Biol.*, **108**, 13.

99. Gilbert, T., Le Bivic, A., Quaroni, A., and Rodriguez-Boulan, E. (1991) Microtubular

organization and its involvement in the biogenetic pathways of plasma membrane proteins in Caco-2 intestinal epithelial cells. *J. Cell Biol.*, **113**, 275.

100. Rindler, M. J., Ivanov, I. E., and Sabatini, D. D. (1987) Microtubule-acting drugs lead to the nonpolarized delivery of the influenza hemagglutinin to the cell surface of polarized Madin–Darby canine kidney cells. *J. Cell Biol.*, **104**, 231.

101. van der Sluijs, P., Bennett, M. K., Antony, C., Simons, K., and Kreis, T. E. (1990) Binding of exocytic vesicles from MDCK cells to microtubules *in vitro*. *J. Cell Sci.*, **95**, 545.

102. Lafont, F., Burkhardt, J. K., and Simons, K. (1994) Involvement of microtubule motors in basolateral and apical transport in kidney cells. *Nature*, **372**, 801.

103. Rothman, J. E. and Warren, G. (1994) Implications of the SNARE hypothesis for intracellular membrane topology and dynamics. *Curr. Biol.*, **4**, 220.

104. Calakos, N., Bennett, M. K., Peterson, K. E., and Scheller, R. H. (1994) Protein–protein interactions contributing to the specificity of intracellular vesicular trafficking. *Science*, **263**, 1146.

105. Sollner, T., Bennett, M. K., Whiteheart, S. W., Scheller, R. H., and Rothman, J. E. (1993) A protein assembly-disassembly pathway *in vitro* that may correspond to sequential steps of synaptic vesicle docking, activation, and fusion. *Cell*, **75**, 409.

106. Block, M. R., Glick, B. S., Wilcox, C. A., Wieland, F. T., and Rothman, J. E. (1988) Purification of an *N*-ethylmaleimide-sensitive protein catalyzing vesicular transport. *Proc. Natl Acad. Sci., USA*, **85**, 7852.

107. Malhotra, V., Orci, L., Glick, B. S., Block, M. R., and Rothman, J. E. (1988) Role of an *N*-ethylmaleimide-sensitive transport component in promoting fusion of transport vesicles with cisternae of the Golgi stack. *Cell*, **54**, 221.

108. Clary, D. O., Griff, I. C., and Rothman, J. E. (1990) SNAPs, a family of NSF attachment proteins involved in intracellular membrane fusion in animals and yeast. *Cell*, **61**, 709.

109. Weidman, P. J., Melancon, P., Block, M. R., and Rothman, J. E. (1989) Binding of an *N*-ethylmaleimide-sensitive fusion protein to Golgi membranes requires both a soluble protein(s) and an integral membrane receptor. *J. Cell Biol.*, **108**, 1589.

110. Sollner, T., Whiteheart, S. W., Brunner, M., Erdjument-Bromage, H., Geromanos, S., Tempst, P., and Rothman, J. E. (1993) SNAP receptors implicated in vesicle targeting and fusion [see comments]. *Nature*, **362**, 318.

111. Whiteheart, S. W., Brunner, M., Wilson, D. W., Wiedmann, M., and Rothman, J. E. (1992) Soluble *N*-ethylmaleimide-sensitive fusion attachment proteins (SNAPs) bind to a multi-SNAP receptor complex in Golgi membranes. *J. Biol. Chem.*, **267**, 12239.

112. Wilson, D. W., Whiteheart, S. W., Wiedmann, M., Brunner, M., and Rothman, J. E. (1992) A multisubunit particle implicated in membrane fusion. *J. Cell Biol.*, **117**, 531.

113. Trimble, W. S., Cowan, D. M., and Scheller, R. H. (1988) VAMP-1: a synaptic vesicle-associated integral membrane protein. *Proc. Natl Acad. Sci., USA*, **85**, 4538.

114. Elferink, L. A., Trimble, W. S., and Scheller, R. H. (1989) Two vesicle-associated membrane protein genes are differentially expressed in the rat central nervous system. *J. Biol. Chem.*, **264**, 11061.

115. Baumert, M., Maycox, P. R., Navone, F., De Camilli, P., and Jahn, R. (1989) Synapto-brevin: an integral membrane protein of 18,000 daltons present in small synaptic vesicles of rat brain. *EMBO J.*, **8**, 379.

116. Sudhof, T. C., Baumert, M., Perin, M. S., and Jahn, R. (1989) A synaptic vesicle membrane protein is conserved from mammals to *Drosophila*. *Neuron*, **2**, 1475.

117. Hay, J. C., Chao, D. S., Kuo, C. S., and Scheller, R. H. (1997) Protein interactions regulating vesicle transport between the endoplasmic reticulum and Golgi apparatus in mammalian cells. *Cell*, **89**, 149.

118. Hay, J. C., Hirling, H., and Scheller, R. H. (1996) Mammalian vesicle trafficking proteins of the endoplasmic reticulum and Golgi apparatus. *J. Biol. Chem.*, **271**, 5671.

119. Mandic, R., Trimble, W. S., and Lowe, A. W. (1997) Tissue-specific alternative RNA splicing of rat vesicle-associated membrane protein-1 (VAMP-1). *Gene*, **199**, 173.

120. McNew, J. A., Sogaard, M., Lampen, N. M., Machida, S., Ye, R. R., Lacomis, L., Tempst, P., Rothman, J. E., and Sollner, T. H. (1997) Ykt6p, a prenylated SNARE essential for endoplasmic reticulum-Golgi transport. *J. Biol. Chem.*, **272**, 17776.

121. Paek, I., Orci, L., Ravazzola, M., Erdjument, B. H., Amherdt, M., Tempst, P., Sollner, T. H., and Rothman, J. E. (1997) ERS-24, a mammalian v-SNARE implicated in vesicle traffic between the ER and the Golgi. *J. Cell Biol.*, **137**, 1017.

122. Nagahama, M., Orci, L., Ravazzola, M., Amherdt, M., Lacomis, L., Tempst, P., Rothman, J. E., and Sollner, T. H. (1996) A v-SNARE implicated in intra-Golgi transport. *J. Cell Biol.*, **133**, 507.

123. Sutton, R. B., Fasshauer, D., Jahn, R., and Brunger, A. T. (1998) Crystal structure of a SNARE complex involved in synaptic exocytosis at 2.4 Å resolution. *Nature*, **395**, 347

124. Bennett, M. K., Calakos, N., and Scheller, R. H. (1992) Syntaxin: a synaptic protein implicated in docking of synaptic vesicles at presynaptic active zones. *Science*, **257**, 255.

125. Bennett, M. K., Garcia-Arraras, J. E., Elferink, L. A., Peterson, K., Fleming, A. M., Hazuka, C. D., and Scheller, R. H. (1993) The syntaxin family of vesicular transport receptors. *Cell*, **74**, 863.

126. Bock, J. B., Lin, R. C., and Scheller, R. H. (1996) A new syntaxin family member implicated in targeting of intracellular transport vesicles. *J. Biol. Chem.*, **271**, 17961.

127. Bock, J. B., Klumperman, J., Davanger, S., and Scheller, R. H. (1997) Syntaxin 6 functions in trans-Golgi network vesicle trafficking. *Mol. Biol. Cell*, **8**, 1261.

128. Darsow, T., Rieder, S. E., and Emr, S. D. (1997) A multispecificity syntaxin homologue, Vam3p, essential for autophagic and biosynthetic protein transport to the vacuole. *J. Cell Biol.*, **138**, 517.

129. Hui, N., Nakamura, N., Sonnichsen, B., Shima, D. T., Nilsson, T., and Warren, G. (1997) An isoform of the Golgi t-SNARE, syntaxin 5, with an endoplasmic reticulum retrieval signal. *Mol. Biol. Cell*, **8**, 1777.

130. Jagadish, M. N., Tellam, J. T., Macaulay, S. L., Gough, K. H., James, D. E., and Ward, C. W. (1997) Novel isoform of syntaxin 1 is expressed in mammalian cells. *Biochem J.*, **321**, 151.

131. Brennwald, P., Kearns, B., Champion, K., Keranen, S., Bankaitis, V., and Novick, P. (1994) Sec9 is a SNAP-25-like component of a yeast SNARE complex that may be the effector of Sec4 function in exocytosis. *Cell*, **79**, 245.

132. Calakos, N. and Scheller, R. H. (1996) Synaptic vesicle biogenesis, docking, and fusion: a molecular description. *Physiol. Rev.*, **76**, 1.

133. Chapman, E. R., An, S., Barton, N., and Jahn, R. (1994) SNAP-25, a t-SNARE which binds to both syntaxin and synaptobrevin via domains that may form coiled coils. *J. Biol. Chem.*, **269**, 27427.

134. Veit, M., Sollner, T. H., and Rothman, J. E. (1996) Multiple palmitoylation of synapto-tagmin and the t-SNARE SNAP-25. *FEBS Lett.*, **385**, 119.

135. Mandon, B., Chou, C. L., Nielsen, S., and Knepper, M. A. (1996) Syntaxin-4 is localized to the apical plasma membrane of rat renal collecting duct cells: possible role in aquaporin-2 trafficking. *J. Clin. Invest.*, **98**, 906.

136. Mandon, B., Nielsen, S., Kishore, B. K., and Knepper, M. A. (1997) Expression of syntaxins in rat kidney. *Am. J. Physiol.*, **273**, F718.

137. Nielsen, S., Marples, D., Birn, H., Mohtashami, M., Dalby, N. O., Trimble, M., and Knepper, M. (1995) Expression of VAMP-2-like protein in kidney collecting duct intracellular vesicles. Colocalization with Aquaporin-2 water channels. *J. Clin. Invest.*, **96**, 1834.

138. Braun, J. E., Fritz, B. A., Wong, S. M., and Lowe, A. W. (1994) Identification of a vesicle-associated membrane protein (VAMP)-like membrane protein in zymogen granules of the rat exocrine pancreas. *J. Biol. Chem.*, **269**, 5328.

139. Gaisano, H. Y., Sheu, L., Foskett, J. K., and Trimble, W. S. (1994) Tetanus toxin light chain cleaves a vesicle-associated membrane protein (VAMP) isoform 2 in rat pancreatic zymogen granules and inhibits enzyme secretion. *J. Biol. Chem.*, **269**, 17062.

140. Gaisano, H. Y., Ghai, M., Malkus, P. N., Sheu, L., Bouquillon, A., Bennett, M. K., and Trimble, W. S. (1996) Distinct cellular locations of the syntaxin family of proteins in rat pancreatic acinar cells. *Mol. Biol. Cell*, **7**, 2019.

141. Gaisano, H. Y., Sheu, L., Grondin, G., Ghai, M., Bouquillon, A., Lowe, A., Beaudoin, A., and Trimble, W. S. (1996) The vesicle-associated membrane protein family of proteins in rat pancreatic and parotid acinar cells. *Gastroenterology*, **111**, 1661.

142. Gaisano, H. Y., Sheu, L., Wong, P. P., Klip, A., and Trimble, W. S. (1997) SNAP-23 is located in the basolateral plasma membrane of rat pancreatic acinar cells. *FEBS Lett.*, **414**, 298.

143. Low, S. H., Chapin, S. J., Weimbs, T., Komuves, L. G., Bennett, M. K., and Mostov, K. E. (1996) Differential localization of syntaxin isoforms in polarized Madin–Darby canine kidney cells. *Mol. Biol. Cell*, **7**, 2007.

144. Delgrossi, M. H., Breuza, L., Mirre, C., Chavrier, P., and Le, B. A. (1997) Human syntaxin 3 is localized apically in human intestinal cells. *J. Cell Sci.*, **110**, 2207.

145. Fujita, H., Tuma, P. L., Finnegan, C. M., Locco, L., and Hubbard, A. L. (1998) Endogenous syntaxins 2, 3 and 4 exhibit distinct but overlapping patterns of expression at the hepatocyte plasma membrane. *Biochem. J.*, **329**, 527.

146. Galli, T., Zahraoui, A., Vaidyanathan, V. V., Raposo, G., Tian, J. M., Karin, M., Niemann, H., and Louvard, D. (1998) A novel tetanus neurotoxin-insensitive vesicle-associated membrane protein in SNARE complexes of the apical plasma membrane of epithelial cells. *Mol. Biol. Cell*, **9**, 1437.

147. Low, S. H., Roche, P. A., Anderson, H. A., van Italie, S., Zhang, M., Mostov, K. E., and Weimbs, T. (1998) Targeting of SNAP-23 and SNAP-25 in polarized epithelial cells. *J. Biol. Chem.*, **273**, 3422.

148. Weimbs, T., Low, S. H., Chapin, S. J., and Mostov, K. E. (1997) Apical targeting in polarized epithelial cells: there's more afloat than rafts. *Trends Cell Biol.*, **7**, 393.

149. Ikonen, E., Tagaya, M., Ullrich, O., Montecucco, C., and Simons, K. (1995) Different requirements for NSF, SNAP, and Rab proteins in apical and basolateral transport in MDCK cells. *Cell*, **81**, 571.

150. Link, E., Edelmann, L., Chou, J. H., Binz, T., Yamasaki, S., Eisel, U., Baumert, M., Sudhof, T. C., Niemann, H., and Jahn, R. (1992) Tetanus toxin action: inhibition of neurotransmitter release linked to synaptobrevin proteolysis. *Biochem. Biophys. Res. Comm.*, **189**, 1017.

151. McMahon, H. T., Ushkaryov, Y. A., Edelmann, L., Link, E., Binz, T., Niemann, H., Jahn, R., and Sudhof, T. C. (1993) Cellubrevin is a ubiquitous tetanus-toxin substrate homologous to a putative synaptic vesicle fusion protein [see comments]. *Nature*, **364**, 346.

152. Fiedler, K., Lafont, F., Parton, R. G., and Simons, K. (1995) Annexin XIIIb: a novel

epithelial specific annexin is implicated in vesicular traffic to the apical plasma membrane. *J. Cell Biol.*, **128**, 1043.

153. Lafont, F., Verade, P., Galli, T., Wimmer, C., Louvard, D., and Simons, K. (1999) Raft association of SNAP receptors actin in apical trafficking in Madin–Darby canine kidney cells. *Proc. Natl Acad. Sci., USA*, **96**, 3734.

154. Grindstaff, K. K., Bacallao, R. L., and Nelson, W. J. (1998) Apiconuclear organization of microtubules does not specify protein delivery from the trans-Golgi network to different membrane domains in polarized epithelial cells. *Mol. Biol. Cell*, **9**, 685.

155. Wang, A. Z., Ojakian, G. K., and Nelson, W. J. (1990) Steps in the morphogenesis of a polarized epithelium. I. Uncoupling the roles of cell–cell and cell–substratum contact in establishing plasma membrane polarity in multicellular epithelial (MDCK) cysts. *J. Cell Sci.*, **95**, 137.

156. Wang, A. Z., Ojakian, G. K., and Nelson, W. J. (1990) Steps in the morphogenesis of a polarized epithelium. II. Disassembly and assembly of plasma membrane domains during reversal of epithelial cell polarity in multicellular epithelial (MDCK) cysts. *J. Cell Sci.*, **95**, 153.

157. Wang, A. Z., Wang, J. C., Ojakian, G. K., and Nelson, W. J. (1994) Determinants of apical membrane formation and distribution in multicellular epithelial MDCK cysts. *Am. J. Physiol.*, **267**, C473.

158. Gumbiner, B. M. (1996) Cell adhesion: the molecular basis of tissue architecture and morphogenesis. *Cell*, **84**, 345.

159. Shapiro, L., Fannon, A. M., Kwong, P. D., Thompson, A., Lehman, M. S., Grube, G., Legrand, J.-L., Als-Nielson, J., Colman, D. R., and Hendrickson, W. A. (1995) Structural basis of cell–cell adhesion by cadherins. *Nature*, **372**, 327.

160. Nose, A., Nagafuchi, A., and Takeichi, M. (1988) Expressed recombinant cadherins mediate cell sorting in model systems. *Cell*, **54**, 993.

161. Nagar, B., Overduin, M., Ikura, M., and Rini, J. M. (1996) Structural basis for calcium-induced E-cadherin rigidification and dimerization. *Nature*, **380**, 360.

162. Brieher, W. M., Yap, A. S., and Gumbiner, B. M. (1996) Lateral dimerization is required for the homophilic binding activity of C-cadherin. *J. Cell Biol.*, **135**, 487.

163. Yap, A. S., Brieher, W. M., Pruschy, M., and Gumbiner, B. M. (1997) Lateral clustering of the adhesive ectodomain: a fundamental determinant of cadherin function. *Curr. Biol.*, **7**, 308.

164. Aberle, H., Schwartz, H., and Kemler, R. (1996) Cadherin–catenin complex: protein interactions and their implications for cadherin function. *J. Cell Biochem.*, **61**, 514.

165. Aberle, H., Butz, S., Stappert, J., Weissig, H., Kemler, R., and Hoschuetzky, H. (1994) Assembly of the cadherin–catenin complex *in vitro* with recombinant proteins. *J. Cell Sci.*, **107**, 3655.

166. Ozawa, M., Baribault, H., and Kemler, R. (1989) The cytoplasmic domain of the cell adhesion molecule uvomorulin associates with three independent proteins structurally related in different species. *EMBO J.*, **8**, 1711.

167. Jou, T. S., Stewart, D. B., Stappert, J., Nelson, W. J., and Marrs, J. A. (1995) Genetic and biochemical dissection of protein linkages in the cadherin–catenin complex. *Proc. Natl Acad. Sci., USA*, **92**, 5067.

168. Reynolds, A. B., Daniel, J., McCrea, P. D., Wheelock, M. J., Wu, J., and Zhang, Z. (1994) Identification of a new catenin: the tyrosine kinase substrate p120cas associates with E-cadherin complexes. *Mol. Cell. Biol.*, **14**, 8333.

169. Herrenknecht, K., Ozawa, M., Eckerskorn, C., Lottspeich, F., Lenter, M., and Kemler, R.

(1991) The uvomorulin-anchorage protein alpha catenin is a vinculin homologue. *Proc. Natl Acad. Sci., USA*, **88**, 9156.

170. Ozawa, M. and Kemler, R. (1992) Molecular organization of the uvomorulin-catenin complex. *J. Cell Biol.*, **116**, 989.

171. Rimm, D. L., Koslov, E. R., Kebriaei, P., Cianci, C. D., and Morrow, J. S. (1995) Alpha 1(E)-catenin is an actin-binding and -bundling protein mediating the attachment of F-actin to the membrane adhesion complex. *Proc. Natl Acad. Sci., USA*, **92**, 8813.

172. Knudsen, K. A., Soler, A. P., Johnson, K. R., and Wheelock, M. J. (1995) Interaction of alpha-actinin with the cadherin/catenin cell–cell adhesion complex via alpha-catenin. *J. Cell Biol.*, **130**, 67.

173. Nagafuchi, A. and Takeichi, M. (1988) Cell binding function of E-cadherin is regulated by the cytoplasmic domain. *EMBO J.*, **7**, 3679.

174. Small, V. J., Anderson, K., and Rottner, K. (1996) Actin and the co-ordination of protrusion, attachment and retraction in cell crawling. *Biosci. Reports*, **16**, 351.

175. Hirano, S., Nose, A., Hatta, K., Kawakami, A., and Takeichi, M. (1987) Calcium-dependent cell–cell adhesion molecules (cadherins): subclass specificities and possible involvement of actin bundles. *J. Cell Biol.*, **105**, 2501.

176. Adams, C. L., Chen, Y.-T., Smith, S. J., and Nelson, W. J. (1998) Mechanisms of epithelial cell–cell adhesion and cell compaction revealed by high-resolution tracking of E-cadherin-green fluorescent protein. *J. Cell Biol.*, **142**, 1105.

177. Adams, C. L., Nelson, W. J., and Smith, S. J. (1996) Quantitative analysis of cadherin–catenin–actin reorganization during development of cell–cell adhesion. *J. Cell Biol.*, **135**, 1899.

178. Yap, A. S., Mullin, J. M., and Stevenson, B. R. (1998) Molecular analyses of tight junction physiology: insights and paradoxes. *J. Membr. Biol.*, **163**, 159.

179. Ragsdale, K. G., Phelps, J., and Luby-Phelps, K. (1997) Viscoelestic response of fibroblasts to tension transmitted through adherens junctions. *Biophys. J.*, **73**, 2798.

180. Angres, B., Barth, A., and Nelson, W. J. (1996) Mechanism for transition from initial to stable cell–cell adhesion: kinetic analysis of E-cadherin-mediated adhesion using a quantitative adhesion assay. *J. Cell Biol.*, **134**, 549.

181. Verkhovsky, A. B., Svutkina, T. M., and Borisy, G. G. (1997) Polarity sorting of actin filaments in cytochalsin-treated fibroblasts. *J. Cell Sci.*, **110**, 1693.

182. Gloushankova, N. A., Alieva, N. A., Krendel, M. F., Bonder, E. M., Feder, H. H., Vailiev, J. M., and Gelfand, I. M. (1997) Cell–cell contact chages the dynamics of lammelar activity in nontransformed epitheliocytes but not in their ras-transformed descendents. *Proc. Natl Acad. Sci., USA*, **94**, 879.

183. Nelson, W. J. and Veshnock, P. J. (1986) Dynamics of membrane-skeleton (fodrin) organization during development of polarity in Madin–Darby canine kidney epithelial cells. *J. Cell Biol.*, **103**, 1751.

184. Nelson, W. J. and Veshnock, P. J. (1987) Modulation of fodrin (membrane skeleton) stability by cell–cell contact in Madin–Darby canine kidney epithelial cells. *J. Cell Biol.*, **104**, 1527.

185. Nelson, W. J. (1989) Topogenesis of plasma membrane domains in polarized epithelial cells. *Curr. Opin. Cell Biol.*, **1**, 660.

186. Nelson, W. J. and Hammerton, R. W. (1989) A membrane-cytoskeletal complex containing $Na^+,K^+$-ATPase, ankyrin, and fodrin in Madin–Darby canine kidney (MDCK) cells: implications for the biogenesis of epithelial cell polarity. *J. Cell Biol.*, **108**, 893.

187. Bennett, V. (1990) Spectrin-based membrane skeleton: a multipotential adaptor between plasma membrane and cytoplasm. *Physiol. Rev.*, **70**, 1029.

188. Nelson, W. J., Shore, E. M., Wang, A. Z., and Hammerton, R. W. (1990) Identification of a membrane-cytoskeletal complex containing the cell adhesion molecule uvomorulin (E-cadherin), ankyrin, and fodrin in Madin–Darby canine kidney epithelial cells. *J. Cell Biol.*, **110**, 349.

189. Nelson, W. J. and Veshnock, P. J. (1987) Ankyrin binding to (Na$^+$ + K$^+$)ATPase and implications for the organization of membrane domains in polarized cells. *Nature*, **328**, 533.

190. Beck, K. A., Buchanan, J. A., and Nelson, W. J. (1997) Golgi membrane skeleton: identification, localization and oligomerization of a 195 kDa ankyrin isoform associated with the Golgi complex. *J. Cell Sci.*, **110**, 1239.

191. McNeill, H., Ozawa, M., Kemler, R., and Nelson, W. J. (1990) Novel function of the cell adhesion molecule uvomorulin as an inducer of cell surface polarity. *Cell*, **62**, 309.

192. Marrs, J. A., Andersson-Fisone, C., Jeong, M. C., Cohen-Gould, L., Zurzolo, C., Nabi, I. R., Rodriguez-Boulan, E., and Nelson, W. J. (1995) Plasticity in epithelial cell phenotype: modulation by expression of different cadherin cell adhesion molecules. *J. Cell Biol.*, **129**, 507.

193. De Camilli, P. and Takei, K. (1996) Molecular mechanisms in synaptic vesicle endocytosis and recycling. *Neuron*, **16**, 481.

194. Govindan, B. and Novick, P. (1995) Development of cell polarity in budding yeast. *J. Exp. Zool.*, **273**, 401.

195. Galli, T., Garcia, E. P., Mundigl, O., Chilcote, T. J., and De Camilli, P. (1995) v- and t-SNAREs in neuronal exocytosis: a need for additional components to define sites of release. *Neuropharmacology*, **34**, 1351.

196. Garcia, E. P., McPherson, P. S., Chilcote, T. J., Takei, K., and De Camilli, P. (1995) rbSec1A and B colocalize with syntaxin 1 and SNAP-25 throughout the axon, but are not in a stable complex with syntaxin. *J. Cell Biol.*, **129**, 105.

197. Rothman, J. E. and Sollner, T. H. (1997) Throttles and dampers: controlling the engine of membrane fusion. *Science*, **276**, 1212.

198. Adams, A. E. and Pringle, J. R. (1984) Relationship of actin and tubulin distribution to bud growth in wild-type and morphogenetic-mutant *Saccharomyces cerevisiae*. *J. Cell Biol.*, **98**, 934.

199. Garcia, E. P., Gatti, E., Butler, M., Burton, J., and De Camilli, P. (1994) A rat brain Sec1 homologue related to Rop and UNC18 interacts with syntaxin. *Proc. Natl Acad. Sci., USA*, **91**, 2003.

200. Hata, Y., Slaughter, C. A., and Sudhof, T. C. (1993) Synaptic vesicle fusion complex contains unc-18 homologue bound to syntaxin. *Nature*, **366**, 347.

201. Hata, Y. and Sudhof, T. C. (1995) A novel ubiquitous form of Munc-18 interacts with multiple syntaxins. Use of the yeast two-hybrid system to study interactions between proteins involved in membrane traffic. *J. Biol. Chem.*, **270**, 13022.

202. Pevsner, J., Hsu, S. C., and Scheller, R. H. (1994) n-Sec1: a neural-specific syntaxin-binding protein. *Proc. Natl Acad. Sci., USA*, **91**, 1445.

203. Tellam, J. T., McIntosh, S., and James, D. E. (1995) Molecular identification of two novel Munc-18 isoforms expressed in non-neuronal tissues. *J. Biol. Chem.*, **270**, 5857.

204. Riento, K., Jantti, J., Jansson, S., Hielm, S., Lehtonen, E., Ehnholm, C., Keranen, S., and Olkkonen, V. M. (1996) A sec1-related vesicle-transport protein that is expressed predominantly in epithelial cells. *Eur. J. Biochem.*, **239**, 638.

205. Pevsner, J., Hsu, S. C., Braun, J. E., Calakos, N., Ting, A. E., Bennett, M. K., and Scheller, R. H. (1994) Specificity and regulation of a synaptic vesicle docking complex. *Neuron*, **13**, 353.

206. Tellam, J. T., Macaulay, S. L., McIntosh, S., Hewish, D. R., Ward, C. W., and James, D. E. (1997) Characterization of Munc-18c and syntaxin-4 in 3T3-L1 adipocytes. Putative role in insulin-dependent movement of GLUT-4. *J. Biol. Chem.*, **272**, 6179.

207. Fujita, Y., Sasaki, T., Fukui, K., Kotani, H., Kimura, T., Hata, Y., Sudhof, T. C., Scheller, R. H., and Takai, Y. (1996) Phosphorylation of Munc-18/n-Sec1/rbSec1 by protein kinase C: its implication in regulating the interaction of Munc-18/n-Sec1/rbSec1 with syntaxin. *J. Biol. Chem.*, **271**, 7265.

208. Hirling, H. and Scheller, R. H. (1996) Phosphorylation of synaptic vesicle proteins: modulation of the alpha SNAP interaction with the core complex. *Proc. Natl Acad. Sci.*, *USA*, **93**, 11945.

209. Okamoto, M. and Sudhof, T. C. (1997) Mints, Munc18-interacting proteins in synaptic vesicle exocytosis. *J. Biol. Chem.*, **272**, 31459.

210. Verhage, M., Kj, D. V., Roshol, H., Burbach, J. P., Gispen, W. H., and Sudhof, T. C. (1997) DOC2 proteins in rat brain: complementary distribution and proposed function as vesicular adapter proteins in early stages of secretion. *Neuron*, **18**, 453.

211. Novick, P. and Brennwald, P. (1993) Friends and family: the role of the Rab GTPases in vesicular traffic. *Cell*, **75**, 597.

212. Goud, B., Salminen, A., Walworth, N. C., and Novick, P. J. (1988) A GTP-binding protein required for secretion rapidly associates with secretory vesicles and the plasma membrane in yeast. *Cell*, **53**, 753.

213. Salminen, A. and Novick, P. J. (1987) A ras-like protein is required for a post-Golgi event in yeast secretion. *Cell*, **49**, 527.

214. Huber, L. A., Pimplikar, S., Parton, R. G., Virta, H., Zerial, M., and Simons, K. (1993) Rab8, a small GTPase involved in vesicular traffic between the TGN and the basolateral plasma membrane. *J. Cell Biol.*, **123**, 35.

215. Chavrier, P., Vingron, M., Sander, C., Simons, K., and Zerial, M. (1990) Molecular cloning of YPT1/SEC4-related cDNAs from an epithelial cell line. *Mol. Cell Biol.*, **10**, 6578.

216. Zahraoui, A., Joberty, G., Arpin, M., Fontaine, J. J., Hellio, R., Tavitian, A., and Louvard, D. (1994) A small rab GTPase is distributed in cytoplasmic vesicles in non polarized cells but colocalizes with the tight junction marker ZO-1 in polarized epithelial cells. *J. Cell Biol.*, **124**, 101.

217. Sogaard, M., Tani, K., Ye, R. R., Geromanos, S., Tempst, P., Kirchhausen, T., Rothman, J. E., and Sollner, T. (1994) A rab protein is required for for the assembly of SNARE complexes in the doking of transport vesicles. *Cell*, **78**, 937.

218. Weber, E., Berta, G., Tousson, A., St. John, P., Green, M. W., Gopalokrishnan, U., Jilling, T., Sorscher, E. J., Elton, T. S., Abrahamson, D. R., *et al.* (1994) Expression and polarized targeting of a rab3 isoform in epithelial cells. *J. Cell Biol.*, **125**, 583.

219. Grindstaff, K. K., Yeaman, C., Anandasabapathy, N., Hsu, S.-C., Rodriguez-Boulan, E., Scheller, R. H., and Nelson, W. J. (1998) Sec6/8 complex is recruited to cell–cell contacts and specifies transport vesicle delivery to the basal-lateral membrane in polarized epithelial cells. *Cell*, **93**, 731,

220. TerBush, D. R. and Novick, P. (1995) Sec6, Sec8, and Sec15 are components of a multisubunit complex which localizes to small bud tips in *Saccharomyces cerevisiae*. *J. Cell Biol.*, **130**, 299.

221. TerBush, D. R., Maurice, T., Roth, D., and Novick, P. (1996) The Exocyst is a multiprotein complex required for exocytosis in *Saccharomyces cerevisiae*. *EMBO J.*, **15**, 6483.

222. Finger, F. P., Hughes, T. E., and Novick, P. (1998) Sec3p is a spatial landmark for polarized secretion in budding yeast. *Cell*, **92**, 559.

223. Finger, F. P. and Novick, P. (1998) Spatial regulation of exocytosis: lessons from yeast. *J. Cell Biol.*, **142**, 609.

224. Guo, W., Roth, D., Gatti, E., De Camilli, P., and Novick, P. (1997) Identification and characterization of homologues of the Exocyst component Sec10p. *FEBS Lett.*, **404**, 135.

225. Hazuka, C. D., Hsu, S. C., and Scheller, R. H. (1997) Characterization of a cDNA encoding a subunit of the rat brain rsec6/8 complex. *Gene*, **187**, 67.

226. Kee, Y., Yoo, J. S., Hazuka, C. D., Peterson, K. E., Hsu, S. C., and Scheller, R. H. (1997) Subunit structure of the mammalian exocyst complex. *Proc. Natl Acad. Sci., USA*, **94**, 14438.

227. Ting, A. E., Hazuka, C. D., Hsu, S. C., Kirk, M. D., Bean, A. J., and Scheller, R. H. (1995) rSec6 and rSec8, mammalian homologs of yeast proteins essential for secretion. *Proc. Natl Acad. Sci., USA*, **92**, 9613.

228. Hsu, S. C., Hazuka, C. D., Roth, R., Foletti, D. L., Heuser, J., and Scheller, R. H. (1998) Subunit composition, protein interactions, and structures of the mammalian brain sec6/8 complex and septin filaments. *Neuron*, **20**, 1111.

229. Pasdar, M. and Nelson, W. J. (1989) Regulation of desmosome assembly in epithelial cells: kinetics of synthesis, transport, and stabilization of desmoglein I, a major protein of the membrane core domain. *J. Cell Biol.*, **109**, 163.

230. Rodriguez-Boulan, E., Paskiet, K. T., and Sabatini, D. D. (1983) Assembly of enveloped viruses in Madin–Darby canine kidney cells: polarized budding from single attached cells and from clusters of cells in suspension. *J. Cell Biol.*, **96**, 866.

231. Gilbert, T. and Rodriguez-Boulan, E. (1991) Induction of vacuolar apical compartments in the Caco-2 intestinal epithelial cell line. *J. Cell Sci.*, **100**, 451.

232. Haas, A. and Wickner, W. (1996) Homotypic vacuole fusion requires Sec17p (yeast alpha-SNAP) and Sec18p (yeast NSF). *EMBO J.*, **15**, 3296.

233. Mayer, A., Wickner, W., and Haas, A. (1996) Sec18p (NSF)-driven release of Sec17p (alpha-SNAP) can precede docking and fusion of yeast vacuoles. *Cell*, **85**, 83.

234. Nichols, B. J., Ungermann, C., Pelham, H. R., Wickner, W. T., and Haas, A. (1997) Homotypic vacuolar fusion mediated by t- and v-SNAREs [see comments]. *Nature*, **387**, 199.

235. Ungermann, C., Nichols, B. J., Pelham, H. R., and Wickner, W. (1998) A vacuolar v-t-SNARE complex, the predominant form *in vivo* and on isolated vacuoles, is disassembled and activated for docking and fusion. *J. Cell Biol.*, **140**, 61.

236. Calhoun, B. C. and Goldenring, J. R. (1997) Two Rab proteins, vesicle-associated membrane protein 2 (VAMP-2) and secretory carrier membrane proteins (SCAMPs), are present on immunoisolated parietal cell tubulovesicles. *Biochem. J.*, **325**, 559.

237. Peng, X. R., Yao, X., Chow, D. C., Forte, J. G., and Bennett, M. K. (1997) Association of syntaxin 3 and vesicle-associated membrane protein (VAMP) with $H^+/K^{(+)}$-ATPase-containing tubulovesicles in gastric parietal cells. *Mol. Biol. Cell*, **8**, 399.

238. Walch-Solimena, C., Blasi, J., Edelmann, L., Chapman, E. R., Fischer von Mollard, G., and Jahn, R. (1995) The t-SNAREs syntaxin 1 and SNAP-25 are present on organelles that participate in synaptic vesicle recycling. *J. Cell Biol.*, **128**, 637.

239. Vega-Salas, D. E., Salas, P. J., and Rodriguez-Boulan, E. (1987) Modulation of the expression of an apical plasma membrane protein of Madin–Darby canine kidney epithelial cells: cell–cell interactions control the appearance of a novel intracellular storage compartment. *J. Cell Biol.*, **104**, 1249.

240. Ojakian, G. K., Nelson, W. J., and Beck, K. A. (1997) Mechanisms for *de novo* biogenesis of an apical membrane compartment in groups of simple epithelial cells surrounded by extracellular matrix. *J. Cell Sci.*, **110**, 2781.

241. Vega-Salas, D. E., Salas, P. J., and Rodriguez-Boulan, E. (1988) Exocytosis of vacuolar apical compartment (VAC): a cell–cell contact controlled mechanism for the establishment of the apical plasma membrane domain in epithelial cells. *J. Cell Biol.*, **107**, 1717.

242. Diamond, J. M. (1977) Twenty-first Bowditch lecture. The epithelial junction: bridge, gate, and fence. *Physiologist*, **20**, 10.

243. Dragsten, P. R., Blumenthal, R., and Handler, J. S. (1981) Membrane asymmetry in epithelia: is the tight junction a barrier to diffusion in the plasma membrane? *Nature*, **294**, 718.

244. van Meer, G. and Simons, K. (1986) The function of tight junctions in maintaining differences in lipid composition between the apical and the basolateral cell surface domains of MDCK cells. *EMBO J.*, **5**, 1455.

245. van Meer, G., Gumbiner, B., and Simons, K. (1986) The tight junction does not allow lipid molecules to diffuse from one epithelial cell to the next. *Nature*, **322**, 639.

246. Furuse, M., Hirase, T., Itoh, M., Nagafuchi, A., Yonemura, S., Tsukita, S., and Tsukita, S. (1993) Occludin: a novel integral membrane protein localizing at tight junctions [see comments]. *J. Cell Biol.*, **123**, 1777.

247. Furuse, M., Itoh, M., Hirase, T., Nagafuchi, A., Yonemura, S., Tsukita, S., and Tsukita, S. (1994) Direct association of occludin with ZO-1 and its possible involvement in the localization of occludin at tight junctions. *J. Cell Biol.*, **127**, 1617.

248. Haskins, J., Gu, L., Wittchen, E. S., Hibbard, J., and Stevenson, B. R. (1998) ZO-3, a novel member of the MAGUK protein family found at the tight junction, interacts with ZO-1 and occludin. *J. Cell Biol.*, **141**, 199.

249. Itoh, M., Yonemura, S., Nagafuchi, A., Tsukita, S., and Tsukita, S. (1991) A 220-kD undercoat-constitutive protein: its specific localization at cadherin-based cell–cell adhesion sites. *J. Cell Biol.*, **115**, 1449.

250. Itoh, M., Nagafuchi, A., Yonemura, S., Kitani-Yasuda, T., Tsukita, S., and Tsukita, S. (1993) The 220-kD protein colocalizing with cadherins in non-epithelial cells is identical to ZO-1, a tight junction-associated protein in epithelial cells: cDNA cloning and immunoelectron microscopy. *J. Cell Biol.*, **121**, 491.

251. Jesaitis, L. A. and Goodenough, D. A. (1994) Molecular characterization and tissue distribution of ZO-2, a tight junction protein homologous to ZO-1 and the *Drosophila* discs-large tumor suppressor protein. *J. Cell Biol.*, **124**, 949.

252. Willott, E., Balda, M. S., Fanning, A. S., Jameson, B., Van, I. C., and Anderson, J. M. (1993) The tight junction protein ZO-1 is homologous to the *Drosophila* discs-large tumor suppressor protein of septate junctions. *Proc. Natl Acad. Sci., USA*, **90**, 7834.

253. Fanning, A. S. and Anderson, J. M. (1996) Protein–protein interactions: PDZ domain networks. *Curr. Biol.*, **6**, 1385.

254. Fanning, A. S. and Anderson, J. M. (1998) PDZ domains and the formation of protein networks at the plasma membrane. *Curr. Top. Microbiol. Immunol.*, **228**, 209.

255. Jou, T. S., Schneeberger, E. E., and Nelson, W. J. (1998) Structural and functional regulation of tight junctions by RhoA and Rac1 small GTPases. *J. Cell Biol.*, **142**, 101.

256. Louvard, D. (1980) Apical membrane aminopeptidase appears at site of cell–cell contact in cultured kidney epithelial cells. *Proc. Natl Acad. Sci., USA*, **77**, 4132.

# 5 | Cell polarity in algae and vascular plants

JOHN E. FOWLER

## 1. Introduction

Due to the presence of the plant cell wall, the morphogenesis and physiology of plant cells and animal cells are under different constraints. Plant cells are immobile, and fixed in relation to their neighbours; morphogenetic changes in the plant cannot occur by the cellular migration mechanisms common during animal development. Furthermore, modification of plant cell shape must occur through modification of the surrounding cell wall. Thus, mechanisms that allow co-ordination and expression of morphogenetic changes in plants are coupled with mechanisms that regulate growth, both by expansion and division of cells.

Although most plant cells grow by delocalized expansion along an axis or multiply through approximately symmetrical divisions, asymmetrical growth, oriented by directional cues, is apparent in many cell types. The morphological expression of cell polarity is associated both with cell expansion at a localized site (tip growth; see Fig. 1A) and with asymmetrical division (see Fig. 1B). It is worth noting that asymmetrical divisions are often accompanied by the establishment of two alternative developmental pathways for the daughter cells. Thus, mechanisms that establish asymmetrical division may also affect the partitioning of molecules that influence cell fate. Finally, polarity may exist in terminally differentiated cells in the form of asymmetries in the plasma membrane that aid in carrying out physiological functions (Fig. 1C). Although not associated with morphological changes, such localized plasma membrane asymmetries could be established and maintained by mechanisms similar to those that establish morphological asymmetries.

Evidence suggests that plant cell polarity is established through pathways analogous to the hierarchy proposed for yeast and mammalian cells (1, 2). The reception of a directional signal allows the subsequent definition of a distinct cortical site oriented by that signal. The cortical site can then be modified by interactions with other cortical components, leading to its stabilization and/or the acquisition of specific functions at that site. Finally, components at the modified site influence cellular processes and generate the required cellular asymmetries to produce a polarized cell (see Fig. 1). In comparison to non-plant eukaryotes, however, the plant cell wall may

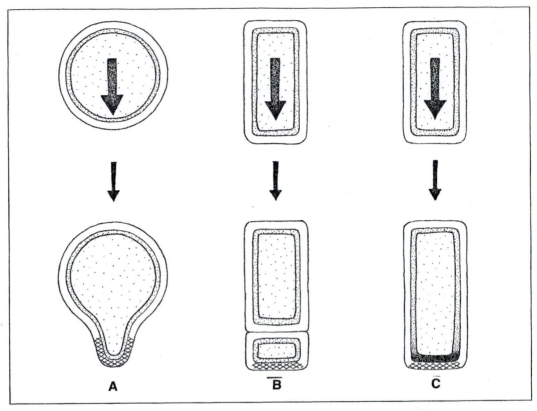

**Fig. 1** Cell polarity in plant cells; the plasma membrane and cell wall are thickened for emphasis. Directional cues can position or direct (A) localized tip growth; (B) asymmetrical division; or (c) localized specializations in the plasma membrane (concentrated dots). Note that the cell wall surrounding all plant cells could also have regions that are locally distinct (cross-hatches), e.g. at the growing tip, in one of the two daughter cells, or adjacent to the plasma membrane specialization.

have a unique and important role in this process, as it provides a structure that may also be locally distinct. This chapter will discuss some of the best-characterized examples of cell polarity in plants with these themes in mind.

## 2. Polar growth in brown algal zygotes

### 2.1 Embryogenesis in fucoid algae

Zygotes of fucoid marine brown algae (from the genera *Fucus* and *Pelvetia*) are useful experimental models for studying the establishment of cell polarity in response to extracellular cues (2–6). Populations of apolar eggs, fertilized in culture, proceed synchronously through embryogenesis for several days. In the first 24 hours after fertilization (AF), zygotes establish a polar growth axis oriented with respect to an external gradient, and grow from a localized site before dividing asymmetrically,

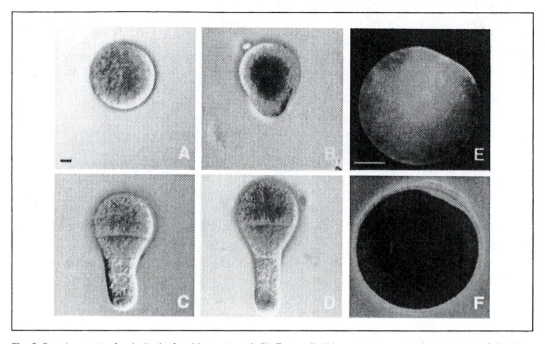

**Fig. 2** Development of polarity in fucoid zygotes. (A–D) *Fucus distichus* zygotes, at various stages of development. Unilateral light was applied from approximately 12 o'clock. Bar = 10 μm. (A) 3.5 h after fertilization (AF), prior to rhizoid tip germination. (B) 13 h AF, the rhizoid tip has germinated. (C) 20 h AF, the zygote has divided asymmetrically, producing the tip-growing rhizoid cell and spherical thallus cell. (D) 28 h AF, the second round of division has occurred. The first division of the rhizoid cell is perpendicular to the growth axis, whereas the first division of the thallus cell is perpendicular to the first division plane. (E–F) Early markers of an aligned, labile cell polarity, 4 h AF in a single living *Pelvetia compressa* zygote. Unilateral light was applied from approximately 7 o'clock; the presumptive rhizoid growth site is at approximately 1 o'clock. Bar = 25 μm. (E) Polar localization of the fluorescent lipid probe FM4-64 (Molecular Probes), likely marking asymmetrical vesicle transport (144). (F) Polar adhesive secretion, visualized by application of a halo of fluorescent microspheres. The appearance of a space between the zygote's plasma membrane and cell wall ('cortical clearing') is also associated with an aligned axis; this may be due to intensive secretory activity at this site.

perpendicular to the growth axis (Fig. 2). This division produces two distinct cell types: the tip-growing rhizoid cell and the spherical thallus cell. The rhizoid cell is specialized to secrete an adhesive material that attaches the developing zygote to the underlying substrate; this attachment probably serves to prevent the free-living embryos from drifting out of the tidal pools that they inhabit. The zygotic axis also defines the developmental axis of the organism. Thus, the establishment of zygotic polarity has important physiological and developmental consequences.

The ability to obtain synchronized populations of zygotes, and to readily manipulate the orientation of the polar axis via application of a unilateral light vector, has provided significant advantages for dissecting the cellular mechanisms that produce the polar two-celled embryo. A classic bioassay provided some of the initial evidence that establishment of polarity occurs in several distinct phases, not necessarily coinciding with obvious morphological changes (7) (see Fig. 3). Using this assay, the

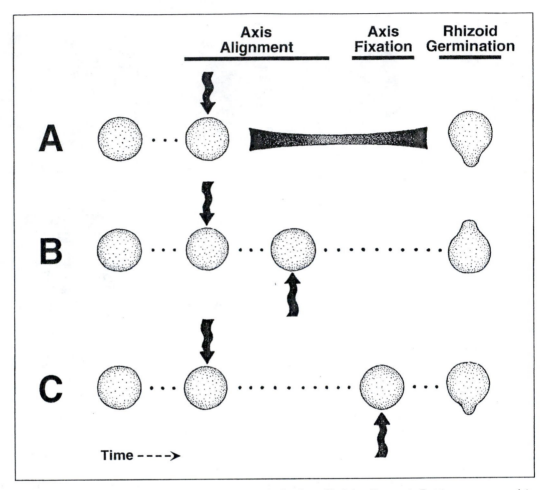

**Fig. 3** Assays for polar axis alignment and polar axis fixation. (A) Axis alignment. Zygotes are exposed to unilateral light (arrow) and then placed in the dark. Rhizoid growth that is aligned with the light vector (from the pole opposite the light) indicates that the axis was aligned at the time the zygote was placed in the dark. (B) Axis realignment. A second light vector (lower arrow) can alter the position of the originally aligned axis. (C) Axis fixation. Beyond a specific time point in zygote development, a second unilateral light vector does not alter the position of a previously aligned axis, indicating that axis fixation has occurred in the zygote.

establishment of zygotic polarity can be divided into four phases: an initial phase that is unresponsive to unilateral light, a phase in which the zygote is continually responsive to the newly applied unilateral light (originally called 'axis formation'), a phase in which the aligned axis is no longer responsive to the light vector (called 'axis fixation'), and a final phase characterized by morphological expression of the axis (i.e. growth of the rhizoid tip opposite the incident light and asymmetrical division). Recent experiments (see below) suggest that the period previously referred to as 'axis formation' should perhaps more fittingly be called 'axis alignment' (8, 9), since a rudimentary axis may form prior to or coincident with the light-sensitive period.

Evidence suggests that establishment of the polar axis is a stepwise process in which cellular components (e.g. F-actin, ion channels, the secretory apparatus) are recruited to the cortex at the future site of rhizoid growth (2, 4, 6). Researchers have concentrated on identifying the localized components of this 'polarity pathway', the temporal phases with which their localization is associated, and the relationships between the various components. For example, polarized adhesive deposition (8) and localization of the lipid probe FM 4–64 (a putative marker for asymmetrical vesicle transport (144)) occur within the period of axis alignment in *Pelvetia compressa* (see Fig. 2E–F). However, an important caveat is that various research groups have used different species as models (e.g. *Fucus distichus*, *Fucus serratus*, *P. compressa*), and although similar mechanisms for establishing polarity appear to exist in each, subtle distinctions between the species are becoming apparent with more extensive research. For example, it is clear that although the general order of phases and some localization events are conserved between the two species, *P. compressa* proceeds more quickly through the phases of axis establishment (germination at ~10 hours AF) than *F. distichus* (germination at ~14 hours AF) (8). Whether observed differences between each species are due to significant differences in mechanisms for the establishment of polarity is an unanswered question.

## 2.2 Directional cues and axis alignment

During the initial phase of axis alignment, the zygote receives the extracellular orienting signal, transduces that signal to cytoplasmic and membrane components, and marks the cortical site that will produce the rhizoid tip. Brown algal zygotes can interpret a wide variety of cues, including unidirectional light, ion and temperature gradients, flow of sea water, and proximity to another zygote, to orient their growth (10). However, zygotes are still able to polarize in the absence of any of these influences; an initial, default directional cue may be provided by sperm entry (11). How such a cue might be transduced to influence the later events in axis establishment is unknown.

The best studied of the orienting vectors is light, which is probably received by a blue/UV light photoreceptor at the cell cortex (12). Photopolarization is likely to involve localized electron transport activity at the plasma membrane of the presumptive rhizoid pole, visualized using hexacyanoferrate (HCF) reduction as a marker (13). However, this electron transport intermediate is probably specific to transduction of the light signal, since inhibition of redox transport blocks orientation of growth by a light vector, but does not affect development of polarity that is oriented by the 'group effect' (i.e. growth of neighbouring zygotes toward each other) (13). Since exposure to blue light increases cyclic guanosine monophosphate (cGMP) levels in *P. fastigiata* zygotes, and the guanylyl cyclase inhibitor LY 83583 inhibits both the increase in [cGMP] and photopolarization, cGMP may also be an intermediate in the light transduction pathway (14). A speculative model has been proposed in which local dissociation of calcium channels from cytoskeletal 'tethers' at the more highly lit presumptive thallus pole is triggered by a light-activated localized cortical increase in cGMP. This release of calcium channels at the thallus pole results in a greater concentration of such channels at the rhizoid pole (where channels remain tethered), initiating a cytoplasmic $Ca^{2+}$ gradient at the presumptive rhizoid cortex (see below) (6).

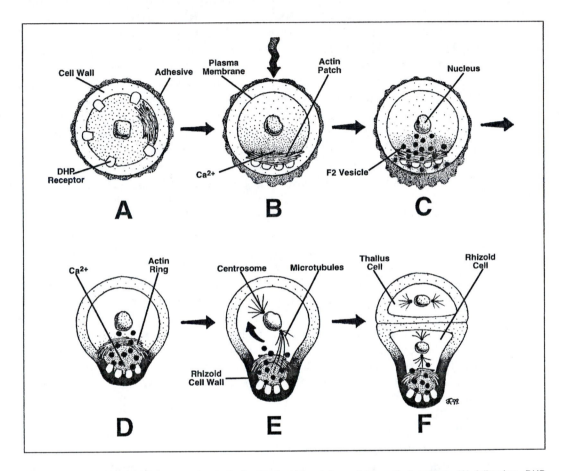

**Fig. 4** A model for establishment of polarity in the fucoid zygote and two-celled embryo. (A) Adhesion. DHP receptors are distributed uniformly throughout the plasma membrane, and the actin patch is not oriented with respect to light. A uniform layer of adhesive is secreted, which attaches the zygote to the substrate. (B) Axis alignment. The actin patch, DHP receptors and a $Ca^{2+}$ gradient (small dots in the cytoplasm) localize at a cortical site on the side of the zygote opposite the light vector (wavy arrow). This cortical site becomes a target site for secretion, and adhesive is preferentially deposited on the shaded side of the zygote. (C) Axis fixation. F2 vesicles appear and are targeted to the cortical site. Deposition of the contents of this vesicle population into the cell wall stabilizes the position of the cortical target site (perhaps by forming transmembrane connections). (D) Tip germination. F2 vesicle secretion continues to be targeted to the site at the rhizoid apex, producing a cell wall with distinct components. The actin patch becomes a subapical ring. (E) Nuclear rotation. Microtubules ftom one centrosome may interact with the cortical structure in the rhizoid to provide positional information for centrosomal (and thus spindle) alignment. (F) Polarized, two-celled embryo. Targeted secretion and tip growth continues in the rhizoid, influenced by the differentiated rhizoid cell wall. Each cell type also exhibits a characteristic division orientation.

It is not known which cues or events initiate the axis alignment phase (i.e. the development of competence to photopolarize). Interestingly, the deposition of an adhesive layer, uniformly distributed outside the cell wall, is one of the first physiological events apparent in *P. compressa* zygotes; deposition of this layer coincides with both adhesion of the zygote to the substrate and initiation of axis alignment (see Fig. 4A) (8). Blocking secretion during the early phases of polarity establishment by application of the drug brefeldin A (BFA) prevents adhesive deposition, adhesion, and photopolarization (8); however, it is unclear whether a causal relationship exists between adhesion and photopolarization. Early BFA treatment also blocks the light-oriented positioning of an early marker of axis alignment, an actin patch (see below), but does not block the formation of the patch, suggesting that BFA interrupts the signal transduction pathway that aligns the axis, but not the one that generates the axis (9). One model to explain the dependence of photopolarization on secretion is that vesicle transport inserts molecules necessary for photopolarization into the plasma membrane or cell wall (8) .

The various signals that orient the axis are probably perceived by distinct mechanisms and receptors (e.g. light receptors versus 'group effect' receptors). Other cellular components, such as actin, are likely to be involved in a common pathway for establishing polarity downstream of the convergence of these signal detection mechanisms; however, this hypothesis has not been rigorously tested for all known orienting stimuli. Two such components, the actin cytoskeleton and internal $[Ca^{2+}]$, appear to play important roles in alignment of the labile axis.

## 2.2.1  Actin and axis alignment

Treatment of zygotes with the F-actin inhibitors cytochalasin B/D (CB/CD) or latrunculin B (LatB) during the period of axis alignment prevents the orientation of polar growth with respect to unilateral light, resulting in random orientation of rhizoids; however, such drugs do not randomize the position of an axis once it has been aligned (7, 8). Drug inhibition of F-actin does not prevent deposition of the uniform adhesive that is blocked by BFA treatment, suggesting that the actin cytoskeleton's role in photopolarization is distinct from the role of secretion (8) .

Exciting new results suggest that a localized actin patch is likely to be the structure compromised by drug treatment during axis alignment (9). Previous studies using aldehyde-based fixation methods had suggested that F-actin was distributed uniformly throughout the zygotic cortex during axis alignment, but localized in a patch at the presumptive rhizoid coincident with axis fixation (15, 16). However, new methods of assaying F-actin distribution in methanol-fixed and living *P. compressa* zygotes indicate that a patch of F-actin localizes to the presumptive rhizoid pole during axis alignment (9). The F-actin patch is initially detectable at a time roughly coincident with the beginning of axis alignment. However, the original position of the patch is not at the presumptive rhizoid pole (specified by an incident light vector); it becomes localized to that site within about 1 hour after its appearance. Furthermore, elegant experiments demonstrate that the position of the actin patch is responsive to a new light vector (i.e. changing the position of the unilateral light changes the position of the actin patch), thus clearly associating the actin patch with the labile axis (9). Interestingly, repositioning appears to occur by *de novo* assembly of a new patch and disassembly of the old patch: during actin repositioning, a small percentage of zygotes were observed with two patches at opposite poles (corresponding to the two light vectors); no zygotes were observed with patches at intermediate sites between the poles (9). The ability of zygotes to form an axis *de novo* is also supported by the observation that plane-polarized light induces zygotes to form two rhizoids (on opposite poles) during the initial cell cycle (12).

Localized accumulation of actin also predicts the site of growth in the budding yeast *S. cerevisiae* (17, see also Chapter 2 in this volume), and is required for the orientation of growth in response to an extracellular gradient (18). The localized actin is important in yeast for the recruitment of several proteins to the gradient-specified growth site (18). The actin patch in fucoid zygotes may serve an analogous function, recruiting various proteins in an ordered fashion to amplify the cortical asymmetry marked by the patch. Circumstantial evidence to support this idea comes from experiments investigating the sensitivity of axis realignment to F-actin inhibitors. Realignment is blocked by CD or LatB treatments when the zygote is exposed to a second, opposing unilateral light signal; surprisingly, the same treatments did not block realignment if the zygotes were initially exposed to uniform, rather than unilateral, light (8). This suggests that not all alignment processes are the same; rather, some are more sensitive to F-actin inhibition than others. Thus, drug sensitivity is dependent on both the time of treatment (i.e. is there a previously established axis?) and on the 'type' of axis previously established (i.e. was it aligned by unilateral light or not?). Based on these results, Hable and Kropf suggest that the axis is dynamic, and that the developmental stage of the axis can influence the mechanism by which it is realigned (8). It is unknown which molecular differences distinguish various developmental stages of the axis.

### 2.2.2 The role of $Ca^{2+}$ in axis alignment

One probable component of the developing axis, which may function in conjunction with the actin patch, is a cytosolic $Ca^{2+}$ gradient, with its highest concentration at the cortical site corresponding to the presumptive rhizoid pole (see Fig. 4B) (6). The presence of a cortical $Ca^{2+}$ gradient at the presumptive rhizoid pole has been demonstrated, despite technical difficulties presented by the optical properties of the zygote (19, 20); however, relocalization of the gradient in response to an altered light vector has not been confirmed. Dispersal of cytoplasmic $Ca^{2+}$ gradients by injection of BAPTA calcium buffers blocks rhizoid tip growth and interferes with cell division, indicating that an early event in the establishment of polarity depends upon such a gradient (21); the exact cellular process affected is unclear.

Other experiments suggest that cytoplasmic $Ca^{2+}$, arising in part from $Ca^{2+}$ influx through plasma membrane channels, plays a role in photopolarization. An inward transmembrane current (a portion of which is carried by $Ca^{2+}$) is detectable at the presumptive rhizoid pole during axis alignment, and the location of influx is responsive to axis realignment by changing the incident light vector (22); furthermore, this

current is sensitive to CD, suggesting a dependence on the localized actin patch (15). The use of $Ca^{2+}$-free artificial seawater medium during exposure to a light vector also reduces photopolarization when zygotes are stimulated with less than maximally polarizing light (23); however, lack of $Ca^{2+}$ in the medium does not appreciably affect photopolarization when zygotes are maximally stimulated by a light vector (24, 25). One explanation for these observations is that $Ca^{2+}$ influx may amplify the cellular asymmetry at the cortical site that is marked by the actin patch, and such amplification is only necessary under less than maximal light regimes (6).

Formation of the $Ca^{2+}$ gradient may be aided by molecules in the plasma membrane that have been visualized by their binding to fluorescently tagged dihydro-pyridine (DHP) *in vivo*. These DHP receptors localize, in an actin-dependent process, to the rhizoid pole in response to light, and realign in response to a second light vector (26). Thus, localization of DHP receptors, like localization of the actin patch, is a marker for an aligned axis. DHP binds to $Ca^{2+}$ channels in mammalian cells (27), and could thus identify analogous algal channels; however, conflicting data exist on the effects that various calcium channel inhibitors have on polarization of (23, 25) and further experiments are needed to address the possible significance of these observations.

### 2.2.3 Targets of actin and $Ca^{2+}$

A likely interactor with the cortical $Ca^{2+}$ is calmodulin, although its exact role in axis alignment is unclear. Pharmacological inhibition of calmodulin function during axis alignment has been reported to inhibit photopolarization (25) or to increase zygote sensitivity to a submaximal light regime (23). More experiments will be necessary to deduce the exact role(s) of this multifunctional protein in the establishment of polarity.

One actin-dependent process during axis alignment is a shift from a uniform pattern of adhesive secretion to a pattern of secretion limited to the rhizoid pole (see Fig. 2F) (8). In *P. compressa*, this shift occurs approximately 1 hour after photopolarization, is sensitive to F-actin inhibitors (8), and can be redirected by a new light vector (4). Interestingly, high concentrations of LatB completely block this secretory event, whereas lower concentrations of LatB (which also block photopolarization) merely delocalize the secretion, resulting in continued uniform deposition of adhesive. This experiment indicates that distinct processes in the zygote (secretion versus localization of secretion) are differentially sensitive to actin inhibitors (8). $Ca^{2+}$ regulates secretion in vascular plant cells (28), so it might also influence this localized secretory process in fucoid zygotes.

A model of axis alignment, modified from an earlier version (2), is shown in Fig. 4A–B. Prior to axis alignment, the actin patch and molecules in the plasma membrane are distributed independently of environmental cues. During axis alignment, a signal transduction pathway using light (or other) receptors generates a specialized complex at an oriented cortical site. Newly positioned cortical sites can be formed upon redirection of the environmental cue, probably by reassembly of components at a new site and disassembly of the initial complex. One of the primary functions of the

complex is to target the secretion of adhesive to the presumptive rhizoid pole. Undoubtedly, there are many as yet unidentified components of this complex, as well as some unclear relationships between the known components. For example, are the localization of the actin patch and the formation of the $Ca^{2+}$ gradient causally linked?

## 2.3 Axis fixation and polar growth

Axis fixation is an event that permanently positions the aligned axis. One current view is that fixation involves modification of the rhizoid pole cortical complex to stabilize its position (2, 5). Several correlative localization experiments and inhibitor studies support the notion that a specific secretory event serves to stabilize the cortical complex (see Fig. 4C). An alternative view is discussed by Robinson *et al.* (6).

An uncompromised actin cytoskeleton is necessary for axis fixation, possibly due to its role in localized secretion. Treatment of zygotes with CB during the time when axis fixation would normally occur prolongs axis alignment, i.e. the period during which the zygote can be reoriented by a new light vector is extended due to a block in axis fixation (see Fig. 3) (7).

A distinct population of vesicles becomes detectable in *F. distichus* during axis fixation due to the sulphation of a fucoidan polysaccharide, F2, in the Golgi apparatus (29, 30). A staining assay specific for the sulphated F2 shows that F2-containing vesicles are specifically targeted to the presumptive rhizoid pole, and that the vesicle contents are incorporated into the zygotic cell wall at the presumptive site for rhizoid growth (31). The initiation of this secretory pattern coincides with axis fixation. Furthermore, blocking the formation and secretion of these vesicles by BFA treatment prevents axis fixation (31), suggesting that the deposition of the contents of these vesicles into the cell wall is a necessary to stabilize the axis.

Other experiments support the contention that the cell wall is required for axis fixation. Enzymatic removal of the zygotic cell wall also blocks axis fixation (32). Moreover, the generation by laser microsurgery of wall-less protoplasts from rhizoid and thallus cells leads to loss of polarity and a renewal of the protoplasts' ability to orient their growth to an incident light vector (33, 34). The nature of the cell-wall components needed for axis fixation is unknown; so far, the only other known molecule in the F2-containing vesicles is VnF, a protein recognized by an antibody against animal vitronectin (an ECM adhesive glycoprotein) (35).

The hypothesized existence of an axis-stabilizing complex (ASC) at the fixed rhizoid pole could explain the roles of the cell wall and actin in axis fixation (36). The putative ASC, which is analogous to focal adhesions in mammalian cells (37), consists of, at a minimum: actin, cell-wall molecules, and transmembrane linker proteins that anchor the aligned cortical complex at a fixed position via interactions with the cell wall. According to this model, axis fixation is triggered by the localization (via targeted secretion) of component(s) necessary for ASC formation. Interestingly, the actin patch that is associated with the aligned axis becomes noticeably larger as germination nears, supporting the idea that the axis-associated cortical complex is

modified over time (9); however, it is not known whether expansion of F-actin correlates with axis fixation. Also, the rhizoid pole-localized actin patch becomes resistant to dispersal after aldehyde treatment coincident with axis fixation, suggesting that the actin structure becomes more stable upon axis fixation (16). However, no transmembrane linker proteins have been isolated from this complex, and a test of the model awaits molecular identification of putative ASC components.

Germination (i.e. appearance of the rhizoid tip) and further polar growth are associated with several localized processes (see Fig. 4D). F2 vesicle secretion continues to be targeted to the axis-associated cortical site, resulting in specific deposition of the contents of those vesicles into the nascent rhizoid cell wall (31) . Solid evidence indicates that a tip-high $Ca^{2+}$ gradient remains associated with the cortex and aids growth at the rhizoid tip (19, 38), as do DHP receptors (26). Calmodulin localizes to the germinating tip (25), and is necessary for rhizoid growth (20). A pH gradient, most acidic at the tip, is also necessary for continued tip growth (39). An actin-dependent gradient of $poly(A)^+$ RNA is formed in the polarized cell, with its highest concentration at the thallus pole (40). Adhesion sites between the plasma membrane and the cell wall, visualized by plasmolysis and containing F-actin, become concentrated near the germinated rhizoid tip (41). The actin patch undergoes a dramatic rearrangement into a subapical ring coincident with germination, and it remains a relatively constant distance from the tip during rhizoid growth; this ring could aid in defining the region of growth and secretion at the tip (9). Finally, small GTPases of the Rho and Rab families (which regulate the actin cytoskeleton and vesicle transport, respectively, in both yeast and mammalian cells) localize asymmetrically in the growing rhizoid (J.E. Fowler, Z. Vejlupkova, and R.S. Quatrano, in preparation), and could have a role in controlling the spatial distribution of either F-actin or secretory vesicles.

Each of these components is likely to have a specific role in sustaining the continued targeting of secretion and growth to the rhizoid tip; the exact relationships between these elements, and their functions, remain to be clarified. In summary, the cortical site defined earlier in development by axis alignment and fixation remains a localized and specialized domain that directs further polar growth. However, the specific elements associated with the cortical site change during the establishment of polarity, and continued modification of the domain and its associated elements (including its interactions with the cell wall) are likely necessary to allow the growth site to shift as the rhizoid elongates.

# 3. Asymmetrical division and cell fate in brown algal zygotes

The polarity established by the axis alignment/fixation process serves to orient not only polar growth, but also the alignment of the division plane and the distinct fates of the two daughter cells. One general model to explain differences in daughter-cell fates is the differential distribution of fate-directing molecules into the two daughter cells (42). Evidence suggests that this mechanism may operate in fucoid zygotes, with some of the important fate-directing information deposited (possibly via secretion) into the differentiated cell walls of the rhizoid and thallus cells. Furthermore, secretion is necessary for the alignment of the division plane, suggesting that cell-wall components aid in positioning the plane to assist in the correct segregation of fate-directing components (see Fig. 4E–4F) (2, 4, 43).

## 3.1 Rotation of the centrosomes and cell division

The orientation of the division plane in fucoid zygotes is determined by a mechanism similar to that of animal cells, rather than of vascular plant cells, which choose their division plane prior to prophase. Although the exact mechanism of fucoid cyto-kinesis is unclear, the orientation of the zygotic division plane is determined by the orientation of the spindle, which depends upon the alignment of the two centro-

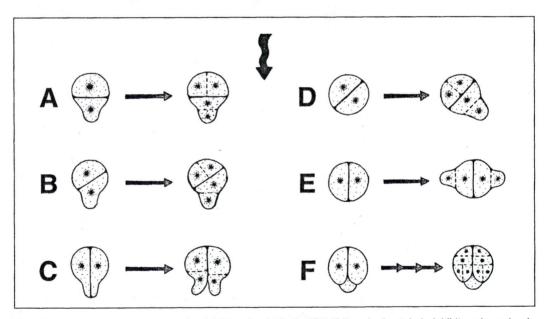

**Fig. 5** Differential effects of a secretion inhibitor (brefeldin A; BFA) (31) and microtubule inhibitors (nocodazole and oryzalin) (45) on division plane orientation, and how subsequent development in fucoid zygotes is influenced by division plane orientation. The dashed lines represent divisions subsequent to the first; the axis-orienting light vector (wavy arrow) is from 12 o'clock. (A) Untreated zygote. (B, C) Zygotes in which BFA was washed out prior to the zygotic cell division. The zygotes establish properly aligned, growing rhizoid tips prior to division, but orientation of the division plane is random, uncoupled from growth polarity. Whether the two daughter cells adopt thallus and rhizoid fates (B) or reiterate the zygotic division pattern (C) depends on whether the orientation of the division plane is parallel to the light vector. (D, E) Zygotes in which BFA was washed out and zygotes were removed from unilateral light after the initial division. BFA prevents growth of the rhizoid until after it is removed. As in (B, C), the fates of the daughter cells depend on the orientation of the division plane with respect to light; however, the axis was not fixed when BFA was washed out, resulting in a labile rhizoid growth site (D) or sites (E). (F) Zygote treated with a pulse of nocodazole or oryzalin with a division plane parallel to the growth axis. The division plane bifurcates, producing three cells, but never bisects the growing rhizoid tip (compare to C). Three-celled embryos that lack a rhizoid nucleus do not produce any tip-growing rhizoid cells, even after several rounds of division.

somes on opposite sides of the nucleus (44). The two centrosomes are aligned randomly until late in the cell cycle, and become aligned just prior to mitosis by rotation of the nucleus and its associated centrosomes (45). This alignment with the growth axis can be perturbed by short treatments that inhibit either F-actin (CD) or microtubules (oryzalin and nocodazole) (44, 45). Microtubules that extend from the centrosome nearest the rhizoid toward the rhizoid tip have been detected, and the nucleus often appears misshapen during rotation; this has led to the proposal that force exerted on the centrosomes rotates the nucleus to align the centrosomes, perhaps via microtubules that extend to the cell cortex (see Fig. 4E) (44, 45). Similar mechanisms for positioning the spindle appear to act in the *C. elegans* zygote (46) and in budding yeast (17; Chapter 2 of this volume).

A pulse of BFA can also lead to misalignment of the division plane in *F. distichus* zygotes (see Fig. 5) (31). This result suggests that proper rotation of the nucleus depends on the deposition of components into the cell wall (31). The cortical site modified during axis fixation by targeted secretion could interact with microtubules that extend from the centrosomes; this connection could aid in providing some of the force and positional information necessary for proper alignment of the centrosomes (4, 5). Interestingly, induction of division plane misalignment by microtubule inhibitors in *P. compressa* leads to different cellular morphologies than misalignment induced by BFA in *F. distichus* (45). Oryzalin- or nocodazole-misaligned division planes that are approximately parallel to the growth axis significantly inhibit elongation of the rhizoid, and such planes occasionally bifurcate in the growing rhizoid (see Fig. 5F); bifurcated division planes were not reported in response to BFA treatment (compare to Fig. 5C). It is unclear whether these distinct pharmacological effects are due to mechanistic differences between the two species, or due to differences in the cellular sensitivity to the secretory inhibitor versus microtubule inhibitors (45).

## 3.2 Asymmetrical distribution of fate-directing information in the cell wall

The normally aligned division plane in fucoid zygotes causes the asymmetry in the surrounding cell wall (marked by the presence or absence of F2) to be segregated into two cells (see Fig. 4F). The importance of asymmetry in the cell wall has been clearly shown by Berger *et al.* (47), using microsurgical techniques to demonstrate that the embryonic cell wall can direct and maintain the rhizoid and thallus cell fates. When either the thallus or the rhizoid cell is ablated, and the remaining cell retains its associated cell wall, the cell continues to divide, and maintains its original differentiated cell-type characteristics (e.g. tip growth for the rhizoid cell). However, if a daughter cell from such a dissected embryo comes in contact with cell-wall material from the other cell type, that daughter cell adopts characteristics of that other cell type. Thus, contact with the cell wall of a specific cell type (rhizoid or thallus) directs the contacting cell to adopt that cell fate (47). Furthermore, rhizoid-cell protoplasts that lack all cell-wall material appear to reiterate the axis establishment process carried out by the zygote; this suggests that the cell wall also contains factors necessary for maintaining differentiated cell fates (33).

Examination of embryos derived from BFA-treated zygotes highlights the importance of the aligned axis (and its specialized cortical site) in distributing factors that influence cell differentiation (5, 31). Under the conditions used in these experiments (see Fig. 5), BFA treatment does not block axis alignment, but does lead to random orientation of the division plane. In BFA-treated cells in which the division plane is somewhat perpendicular to the incident light vector, the two daughter cells of the first division adopt the normal thallus and rhizoid cell fates (see Fig. 5B, 5D). However, if the division plane is approximately parallel to the light vector, a high percentage of the two-celled embryos reiterate the developmental characteristics of the original zygote in both daughter cells, producing two rhizoid tips (see Fig. 5C, 5E) (31). The proposed explanation for these observations is that a division plane parallel to the light vector will bisect and partition the cortical site into both daughter cells, rather than into only one cell, leading to the formation of a growing rhizoid tip in both daughters. Thus, the cortical site formed during the process of axis establishment is crucial for establishing the developmental polarity of the embryo (31). Because the cortical site directs targeted secretion, and has been shown to be capable of establishing cell-wall asymmetry (see Fig. 4F), it has been speculated that targeted secretion could distribute the cell-wall-associated factor discovered by Berger *et al.* that directs the rhizoid cell fate (5).

In conclusion, the fucoid zygote model has been extremely useful for assembling a coherent working framework for the establishment of polarity during embryogenesis (see Fig. 4) (2). The ability to analyse each of the steps in the process of establishing polarity has resulted in a general model, which progresses from the initial directional signal to the oriented and differentiated two-celled zygote. However, there are clearly many gaps in this model; it would be useful to isolate and test the function of more of the specific molecular components of the cell-polarity machinery. It is hoped that some of these proteins (e.g. small GTPases, calmodulin) can be isolated from fucoid zygotes based on their homology to polarity genes discovered in vascular plants (see below), yeast (17; Chapter 2 in this volume), or animals (37; Chapters 4 and 7 in this volume). None the less, some of the general lessons learned from *Fucus* and *Pelvetia* are also likely to be useful in studies of polarity in vascular plants. For example, do localized cortical actin structures aid in defining the position of signalling sites at the cell cortex? Do ion gradients (e.g. $Ca^{2+}$) operate as components of localized cortical domains? Is targeted vesicle secretion important for stabilizing the position of cortical complexes? Do informational molecules that are asymmetrically distributed in plant cell walls affect developmental decisions?

## 4. Directed polar growth in vascular plant cells: pollen tubes

The pollen tube exemplifies polar growth (analogous to rhizoid elongation), and is the most rapidly growing plant cell known, with growth rates up to 1 cm $h^{-1}$. The

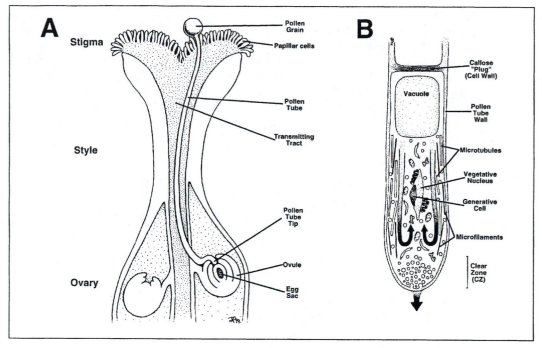

**Fig. 6** Polar growth of the pollen tube from the pollen grain to the egg sac. (A) Three distinct types of signals appear to orient pollen tube growth in the three distinct regions of the carpel (stigma, style, and ovary). (B) The pollen tube vegetative cell at the pollen tube tip. Interior arrows indicate the direction of cytoplasmic streaming; external arrow indicates the site and orientation of tube growth.

tube provides the means by which the sperm nuclei from the pollen grain are transported to the ovule, where double fertilization occurs. To accomplish this, the tube germinates from the pollen grain, following its hydration on the stigma, and grows through the transmitting tract of the style. It finally leaves the transmitting tract in the ovary and grows to the female gametophyte in an ovule (see Fig. 6A) (48, 49). Throughout this process, the pollen tube maintains its polarized organization and restricts its growth to the tip of the tube.

## 4.1 Polar organization of the pollen tube and tip

The pollen tube consists of two or three haploid cells: the vegetative cell and one or two generative cells. The vegetative cell, which carries the generative cell(s), exhibits a stereotypical polarity with respect to the tube tip (see Fig. 6B). At its base, the vegetative cell synthesizes periodic 'plugs' of callose wall as it grows; these callose walls presumably confine the cell to the tip of the tube as the tube's wall elongates apically. A vacuole is usually located near the tube base, and the size of the vacuole may be regulated to aid in confining the cytoplasm to the tube tip. Actin micro-filaments and microtubules extend approximately parallel to the direction of growth along the tube cortex, but do not appear to enter the apex of the tip. The tip apex is occupied by the clear zone (CZ), which contains an aggregation of vesicles that will ultimately be secreted at a localized site at the apex. Just basal to the CZ is a region occupied by several types of organelles (secretory vesicles, mitochondria, endoplasmic reticulum (ER)), which undergo cytoplasmic streaming; the cortical actin filaments are thought to be necessary for this streaming, which transports the vesicles and other organelles that are required for cell growth toward the apex (2, 48–50).

Research has focused on the structure, physiology, and signalling mechanisms of the tip and the CZ, as those portions of the tube that support polar growth and can respond to external signals that orient the growth (2, 48–50). Presumably, the cues that establish the orientation of the tip are transduced to the remainder of the vegetative cell to control the cytoplasmic organization. Microtubules may play an important role in this process, because the appropriate localization and movement of the vacuole and the generative cells fail to occur in the presence of microtubule inhibitors (51).

## 4.2 Orientation of pollen tube growth by external signals

For the pollen tube to reach its final destination successfully, external signals must orient its polar growth to guide it to the ovule. Distinct signals appear to operate during different phases of pollen tube growth (see Fig. 6A).

### 4.2.1 Pollen-tube germination and penetration of the stigma

After landing on the papillar cells of the stigma, pollen grains become hydrated and the pollen tube germinates, then penetrates the cuticle of the papillar cell. Hydration is a necessary cue for tube germination (48); evidence suggests a gradient of water in a lipid mixture may aid in establishing the polarity of the initial tube growth. Many stigmas secrete a lipid-rich exudate, through which the pollen tube grows before penetration. In transgenic stigmaless tobacco plants, pollen tubes do not penetrate and grow into the style (52); however, if the stylar surfaces of stigmaless plants (or of sterile, lipid-synthesis defective *Arabidopsis thaliana* mutants) are supplemented with either purified *Petunia* exudate, or with the purified *cis*-unsaturated triacylglyceride trilinolein, pollen tubes will penetrate and fertilize the ovules (53). Surprisingly, application of lipids onto leaves stripped of their cuticle also promotes pollen germination and polar growth into the intercellular spaces of the leaf (53). A minimal *in vitro* culture system has been established in which tobacco pollen grains placed in olive oil will hydrate, germinate, and grow directionally toward a nearby drop of aqueous medium containing only polyethylene glycol (PEG) (54). The authors hypothesize that a gradient of water diffusing from an aqueous source (i.e. the surface of the stigma) toward the pollen grain in the lipid exudate is a spatial cue that could orient the initial pollen tube growth *in vivo*. This model is supported by the observation that pollen tubes that germinate on a stigma under conditions of high relative humidity show a high incidence of growth into the moist air rather than towards the stigma (54). However, other stigma-derived molecules could also aid in orienting *in vivo*-grown tubes.

## 4.2.2   Signals in the transmitting tract

The pollen tube grows through the extracellular space of the transmitting tract, and its growth in some plant species is most likely aided by molecules in the specialized cell-wall components of this tissue. One group of transmitting-tract cell-wall components that may aid in growth orientation is that of the arabinogalactan proteins (AGPs), such as those encoded by the *transmitting tract specific* (*TTS*) genes of tobacco (*Nicotiana tabacum*) (55, 56). AGPs are a large class of secreted proteoglycans that are incorporated into the cell wall, and are also associated with the cell wall–plasma membrane interface (57). The TTS proteins vary in their degree of glycosylation, and are distributed in a gradient from the top to the bottom of the style, with the most highly glycosylated TTS proteins at the base of the style, closest to the ovary (56). Reduction of TTS levels via antisense methods reduces successful fertilization events, and purified TTS proteins stimulate the growth rate and attract pollen tubes grown *in vitro*, but only if TTS glycosylation is intact (55). Hence, the glycosylation gradient may provide guidance and nutrients to the pollen tube. Interestingly, stylar TTS proteins are deglycosylated during tube growth (probably by an activity that is associated with the tube itself), and TTS proteins are incorporated into the tube cell wall (56), suggesting that TTS itself could interact with the tube tip to orient growth. However, this role for TTS could be specific to *N. tabacum*, as similar studies in *N. alata* were not able to attribute any attractive or stimulatory role to a likely homologue of TTS-1, galactose-rich style glycoprotein (GarSGP) (58).

The pollen tube itself secretes AGPs, which have been studied using the Yariv phenylglycoside, a reagent that binds to AGPs. The Yariv phenylglycoside inhibits fertilization *in vivo*, and tube growth *in vitro*, suggesting that it inhibits extracellular AGP function by binding to the secreted AGPs (59). *In vitro*, phenylglycoside-inhibited tubes display swollen tips and improper cell-wall assembly at the tip, and appear to delocalize the site of exocytosis away from the extreme tip apex (60). Hence, there may be feedback signalling from the tube cell wall to the cytoplasm of the vegetative cell, which could aid in defining the site of exocytosis at the tip, and thus, the orientation of tube growth.

## 4.2.3   Egg sac-dependent signals orient the pollen tube in the ovary

Genetic evidence supports the hypothesis that the final phase of pollen-tube growth, from the transmitting tract to the egg sac, relies upon guidance from signals that require a functional egg sac. Using specific mutations (61) or chromosomal translocations (62) in *A. thaliana* to eliminate either the ovule and egg sac or the egg sac alone results in a similar phenotype: pollen tubes enter and follow the transmitting tract normally, but fail to exit from the transmitting tract and do not grow toward the defective ovules/egg sacs. Thus, the egg sac apparently either provides a long-range signal that attracts the pollen tube directly, or alternatively, induces the surrounding sporophytic cells to provide orienting information.

Two *Arabidopsis POLLEN–PISTIL INCOMPATIBILITY* (*POP*) genes are likely to provide redundant functions necessary for tube guidance during the final phase of growth. Plants doubly mutant for a recessive *pop2* allele and a dominant *pop3* allele are self-sterile; *pop2/pop3* pollen tubes grow normally through the transmitting tract, but aimlessly in the ovary (63). Because this phenotype is only present when both female and male parents are mutant for both genes, the genes appear to encode molecules (perhaps those involved in adhesion) that can function when present either in the pollen tube or the ovary (63). It is not known whether *POP2/POP3* function provides the actual guidance information, or is only indirectly necessary for guidance.

The molecular mechanisms that transduce any of the putative signals from the female tissue (the stigma, transmitting tract, or egg sac) to the cytoplasm of the tube apex and to the growth apparatus are unknown. It has been speculated that these mechanisms may include receptor-like kinases (e.g. LePRK1, LePRK2; 64), trans-membrane adhesion complexes (65), and/or stretch-activated ion channels (54).

## 4.3 The pollen-tube clear zone (CZ) and CZ cortex

Although the mechanisms that transduce orienting signals to the cytoplasm are unknown, several of the cytoplasmic and cortical components of the tube tip that are likely to transduce or effect those signals have been identified. Our current under-standing suggests that the cortex of the CZ is of critical importance for orienting polar growth, and that the cytoplasmic components of the CZ and tube support that growth (see Fig. 6B). The polar growth is dependent on exocytosis of CZ vesicles, which contain pectin and other cell-wall components, at a cortical site (possibly of minimal size) on the tube apex. Hence, extracellular cues that orient growth must ultimately affect the position of the site for secretion at the CZ cortex; it is unknown whether cytoplasmic organization of the vegetative cell is controlled indirectly by the position of the growth site, or alternatively, whether the extracellular cues affect the tube's cytoplasmic organization independently of their effect on the growth/secretion site. Known components that regulate tip growth in the pollen tube include a tip-high $Ca^{2+}$ gradient as well as members of the Rop subfamily of Rho-GTPases. A functional actin cytoskeleton is also necessary for sustained tip growth, although it is unclear whether its role extends beyond the transport of vesicles to the CZ (i.e. whether it also regulates, for example, CZ organization). Thus, the mechanisms that regulate polarity in pollen tubes have similarities to those in fucoid zygotes (e.g. a $Ca^{2+}$ gradient), as well as to those in yeast and animal cells (e.g. Rho-GTPases) (66).

### 4.3.1 The $Ca^{2+}$ gradient

It is well established that $Ca^{2+}$ is necessary for continued tip growth of the pollen tube *in vitro* (48, 67). Dissipation of the tip-high gradient either by injection of calcium buffers (68) or by treatment with a calcium-channel blocker (69) stops tube growth. High-resolution imaging using fluorescent $Ca^{2+}$ indicator dyes in living pollen tubes indicates that $[Ca^{2+}]$ is tightly coupled to growth rates, since both the $[Ca^{2+}]$ and the growth rate oscillate with the same periodicity (70, 71). Furthermore, $Ca^{2+}$ also provides positional information that governs growth orientation (68, 69, 72–74). The

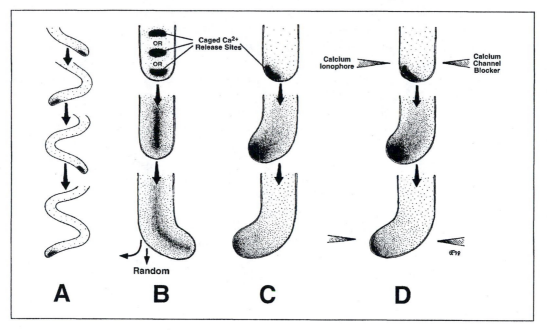

**Fig. 7** A tip-high cytoplasmic $Ca^{2+}$ gradient (small dots) directs pollen tube growth orientation (73). (A) In 'undulating' pollen tubes in culture, oscillating changes in the position of the highest $[Ca^{2+}]$ at the tip apex predicts the change in growth orientation. (B) Release of caged $Ca^{2+}$ by UV laser in any of several positions except immediately adjacent to the CZ cortex is followed by dispersal of $Ca^{2+}$, slowing of tip growth, and random reorientation of tube growth. (C) Release of caged $Ca^{2+}$ on one side of the tube apex, adjacent to the cortex, leads to a transient local increase in $[Ca^{2+}]$ and reorientation of the tube growth toward the side of $Ca^{2+}$ release. (D) Subjecting the growing tube tip to external gradients (by diffusion from a microelectrode) of a $Ca^{2+}$ ionophore (A23187) or a $Ca^{2+}$ channel blocker ($GdCl_3$) attracts or repulses the tube growth, respectively. The cytoplasmic $Ca^{2+}$ gradient again shifts at the tip cortex to predict growth orientation.

region of high cortical $[Ca^{2+}]$ is very limited in size: its position not only predicts the site of future polar growth (73, 74), but is also sufficient to control growth orientation, as demonstrated by manipulation of cytoplasmic $[Ca^{2+}]$, via release of caged $Ca^{2+}$ (see Fig. 7) (73). However, a high cytoplasmic $[Ca^{2+}]$ ectopically placed basal to the CZ does not establish new tube growth at that site (see Fig. 7B), indicating that the cortex of the CZ is somehow specialized to respond to localized increase of $[Ca^{2+}]$.

The location of the $Ca^{2+}$ gradient is likely to be influenced by localized calcium-channel activity in the plasma membrane of the tip. There is a close correlation between the location of extracellular calcium influx into the tip and the site of the cytoplasmic $Ca^{2+}$ gradient, and between the activity of tip-localized $Ca^{2+}$ channels and the presence of the gradient (68, 69). Imposition of an extracellular gradient of calcium-channel blocker laterally across a cultured tube tip causes the internal $Ca^{2+}$ gradient to reposition and further tube growth to be oriented opposite the gradient source (see Fig. 7D) (73), further supporting the idea that channel activity influences the location of the internal gradient. It has been hypothesized that stretch-activated calcium channels would be maximally activated at regions of localized growth, and thus could generate a feedback loop to maintain the gradient at the site of exocytosis (68, 69). However, recent data indicate that the period of maximal $Ca^{2+}$ influx into the tip is not coincident with the period of maximal cytosolic $[Ca^{2+}]$ and maximal tip growth, but rather that maximum influx lags behind; this suggests that secondary stores of $Ca^{2+}$ (either internal or in the cell wall) are involved in generating the gradient (70). Rop-GTPases have also been implicated as potential interactors with the $Ca^{2+}$-signalling apparatus in the pollen tube (see below).

The molecular mechanisms by which high concentrations of $Ca^{2+}$ are able to position the growth site are unknown. However, a fluorescent probe that allows imaging of kinase activity by protein kinase C- and $Ca^{2+}$-dependent calmodulin-independent protein kinase (CDPK)-like molecules has been used to determine whether such kinases might be targets for $Ca^{2+}$. Experiments using this probe suggest that CDPK-like kinase activity is locally enhanced in the CZ near the cortex during the growth reorientation that follows asymmetrial release of caged $Ca^{2+}$ (75). Annexins, which bind to $Ca^{2+}$ and phospholipids, have also been localized to the tip cortex (76). Thus, both kinases and annexins are possible targets of the tip gradient, and may transduce signals in the cytoplasm (CDPK-like kinase), or aid in exocytosis of secretory vesicles (annexins) (28).

### 4.3.2 Rop-GTPases

Members of the Rho family of small GTPases are important for establishing and maintaining cell polarity in yeast and animals (66). Many members of the vascular plant-specific Rop subfamily of the Rho-GTPases have been identified in each of several species, and there is good evidence that some members of the family are crucial regulators of pollen-tube growth (50). The Rop subfamily is most closely related to the Rac subfamily of mammals, and its members show extensive sequence conservation (80% or greater amino-acid identity between members); at least 10 members of the Rop subfamily have been identified in *A. thaliana* (77, 78).

Three members of this family (Rop1At, Rop3At, and Rop5At/At-Rac2) are expressed in mature pollen of *A. thaliana* (78, 79); however, much of the data addressing the function of these proteins in pollen concern an orthologous protein (Rop1Ps) in pea (80). An antibody against Rop1Ps (which cross-reacts with other members of the Rop family) was used to show that Rop proteins localize both to the tube cytoplasm and, most strongly, to the CZ cortex (81). Overexpression of a GFP–At-Rac2 (At-Rac2 is also known as Rop5At) fusion protein in tobacco pollen (79) and of GFP–Rop1At in *A. thaliana* pollen (145) supports the antibody localization results. Interestingly, the expression level of a GFP–Rop1At fusion in *A. thaliana* correlates with both the concentration of GFP–Rop1At present at the apex membrane, and the size of the cortical region to which GFP–Rop1At localizes, i.e. higher-expressing GFP–Rop1At lines show GFP–Rop1At localizing to a larger cortical area in higher quantities. This implies that Rop proteins can affect their own subcellular distribution (145).

Overexpression of constitutively active mutant forms (and to a lesser extent, wild-type forms) of either At-Rac2 in tobacco pollen (79) or Rop1At in *A. thaliana* pollen (145) leads to isotropic, depolarized growth of the pollen (resulting in dramatic, balloon-shaped cells), indicating that Rop-GTPases can play a crucial role in the establishment of the cortical site for growth. Thus, activation of Rop-GTPases and their polar localization are likely important for their function in control of the growth site.

Microinjection of the Rop1Ps antibodies into pea pollen tubes inhibits tube elongation without interrupting cytoplasmic streaming, indicating that inhibition of Rop function *in vivo* specifically affects a process in the tube apex, without affecting the cortical actin filaments necessary for cytoplasmic motility (82). The effect of the injected antibody on the pollen tube is similar to that created by experimental dissipation of the $Ca^{2+}$ gradient; furthermore, the inhibitory effects of antibody injection are enhanced in the presence of either low extracellular $[Ca^{2+}]$, or caffeine treatment (which affects $Ca^{2+}$ signalling in plants), suggesting that Rop may control tube elongation via interactions with the $Ca^{2+}$ gradient (82). Expression of dominant negative mutant forms of both At-Rac2 (79) and Rop1At (145) also inhibits pollen growth, supporting the results of the antibody-injection experiments. Finally, the connection between regulation by Rop and the $Ca^{2+}$ gradient is further strengthened by experiments demonstrating that the inhibition of tube growth by the dominant negative Rop1At is reversed by increasing extracellular $[Ca^{2+}]$, and that injection of Rop1Ps antibody dissipates the $Ca^{2+}$ gradient (145).

Thus, in plant cells, as in yeast and animal cells, Rho family GTPases are important regulators of cell polarity. However, the regulation of pollen-tube polarity by Rop may be primarily via the $Ca^{2+}$ gradient, rather than via the actin cytoskeleton (as seen in non-plant eukaryotes, Chapters 2 and 4) (see Fig. 8). Furthermore, experiments have not yet separated Rop function in establishing tube polarity and aiding polar growth; these two functions may be inextricably linked in the pollen tube.

### 4.3.3 Other components of the CZ: the actin cytoskeleton and the plasma membrane

Although previous data indicated that an actin meshwork was located adjacent to the CZ cortex, more recent studies, using advanced TEM techniques and probes that label actin *in vivo*, indicate that no actin filaments, or very few, are in the CZ or at the CZ cortex (83–85). Instead, cortical actin filaments in the tube proper terminate near the boundary of the CZ and the region of the 'reverse fountain' streaming (see Figs 6B, 8). The accepted role for these filaments is in organelle transport to the CZ (48, 49). Expression in pollen tubes of a GFP–talin fusion protein (talin is a mouse actin-binding protein) suggested that a localized ring of F-actin (perhaps analogous to that observed in *P. compressa* zygotes; 9) may be present at the cortex at the CZ boundary; this structure might aid in defining the CZ region (85), although more evidence is needed to address this possibility.

Potential regulators of actin include the calcium gradient, Rop-GTPases, and profilin. Interestingly, dissipation of the calcium gradient via treatment with caffeine leads to displacement of the actin filament ends closer to the apex, suggesting that the gradient exerts some control over their position (84). $Ca^{2+}$ is known to regulate

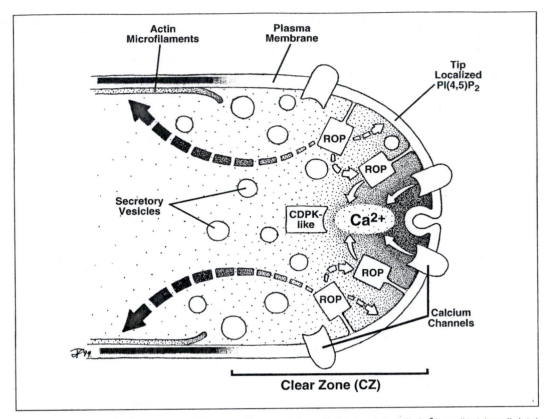

**Fig. 8** A model for molecular interactions at the CZ apex of the pollen tube. A tip-high $Ca^{2+}$ gradient (small dots) aids in defining the site for tip growth and secretion of vesicle contents into the cell wall (the cell wall is not shown). Localized $Ca^{2+}$ channel activity contributes to the $Ca^{2+}$ gradient, possibly via local activation of calcium channels; alternatively, the channels themselves could be localized. The $Ca^{2+}$ gradient regulates a localized $Ca^{2+}$-dependent calmodulin-independent protein kinase (CDPK)-like activity; the function of this kinase activity is unknown. Rop-GTPases also regulate polar growth; active, GTP-bound Rop likely promotes $Ca^{2+}$ influx, although the mechanism by which this occurs is unknown. The tip localization of Rop also appears to be important for restricting growth to the tip; Rop may regulate its own distribution (if indirectly—dashed arrow). Other potential downstream targets of Rop (dashed arrows) are cortical actin filaments that terminate just basal to the clear zone, and phosphatidylinositol (4,5)-bisphosphate ($PI(4,5)P_2$) localized in the CZ plasma membrane.

negatively the actin cytoskeleton (86), and the high tip concentrations could aid in excluding actin from the CZ. Actin cables in an abnormal helical pattern appear at the cortex of the isotropically growing 'balloon' pollen induced by constitutively active At-Rac2 (79), but whether this is a result of direct actin regulation by Rop, or a secondary effect of the change in growth pattern caused by Rop, is unclear. Finally, profilin is an actin-monomer-binding protein that is also a good candidate for a regulator of actin dynamics in pollen (87). The poly-proline binding domain in a maize pollen profilin is important for regulating actin dynamics *in vivo* (88). This is tantalizing, because in animals and yeast, profilin has been shown to interact with the proline-rich motifs present in formin homology (FH) proteins; certain FH proteins are components of actin-regulating complexes that also include Rho family GTPases (e.g. the profilin–Bni1p–Cdc42p complex involved in yeast polarity (89; Chapter 2, this volume)).

A final component possibly associated with pollen-tube polarity is phosphatidylinositol (4,5)-bisphosphate ($PI(4,5)P_2$). Because the pleckstrin homology (PH) domain of mammalian phospholipase C binds $PI(4,5)P_2$ both *in vitro* and *in vivo*, a GFP–PH domain fusion was used to investigate $PI(4,5)P_2$ in tobacco pollen tubes (79). GFP–PH accumulates preferentially at the CZ cortex, and inhibits tube growth, suggesting that $PI(4,5)P_2$ lipids are localized in the CZ plasma membrane, and that they promote growth (79). At-Rac2 associates *in vitro* with a pollen tube phosphatidylinositol kinase activity that generates $PI(4,5)P_2$, and thus could regulate the production of $PI(4,5)P_2$ at the CZ cortex (79); further work is needed to address this prospect.

## 4.4 Pollen-tube versus root-hair growth: similarities and differences

Root hairs are tube-like extensions from certain root epidermal cells, and these hairs play physiological roles in the uptake of water and nutrients from the soil. Their formation occurs via a tip-growing mechanism that has many parallels with that of pollen tubes and the rhizoid tip (2, 50, 90). For example, actin is necessary for their growth (91), and a tip-focused $Ca^{2+}$ gradient is associated with the orientation of their growth (92). Genetic screens in both *A. thaliana* and maize have identified several loci that, when mutant, alter root-hair morphology (93, 94). These genes presumably encode components required for root-hair growth. Genetic analysis of one *A. thaliana* gene, *TIP1*, further emphasizes the commonality between mechanisms used in pollen-tube and root-hair growth, in that mutant plants display defects in both root-hair elongation and pollen-tube growth (95). However, it is worth noting two aspects of root-hair growth that appear to be distinct from those described for the pollen tube.

First, determination of the site of polar growth for root hairs (analogous to axis alignment in fucoid zygotes) is genetically separable from the process of tip growth. Characterization of mutant phenotypes has identified several genes that affect the positioning and initiation of the root-hair growth site, but not hair growth itself (90). For example, the *A. thaliana RHD6* gene appears to act in a signalling process that, in conjunction with the phytohormones auxin and ethylene, positions the site of root-hair initiation at the apical end of root epidermal cells (96). Initiation of the root hair is accompanied by a localized acidification of the cell wall, and can be inhibited reversibly by the application of buffers that prevent acidification; in contrast, tip growth of an initiated root hair is inhibited irreversibly by treatment with the same buffers (97). Thus, a localized change in the cell wall (induced by acidification) may be a necessary event in root-hair initiation; furthermore, initiation of tip growth and tip growth itself are affected differentially by external pH (97).

Secondly, the orientation of root-hair growth in *A. thaliana* appears to be determined by a more stable or constant signal than the one to which pollen tubes respond. Redirection of root-tip growth via manipulation of internal calcium only transiently reorients tip growth; the root hair returns to its original growth orientation after the experimental treatment is removed (92). The microtubule cytoskeleton could be an important component that aids in transmitting or stabilizing this signal because treatment with the microtubule inhibitors taxol and oryzalin leads to various growth defects, such as 'waving' growth (i.e. a zigzagging root hair) or the formation of more than one growing tip (91).

Isolation of mutants important for root-hair growth has already identified two new components of interest: RHD3 in *A. thaliana*, a conserved protein with GTP-binding motifs of unknown function (98), and RTH1 in maize (T.-J. Wen and P.S. Schnable, in preparation), a protein with intriguing similarity to the *S. cerevisiae* Sec3p protein, a subunit of the exocyst complex necessary for targeted secretion (99). With the ability to readily identify and isolate components for root-hair growth via a genetic approach, the prospects for a better mechanistic understanding of how root-hair polarity is initiated and expressed appear favourable.

# 5. Asymmetrical cell division and plant cell fate

The asymmetrical division of the fucoid zygote segregates factors that control the cell fates adopted by the two daughter cells. Similar processes appear to be important in many phases of vascular plant development (see Fig. 1B) in that asymmetrical divisions often produce daughters of different cell types (100, 101). Several studies have reported mutations that concomitantly affect both the division itself and the daughter-cell fates (e.g. *scarecrow* in the *A. thaliana* root (102); *gemini pollen1* in the microspore (103); and *gnom* in the zygote (104)). Thus, these analyses support the idea that, as in fucoid zygotes, the spatial control of cell division plays some role in the differential inheritance of determinants that affect cell fate. Although no comprehensive model has yet emerged, data suggest that the cytoskeleton, the cell wall, and secretion have important functions in generating these asymmetries.

## 5.1  The mechanism of asymmetrical cell division

Plant cytokinesis differs from animal cytokinesis in two fundamental ways: first, in plants the placement of the division plane is determined prior to mitotic prophase by a process that marks the cell cortex; and, secondly, the cytokinetic apparatus is the phragmoplast, a structure that builds the division plane centrifugally from the cell interior to the cortex (2, 105). The process that marks the cortex is associated with F-actin and microtubule rings that appear prior to prophase at the future division site (i.e. where the new cell wall will fuse with the existing wall). These cytoskeletal structures are called the pre-prophase bands (PPBs), and the apparent default mechanism for positioning the PPBs places them so as to produce daughter cells of approximately equal size. Asymmetrical division appears to be associated with a specialized process that causes the division-site apparatus to be positioned at a site distinct from that of the default arrangement.

Development of the stomatal complex in the epidermis of monocot leaves occurs through a predictable pattern of asymmetrical divisions (2). The small stomatal subsidiary cell (SC) is formed by an asymmetrical division of the subsidiary mother cell (SMC), and is flanked by its sister cell and by a cell from a different lineage, the guard mother cell (GMC). A signal that induces and positions the division plane that creates the SC apparently originates in the GMC, as the SC-generating division occurs only after formation of the GMC and is oriented such that the smaller SC is always adjacent to the GMC. Prior to the asymmetrical division of the SMC, its PPBs appear at a position close the GMC (106, 107). Thus the putative signal originating from the GMC appears to shift the site of PPB formation away from its default site toward the GMC. Appearance of the shifted PPB is preceded by accumulation of cytoplasm in the vacuolated SMC at the cortex adjacent to the GMC, migration of the SMC nucleus adjacent to that site, and accumulation of an F-actin patch at the same cortical site (revealed by injection of rhodamine phalloidin into living *Tradescantia virginiana* cells) (107, 108). This oriented accumulation of actin persists throughout mitosis. Centrifugation experiments that shift nuclei away from the GMC-adjacent cortical site indicate that migration back to that site requires an uncompromised actin cytoskeleton (108); however, it is not known whether the asymmetrical placement of the PPBs is induced by the asymmetrically placed nucleus, or is independent of nuclear movement. Treatment with CD leads to the formation of misoriented cell walls in many plant-cell divisions (including in the stomatal complex); however, it is unclear which F-actin structures are affected by such treatments, and it is therefore difficult to assign specific roles to specific actin structures. Interestingly, local application of pressure to the SMC leads to cytoplasmic and nuclear migration to the pressure point; one proposed mechanism to mediate this movement is through an ion flux produced by stretch-activated channels in the plasma membrane (108). Whether a similar mechanism aids in orienting the asymmetrical SMC division is unknown.

## 5.2 Potential connections between targeted secretion, the cell wall, and plant cell fate

Evidence indicates that cell-wall asymmetries in vascular plants can be partitioned by cell division, and that those asymmetries are correlated with distinct developmental fates and functions (109). Carrot embryogenic suspension cultures contain cell types that differ in morphology and in competence to generate somatic embryos. Two molecularly distinguishable cell types, called state B and state C, are both competent to generate embryos, but differ in the presence of an AGP epitope in their cell walls that is recognized by both the monoclonal antibody JIM8 and the Yariv phenylglycoside (110). State B cells (which are JIM8-positive) appear to divide asymmetrically, producing one state C cell (which is JIM8-negative and embryogenic) and one state F cell (which is JIM8-positive, but not embryogenic) (109). Hence, the presence or absence of an epitope in the cell wall of the daughters of a state B cell correlates with their developmental competence. Immunofluorescent detection of JIM8 in dividing cells from these cultures suggests that the JIM8 cell-wall epitope becomes asymmetrically distributed in state B cells, and is partitioned asymmetrically into only one of the two daughters (the state F cell) by an oriented division plane. Further experiments with sorted populations of JIM8(+) and JIM8(-) cells from this culture indicate that the JIM8(-) population (containing the embryogenesis-competent state C cells) do not produce embryos unless cultured in media that was previously conditioned by growth of JIM8(+) cells (109). The authors concluded that at least some of the JIM8(+) cells produce a soluble signal that causes state C cells to proceed through somatic embryogenesis, whereas cells that lack the JIM8 epitope do not produce this signal.

Genetic and molecular analysis of the *GNOM/EMB30* gene of *A. thaliana* has hinted at a further connection between secretion, cell polarity, and cell fate in vascular plants. The allelic *gnom* and *emb30* mutations cause a range of morphological abnormalities, including irregular cell shapes (104, 111). The normally asymmetrical zygotic division in mutant plants often appears to be symmetrical (104), as do other typically asymmetrical divisions (111). Embryonic polarity in *gnom* mutants (as assayed by the expression pattern of the apical marker *AtLTP1*) is variable in its orientation (112). Thus, *GNOM/EMB30* appears to be important in a cellular process that can promote cell asymmetry and aid in proper orientation of embryonic patterning. The GNOM/EMB30 protein contains the conserved Sec7 domain which, based on analysis of mutant alleles, is likely necessary for its function (111, 113). The Sec7 domains of several budding yeast and mammalian proteins have guanine nucleotide exchange activity, a positive regulatory activity directed toward Arf1, which is a conserved small GTPase that is necessary for formation of vesicles in the ER and Golgi (114). Intriguingly, the exchange activity of several Sec7 domain proteins is specifically inhibited by BFA; inhibition of Sec7 domain proteins in yeast is the primary mechanism by which the drug blocks secretion (115). Thus, the morphologically similar effects (e.g. misplaced division planes and incorrect cell-fate specification) elicited by *gnom/emb30* mutations in *A. thaliana* cells and by BFA-treatment of *F. distichus* zygotes may be due to a similar mechanism: inhibition of secretion. Further phenotypic characterization of the *gnom/emb30* mutants at the cellular level will be necessary to address this speculative connection more fully.

In summary, much analysis remains before we have a clear understanding of the polar process of asymmetrical cell division in vascular plants, and its role in development. The shards of evidence so far do suggest that many factors important in the establishment of polarity in fucoid zygotes (e.g. actin, the cell wall, secretion) also have roles in vascular plants, and, based on these similarities, a speculative general model for asymmetrical division in vascular plants has been proposed (2). In this model, a cortical complex and its associated cytoskeletal structures, positioned by a directional cue, direct targeted secretion to produce a locally differentiated cell wall. Interactions between cell-wall components and the cortical complex can stabilize the position of the complex, modify complex function such that it can affect other cellular processes (e.g. nuclear migration, division orientation), and/or direct developmental decisions following partitioning of the localized complex/distinct cell wall into a single daughter cell.

# 6. Cell polarity and polar auxin transport

Cell polarity, expressed as asymmetries associated with the plasma membrane, may also be necessary for physiological functions of differentiated cells (e.g. mammalian epithelial cells; 1) (see Fig. 1C). However, research on plant cell polarity has mainly concentrated on the morphological polarity exhibited by tip-growing cells. The existence of molecular asymmetry in the plasma membrane in two types of non-tip-growing plant cells has recently been demonstrated, and will provide new tools with which to study plant cell polarity. This dramatic discovery provides direct molecular evidence in support of a long-standing model for the polar transport of auxin (116–120).

## 6.1 Polar transport of auxin is important for auxin-mediated responses

Classic physiological studies have demonstrated that the phytohormone auxin influences plant development in many ways (121, 122). The most abundant naturally occurring form of auxin is indole-3-acetic acid (IAA), which regulates several important physiological and developmental processes, including cell division and elongation, as well as embryonic patterning and vascular differentiation.

A distinctive feature of many auxin-based responses, in contrast to those of other plant hormones, is the asymmetrical transport of auxin through various plant tissues (see Fig. 9A) (121, 122). This asymmetry can lead to differentials in auxin accumulation, which result in distinct downstream responses. For example, in the shoot, auxin is synthesized primarily in immature cells at the shoot apex, and is then transported downward (basipetally), inducing vascular differentiation and contributing to vascular organization (123).

Plant gravitropism (directed growth in response to gravity) is also dependent on auxin-transport patterns (124). In roots, auxin transport continues in the same direction as in shoots (but acropetally, with reference to the root meristem) through the stele (the central vasculature). However, at the root tip, auxin is redirected outward to the epidermis and basipetally (away from the root meristem) (see Fig. 9A). Physiological experiments indicate that when the root is placed horizontally, cells within the root cap (acropetal to the meristem) sense the altered gravity vector, resulting in a redistribution of auxin to the lower side of the root tip through an unknown mechanism (125). One model to explain gravitropic curvature is that subsequent basipetal transport of auxin from the root tip through the outer cell layers of the root results in higher concentrations on the lower side of the root elongation zone (basipetal to the root meristem) compared with those on the upper side. The

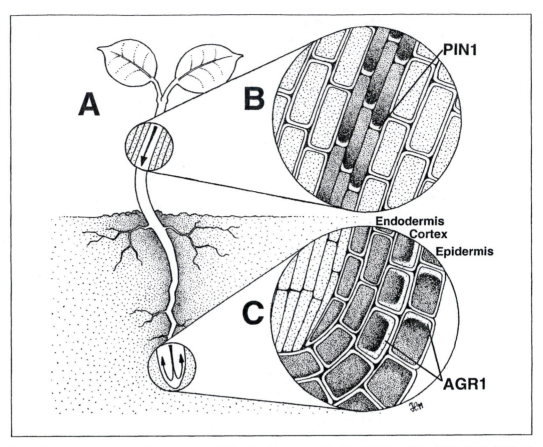

**Fig. 9** Polar transport of auxin in plants could be due to asymmetrical localization of the AGR1/PIN1 family of putative auxin efflux carriers. (A) One stream of auxin transport is associated with the central vasculature in the shoot; transport occurs in a basipetal direction. In the root, auxin is transported acropetally toward the root meristem in the stele (the central vasculature of the root). Unknown mechanisms lead to a shift in the direction of transport near the root tip; in the root cortical tissues near the tip, transport is basipetal (away from the meristem). (B) The PIN1 protein localizes asymmetrically to the plasma membrane at the basal end of parenchymatous cells in the *A. thaliana* shoot xylem. (C) Near the root tip, the AGR1/EIR1/PIN2 protein localizes asymmetrically away from the internal face of cells of the cortex layer, and to the basal end of cells in the epidermal layer.

difference in auxin concentration across the width of the root leads to downward curvature of the root, due to a lateral gradient in growth rate (i.e. more growth occurs on the upper side of the root than on the lower) (126).

## 6.2  The chemiosmotic hypothesis for polar auxin transport

Based on physiological experiments, a model to explain the mechanism of polar auxin transport was proposed independently by Rubery and Sheldrake (127) and by Raven (128). The chemiosmotic model postulates that polar auxin transport occurs via cell-to-cell 'shuttling' of IAA molecules (see Fig. 10). Furthermore, the asymmetry

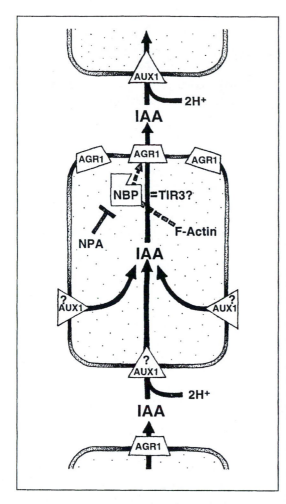

**Fig. 10** The chemiosmotic model of polar auxin (IAA) transport, incorporating some of the candidate molecular components in root epidermal cells. AUX1 is a candidate component of the auxin influx carrier; no data addressing its localization have been published. AGR1 is localized asymmetrically to the basal plasma membrane, and is a candidate component of the auxin efflux carrier. The polarity of auxin transport is provided by the asymmetrical localization of the efflux carrier. The NPA-binding protein (NBP) facilitates polar auxin efflux by interacting with the efflux carrier through another, labile component; the *A. thaliana* *TIR3* gene may encode the NBP. NPA binding by the NBP inhibits auxin efflux. The actin cytoskeleton also interacts with the NBP, and apparently promotes the ability of NBP to regulate the efflux carrier; further details of this interaction are unknown.

of transport is directly dependent on a plasma membrane asymmetry in specialized cells. Experimental tests of this model have supported its basic tenets, and current molecular and genetic approaches are identifying compelling candidate molecules to further support the importance of cell polarity for auxin transport.

The transport of auxin is energy-dependent; specifically, auxin uptake into cells is coupled to the proton motive force present across the plasma membrane (due to the pH difference between the acidic extracellular and neutral cytoplasmic environments) (122). However, the auxin uptake carrier (at least one component of which is probably encoded by the *Arabidopsis AUX1* gene; 129, 130) does not provide directionality to the transport. This does not preclude asymmetrical localization of the uptake carrier, which could, in theory, enhance the directionality of auxin transport (131). The AUX1 protein shows homology to plant amino-acid permeases; its 11 predicted transmembrane segments suggest a role for AUX1 as a transporter, to carry the IAA molecule (which is structurally related to tryptophan) across the plasma membrane (129, 130). The subcellular localization of AUX1 has not yet been determined.

The efflux carrier provides directionality to transport, as was determined following the discovery and use of several auxin-transport inhibitors that block efflux. The most commonly used inhibitors have been 1-*N*-naphthylphthalamic acid (NPA) and 2,3,5-triodobenzoic acid (TIBA); these compounds specifically inhibit both the polar transport of auxin from the apical to the basal end of isolated stem segments, and the tropisms (e.g. gravitropism) effected by auxin (122). Evidence for the asymmetrical localization of the NPA-inhibited efflux activity within each cell has come from studies with both maize and courgettes (zucchini). In both species, two populations of auxin-accumulating plasma-membrane vesicles with distinct sensitivities to NPA could be isolated using dextran density gradients. The authors suggest that each vesicle population corresponds to distinct portions of the cell's plasma membrane (apical versus basal), reflecting in turn the asymmetry of distribution of the NPA-sensitive efflux activity (132).

## 6.3 PIN1 and AGR1 / EIR1 / PIN2 are likely components of the polarly localized auxin efflux carrier

Screening for *Arabidopsis* mutants with defects in auxin-dependent responses (e.g. root gravitropic curvature) or for those that phenocopy the application of auxin-transport inhibitors has provided a successful entrance into the molecular characterization of several components of the polar transport machinery (131). Most exciting has been the isolation of genes encoding a family of proteins that are present in several plant species and that are likely to be auxin efflux channels, due to the near simultaneous convergence of work by four different research groups (116–120). The proteins are believed to be components of the efflux carrier for several reasons. First, mutants in two genes in this family show phenotypes (discussed below) that are consistent with a defect in polar auxin transport. Secondly, the predicted amino-acid sequences of these proteins show limited homology to putative bacterial transporter proteins, and contain approximately 10 predicted transmembrane segments. Finally, *S. cerevisiae* strains that express one of the family members (*AGR1 / EIR1 / PIN2*) were more resistant to the toxic effects of the auxin-related compound 5-fluoro-IAA (116, 118) and retained less radiolabelled IAA (116) than the control strains.

In *Arabidopsis*, at least five members of this family of putative efflux carrier components have been identified, two of which (*PIN1* and *AGR1 / EIR1 / PIN2*) have been characterized functionally by phenotypic analysis. Mutations in the *pin-formed* (*PIN1*) gene lead to a decrease in polar auxin transport, and a shoot phenotype (an inflorescence stem with only a few or defective flowers) that is phenocopied by NPA treatment of wild-type *Arabidopsis* (133, 134). Further analysis of plants carrying different *pin1* alleles revealed defective vascular patterning as well (117). Immuno-detection of PIN1 protein has demonstrated its asymmetrical localization at the basal plasma membrane of elongated cells associated with the shoot vasculature (see

Fig. 9B) (117), as predicted by the chemiosmotic model for basipetal polar auxin transport.

The *AGRAVITROPIC1* (*AGR1*) gene (116, 120) (initially isolated as *AGR* (135) and *WAV6* (136), and subsequently as *EIR1* (118) and *PIN2* (119)) appears to be an efflux carrier that is expressed and functions primarily in the root. *AGR1* is necessary for the root gravitropic response, but not for normal shoot gravitropic response or development. *agr1* mutants retain more preloaded radiolabelled IAA in their roots than do wild-type plants (116), and their growth is less inhibited by TIBA (116, 120), suggesting a defect in the auxin efflux carrier. AGR1 is only expressed in the outer two cell layers of the root, basipetal to the root tip, and, as with PIN1, immuno-localization of the AGR1 protein demonstrated an asymmetrical distribution at the plasma membrane (see Fig. 9C). Furthermore, the subcellular localization again supports the chemiosmotic model and is consistent with the direction of auxin transport demonstrated for the outer regions of the root. In the root cortex, AGR1 localization suggests a role in transporting auxin away from the central stele, whereas the basipetal localization of AGR1 in the root epidermis is consistent with the observed direction of auxin transport basipetally from the root tip to the elongation zone (119).

Thus, in both roots and shoots, cell polarity is evident and physiologically important in the directional transport of auxin. Interestingly, shoot cuttings rooted in an inverted orientation do not alter their original basipetal direction of auxin transport, suggesting that the polarity of the majority of transport components is stable, or 'fixed' (137). The mechanism that establishes and maintains the asymmetrical localization of PIN1 and AGR1 is unknown; however, several models can be imagined. Both proteins contain putative N-terminal signal peptides, suggesting that they reach the plasma membrane through the secretory pathway. Although targeted, asymmetrical secretion of specific vesicle populations has thus far not been demonstrated in non-tip-growing plant cells, it could conceivably aid in localizing the auxin efflux channel. Transporter localization may also be influenced by other cellular factors that have been identified as regulators of polar transport. In the shoot, the lateral transport of auxin necessary to establish differential auxin accumulations triggering gravitropism is thought to occur in response to perception of changes in the gravity vector by specialized endodermal cells (138). It will be interesting to determine whether an AGR1/PIN1 family protein is present in these cells, and whether its localization is altered in response to gravity.

## 6.4 Regulation of the polar auxin efflux carrier

The best characterized regulator of auxin efflux is the NPA-binding protein (NBP) (see Fig. 10). The NPA-binding site resides on a polypeptide distinct from the auxin transporter, and experiments suggest that a short-lived component (which rapidly disappears in the presence of protein synthesis inhibitors) is necessary for the NPA–NBP–carrier interaction that blocks efflux (139). There is still some controversy as to whether the NBP is an integral (140) or a peripheral membrane protein (141);

however, recent evidence suggests that the NBP is on the cytoplasmic face of the membrane and is associated with the actin cytoskeleton both *in vitro* and *in vivo* (142). Furthermore, the association between the NBP and the actin cytoskeleton appears physiologically relevant, as treatment with CD reduces the ability of NPA to inhibit auxin transport in courgette (zucchini) hypocotyls (142).

The physiological function of the NBP (in the absence of NPA) remains a matter of some speculation. Some naturally occurring flavonoid compounds block auxin efflux and compete with NPA for NBP binding; thus, they may serve to regulate auxin transport *in vivo*, via the NPA-binding site (122). Analysis of *Arabidopsis* plants mutant for the *TRANSPORT INHIBITOR RESPONSE3* (*TIR3*) gene suggests that the NBP may promote auxin efflux. *tir3* mutants show reduced sensitivity to NPA, display a reduced level of IAA transport in the inflorescence stem, and show a variety of morphological abnormalities throughout the plant that are likely due to changes in auxin distribution (143). Biochemical analysis shows that isolated microsomes from *tir3* plants have a reduced number of NPA-binding sites compared to wild type; thus, *TIR3* may encode the NBP, or may aid in the functioning of the NBP (143). Molecular characterization of this gene may help address the role of the NBP and the relationship between the actin cytoskeleton and the transport of auxin. For example, does the NBP aid in establishing the asymmetrical localization of the efflux transporter? Or, alternatively, is F-actin necessary for the co-localization of NBP and the efflux transporter, where together they can regulate auxin efflux? The isolation of the *PIN1/AGR1* family provides molecular tools with which to further address how the NBP interacts with the efflux carrier.

## 7. Conclusions

No single experimental system has yet provided a thorough picture of the mechanisms that establish and regulate cell polarity in plants. Many key molecular components that function in these processes have yet to be isolated, and attempting to find common themes in these disparate examples is perhaps premature. However, our current knowledge suggests that some features are important in several or all of the cited examples: association of the actin cytoskeleton with the plasma membrane, a localized $[Ca^{2+}]$ gradient, Rho family GTPases, secretory function, and cell-wall asymmetries. This raises the possibility that common, or at least related, mechanisms exist in the generation of cell polarity in plants. One future challenge will be to determine whether these mechanisms are indeed related; for example, do Rho-GTPases influence the distribution of PIN1/AGR1 family proteins? Another challenge is to define the causal connections between the isolated components; for example, localized $[Ca^{2+}]$, the actin cytoskeleton, and secretion in fucoid zygotes. Finally, identifying new components in the polarity machinery, and establishing genetic approaches to studying polar cells (e.g. in root hairs) should also provide a more thorough understanding of cell polarity and its developmental and physiological roles in plants.

## Acknowledgements

The author would like to give special thanks to Joe Patterson for providing excellent illustrations, and to Zuzana Vejlupkova for assistance. Thanks are also due to Margit Foss, Terri Lomax, and Zhenbiao Yang for critical review of parts or all of the manuscript. Many of the ideas presented in this manuscript were developed in discussions with Ralph Quatrano and Sid Shaw. The author's work is supported by the NSF (9803156-IBN) and the USDA (98–35304–6670).

## References

1. Drubin, D. G. and Nelson, W. J. (1996) Origins of cell polarity. *Cell*, **84**, 335.
2. Fowler, J. E. and Quatrano, R. S. (1997) Plant cell morphogenesis: plasma membrane interactions with the cytoskeleton and cell wall. *Annu. Rev. Cell Dev. Biol.*, **13**, 697.
3. Kropf, D. L. (1992) Establishment and expression of cellular polarity in fucoid zygotes. *Microbiol. Rev.*, **56**, 316.
4. Kropf, D. L. (1997) Induction of polarity in Fucoid zygotes. *Plant Cell*, **9**, 1011.
5. Quatrano, R. S. and Shaw, S. L. (1997) Role of the cell wall in the determination of cell polarity and the plane of cell division in *Fucus* embryos. *Trends Plant Sci.*, **2**, 15.
6. Robinson, K. R., Wozniak, M., Pu, R., and Messerli, M. (1999) Symmetry breaking in the zygotes of the fucoid algae: controversies and recent progress. *Curr. Top. Dev. Biol.*, **44**, 101.
7. Quatrano, R. S. (1973) Separation of processes associated with the differentiation of two-celled *Fucus* embryos. *Dev. Biol.*, **30**, 209.
8. Hable, W. E. and Kropf, D. L. (1998) Roles of secretion and the cytoskeleton in cell adhesion and polarity establishment in *Pelvetia compressa* zygotes. *Dev. Biol.*, **198**, 45.
9. Alessa, L. and Kropf, D. L. (1999) F-actin marks the rhizoid pole in living *Pelvetia compressa* zygotes. *Development*, **126**, 201.
10. Jaffe, L. F. (1968) Localization in the developing *Fucus* egg and the general role of localizing currents. *Adv. Morphol.*, **7**, 295.
11. Knapp, E. (1931) Entwicklungsphysiologische untersuchungen an Fucaceen-Eiren. I. Zur Kenntnis der Polaritat der eier von Cystosira barbata. *Planta*, **14**, 731.
12. Jaffe, L. F. (1958) Tropistic responses of zygotes of the Fucaceae to polarized light. *Exp. Cell Res.*, **15**, 282.
13. Berger, F. and Brownlee, C. (1994) Photopolarization of *Fucus* sp. zygote by blue light involves a plasma membrane redox chain. *Plant Physiol.*, **105**, 519.
14. Robinson, K. R. and Miller, B. J. (1997) The coupling of cyclic GMP and photopolarization of *Pelvetia* zygotes. *Dev. Biol.*, **187**, 125.
15. Brawley, S. H. and Robinson, K. R. (1985) Cytochalasin treatment disrupts the endogenous currents associated with cell polarization in fucoid zygotes: studies of the role of F-actin in embryogenesis. *J. Cell Biol.*, **100**, 1173.
16. Kropf, D. L., Berge, S. K., and Quatrano, R. S. (1989) Actin localization during *Fucus* embryogenesis. *Plant Cell*, **1**, 191.
17. Madden, K. and Snyder, M. (1998) Cell polarity and morphogenesis in budding yeast. *Annu. Rev. Microbiol.*, **52**, 687.
18. Ayscough, K. R. and Drubin, D. G. (1998) A role for the yeast actin cytoskeleton in pheromone receptor clustering and signalling. *Curr. Biol.*, **8**, 927.

19. Berger, F. and Brownlee, C. (1993) Ratio confocal imaging of free cytoplasmic calcium gradients in polarising and polarised *Fucus* zygotes. *Zygote*, **1**, 9.
20. Pu, R. and Robinson, K. R. (1998) Cytoplasmic calcium gradients and calmodulin in the early development of the fucoid alga *Pelvetia compressa*. *J. Cell Sci.*, **111**, 3197.
21. Speksnijder, J. E., Miller, A. L., Weisenseel, M. H., Chen, T. H., and Jaffe, L. F. (1989) Calcium buffer injections block fucoid egg development by facilitating calcium diffusion. *Proc. Natl Acad. Sci., USA*, **86**, 6607.
22. Nuccitelli, R. (1978) Ooplasmic segregation and secretion in *Pelvetia* egg is accompanied by the membrane-generated electrical current. *Dev. Biol.*, **62**, 13.
23. Robinson, K. R. (1996) Calcium and the photopolarization of *Pelvetia* zygotes. *Planta*, **198**, 378.
24. Hurst, S. R., and Kropf, D. L. (1991) Ionic requirements for establishment of an embryonic axis in *Pelvetia* zygotes. *Planta*, **185**, 27.
25. Love, J., Brownlee, C., and Trewavas, A. J. (1997) $Ca^{2+}$ and calmodulin dynamics during photopolarization in *Fucus serratus* zygotes. *Plant Physiol.*, **115**, 249.
26. Shaw, S. L. and Quatrano, R. S. (1996) Polar localization of a dihydropyridine receptor on living *Fucus* zygotes. *J. Cell Sci.*, **109**, 335.
27. Knaus, H. G., Moshammer, T., Friedrich, K., Kang, H. C., Haugland, R. P., and Glossman, H. (1992) *In vivo* labeling of L-type $Ca^{2+}$ channels by fluorescent dihydropyridines: evidence for a functional, extracellular heparin-binding site. *Proc. Natl Acad. Sci., USA*, **89**, 3586.
28. Battey, N. H., James, N. C., Greenland, A. J., and Brownlee, C. (1999) Exocytosis and endocytosis. *Plant Cell*, **11**, 643.
29. Quatrano, R. S. and Crayton, M. A. (1973) Sulfation of fucoidan in *Fucus* embryos. I. Possible role in localization. *Dev. Biol.*, **30**, 29.
30. Brawley, S. H. and Quatrano, R. S. (1979) Sulfation of fucoidin in *Fucus* embryos. 4. Autoradiographic investigations of fucoidin sulfation and secretion during differentiation and the effects of cytochalasin treatment. *Dev. Biol.*, **73**, 193.
31. Shaw, S. L. and Quatrano, R. S. (1996) The role of targeted secretion in the establishment of cell polarity and orientation of the division plane in *Fucus* zygotes. *Development*, **122**, 2623.
32. Kropf, D. L., Kloareg, B., and Quatrano, R. S. (1988) Cell wall is required for fixation of the embryonic axis in *Fucus* zygotes. *Science*, **239**, 187.
33. Berger, F. and Brownlee, C. (1995) Physiology and development of protoplasts obtained from *Fucus* embryos using laser microsurgery. *Protoplasma*, **186**, 63.
34. Bouget, F. Y., Berger, F., and Brownlee, C. (1998) Position dependent control of cell fate in the *Fucus* embryo: role of intercellular communication. *Development*, **125**, 1999.
35. Wagner, V. T., Brian, L., and Quatrano, R. S. (1992) Role of a vitronectin-like molecule in embryo adhesion of the brown alga *Fucus*. *Proc. Natl Acad. Sci., USA*, **89**, 3644.
36. Goodner, B. and Quatrano, R. S. (1993) *Fucus* embryogenensis: A model to study the establishment of polarity. *Plant Cell*, **5**, 1471.
37. Burridge, K. and Chrzanowska-Wodnicka, M. (1996) Focal adhesions, contractility, and signaling. *Annu. Rev. Cell Dev. Biol.*, **12**, 463.
38. Brownlee, C. and Wood, J. W. (1986) A gradient of free $Ca^{2+}$ in growing rhizoid cells of *Fucus serratus*. *Nature*, **320**, 624.
39. Gibbon, B. C. and Kropf, D. L. (1994) Cytosolic pH gradients associated with tip growth. *Science*, **263**, 1419.
40. Bouget, F.-Y., Gerttula, S., and Quatrano, R. S. (1995) Spatial redistribution of poly(A)$^+$ RNA during polarization of the *Fucus* zygote is dependent upon microfilaments. *Dev. Biol.*, **171**, 258.

41. Henry, C. A., Jordan, J. R., and Kropf, D. L. (1996) Localized membrane-wall adhesions in *Pelvetia* zygotes. *Protoplasma*, **190**, 39.

42. Horvitz, H. R. and Herskowitz, I. (1992) Mechanisms of asymmetric cell division: two Bs or not two Bs, that is the question. *Cell*, **68**, 237.

43. Brownlee, C. and Bouget, F. Y. (1998) Polarity determination in Fucus: from zygote to multicellular embryo. *Semin. Cell. Dev. Biol.*, **9**, 179.

44. Allen, V. W. and Kropf, D. L. (1992) Nuclear rotation and lineage specification in *Pelvetia* embryos. *Development*, **115**, 873.

45. Bisgrove, S. R. and Kropf, D. L. (1998) Alignment of centrosomal and growth axes is a late event during polarization of *Pelvetia compressa* zygotes. *Dev. Biol.*, **194**, 246.

46. White, J. and Strome, S. (1996) Cleavage plane specification in C. *elegans*: how to divide the spoils. *Cell*, **84**, 195.

47. Berger, F., Taylor, A., and Brownlee, C. (1994) Cell fate determination by the cell wall in early *Fucus* development. *Science*, **263**, 1421.

48. Taylor, L. P. and Hepler, P. K. (1997) Pollen germination and tube growth. *Annu. Rev. Plant Physiol. Plant Mol. Biol.*, **48**, 461.

49. Franklin-Tong, V. E. (1999) Signaling and the modulation of pollen tube growth. *Plant Cell*, **11**, 727.

50. Yang, Z. (1998) Signaling tip growth in plants. *Curr. Opin. Plant Biol.*, **1**, 525.

51. Joos, U., van Aken, J., and Kristen, U. (1994) Microtubules are involved in maintaining the cellular polarity in pollen tubes of *Nicotiana sylvestris*. *Protoplasma*, **179**, 5.

52. Goldman, M. H., Goldberg, R. B., and Mariani, C. (1994) Female sterile tobacco plants are produced by stigma-specific cell ablation. *EMBO J.*, **13**, 2976.

53. Wolters-Arts, M., Lush, W. M., and Mariani, C. (1998) Lipids are required for directional pollen-tube growth. *Nature*, **392**, 818.

54. Lush, W. M., Grieser, F., and Wolters-Arts, M. (1998) Directional guidance of *Nicotiana alata* pollen tubes *in vitro* and on the stigma. *Plant Physiol.*, **118**, 733.

55. Cheung, A. Y., Wang, H., and Wu, H. M. (1995) A floral transmitting tissue-specific glycoprotein attracts pollen tubes and stimulates their growth. *Cell*, **82**, 383.

56. Wu, H. M., Wang, H., and Cheung, A. Y. (1995) A pollen tube growth stimulatory glycoprotein is deglycosylated by pollen tubes and displays a glycosylation gradient in the flower. *Cell*, **82**, 395.

57. Nothnagel, E. A. (1997) Proteoglycans and related components in plant cells. *Int. Rev. Plant Cytol.*, **174**, 195.

58. Sommer-Knudsen, J., Lush, W. M., Bacic, A., and Clarke, A. E. (1998) Re-evaluation of the role of a transmitting tract-specific glycoprotein on pollen tube growth. *Plant J.*, **13**, 529.

59. Jauh, G. Y. and Lord, E. M. (1996) Localization of pectins and arabinogalactan-proteins in lily (*Lilium longiflorum* L.) pollen tube and style, and their possible roles in pollination. *Planta*, **199**, 251.

60. Roy, S., Jauh, G. Y., Hepler, P. K., and Lord, E. M. (1998) Effects of Yariv phenylglycoside on cell wall assembly in the lily pollen tube. *Planta*, **204**, 450.

61. Hülskamp, M., Schneitz, K., and Pruitt, R. E. (1995) Genetic evidence for a long-range activity that directs pollen tube guidance in *Arabidopsis*. *Plant Cell*, **7**, 57.

62. Ray, S. M., Park, S. S., and Ray, A. (1997) Pollen tube guidance by the female gametophyte. *Development*, **124**, 2489.

63. Wilhelmi, L. K. and Preuss, D. (1996) Self-sterility in *Arabidopsis* due to defective pollen tube guidance. *Science*, **274**, 1535.

64. Muschietti, J., Eyal, Y., and McCormick, S. (1998) Pollen tube localization implies a role in

pollen–pistil interactions for the tomato receptor-like protein kinases LePRK1 and LePRK2. *Plant Cell*, **10**, 319.

65. Lord, E. M. and Sanders, L. C. (1992) Roles for the extracellular matrix in plant development and pollination: a special case of cell movement in plants. *Dev. Biol.*, **153**, 16.

66. Hall, A. (1998) Rho GTPases and the actin cytoskeleton. *Science*, **279**, 509.

67. Feijó, J. A., Malhó, R., and Obermeyer, G. (1995) Ion dynamics and its possible role during *in vitro* pollen germination and tube growth. *Protoplasma*, **187**, 155.

68. Pierson, E. S., Miller, D. D., Callaham, D. A., Shipley, A. M., Rivers, B. A., Cresti, M., and Hepler, P. K. (1994) Pollen tube growth is coupled to the extracellular calcium ion flux and the intracellular calcium gradient: effect of BAPTA-type buffers and hypertonic media. *Plant Cell*, **6**, 1815.

69. Malhó, R., Read, N. D., Trewavas, A. J., and Pais, M. S. (1995) Calcium channel activity during pollen tube growth and reorientation. *Plant Cell*, **7**, 1173.

70. Holdaway-Clarke, T. L., Feijó, J. A., Hackett, G. R., Kunkel, J. G., and Hepler, P. K. (1997) Pollen tube growth and the intracellular cytosolic calicium gradient oscillate in phase while extracellular calcium influx is delayed. *Plant Cell*, **9**, 1999.

71. Messerli, M. and Robinson, K. R. (1997) Tip localized $Ca^{2+}$ pulses are coincident with peak pulsatile growth rates in pollen tubes of *Lilium longiflorum*. *J. Cell Sci.*, **110**, 1269.

72. Jaffe, L. A., Weisenseel, M. H., and Jaffe, L. F. (1975) Calcium accumulations within the growing tips of pollen tubes. *J. Cell Biol.*, **67**, 488.

73. Malhó, R. and Trewavas, A. J. (1996) Localized apical increases of cytosolic free calcium control pollen tube orientation. *Plant Cell*, **8**, 1935.

74. Pierson, E. S., Miller, D. D., Callaham, D. A., van Aken, J., Hackett, G., and Hepler, P. K. (1996) Tip-localized calcium entry fluctuates during pollen tube growth. *Dev. Biol.*, **174**, 160.

75. Moutinho, A., Trewavas, A. J., and Malho, R. (1998) Relocation of a $Ca^{2+}$-dependent protein kinase activity during pollen tube reorientation. *Plant Cell*, **10**, 1499.

76. Blackbourn, H. D., Barker, P. J., Huskisson, N. S., and Battey, N. H. (1992) Properties and partial protein sequence of plant annexins. *Plant Physiol.*, **99**, 864.

77. Winge, P., Brembu, T., and Bones, A. M. (1997) Cloning and characterization of rac-like cDNAs from *Arabidopsis thaliana*. *Plant Mol. Biol.*, **35**,

78. Li, H., Wu, G., Ware, D., Davis, K. R., and Yang, Z. (1998) *Arabidopsis* Rho-related GTPases: differential gene expression in pollen and polar localization in fission yeast. *Plant Physiol.*, **118**, 407.

79. Kost, B., Lemichez, E., Spielhofer, P., Hong, Y., Tolias, K., Carpenter, C., and Chua, N.-H. (1999) Rac homologues and compartmentalized phosphatidylinositol 4,5-bisphosphate act in a common pathway to regulate polar pollen tube growth. *J. Cell Biol.*, **145**, 317.

80. Yang, Z. and Watson, J. C. (1993) Molecular cloning and characterization of rho, a ras-related small GTP-binding protein from the garden pea. *Proc. Natl Acad. Sci., USA*, **90**, 8732.

81. Lin, Y., Wang, Y., Zhu, J.-K., and Yang, Z. (1996) Localization of a Rho GTPase implies a role in tip growth and movement of the generative cell in pollen tubes. *Plant Cell*, **8**, 293.

82. Lin, Y. and Yang, Z. (1997) Inhibition of pollen tube elongation by microinjected anti-Rop1Ps antibodies suggests a crucial role for Rho-type GTPases in the control of tip growth. *Plant Cell*, **9**, 1647.

83. Lancelle, S. A. and Hepler, P. K. (1992) Ultrastructure of freeze-substituted pollen tubes of *Lilium longiflorum*. *Protoplasma*, **167**, 215.

84. Miller, D. D., Lancelle, S. A., and Hepler, P. K. (1996) Actin microfilaments do not form a dense meshwork in *Lilium longiflorum* pollen tube tips. *Protoplasma*, **195**, 123.

85. Kost, B., Spielhofer, P., and Chua, N. H. (1998) A GFP–mouse talin fusion protein labels plant actin filaments *in vivo* and visualizes the actin cytoskeleton in growing pollen tubes. *Plant J.*, **16**, 393.

86. Kohno, T. and Shimmen, T. (1988) Mechanism of Ca$^{2+}$ inhibition of cytoplasmic streaming in lily pollen tubes. *J. Cell Sci.*, **95**, 501.

87. Staiger, C. J., Gibbon, B. C., Kovar, D. R., and Zonia, L. E. (1997) Profilin and actin depolymerizing factor: modulators of actin organization in plants. *Trends Plant Sci.*, **2**, 275.

88. Gibbon, B. C., Zonia, L. E., Kover, D. R., Hussey, P. J., and Staiger, C. J. (1998) Pollen profilin function depends on interaction with proline-rich motifs. *Plant Cell*, **10**, 981.

89. Evangelista, M., Blundell, K., Longtine, M. S., Chow, C. J., Adames, N., Pringle, J. R., Peter, M., and Boone, C. (1997) Bni1p, a yeast formin linking cdc42p and the actin cytoskeleton during polarized morphogenesis. *Science*, **276**, 118.

90. Hülskamp, M., Folkers, U., and Grini, P. E. (1998) Cell morphogenesis in *Arabidopsis*. *Bioessays*, **20**, 20.

91. Bibikova, T. N., Blancaflor, E. B., and Gilroy, S. (1999) Microtubules regulate tip growth and orientation in root hairs of *Arabidopsis thaliana*. *Plant J.*, **17**, 657.

92. Bibikova, T. N., Zhigilei, A., and Gilroy, S. (1997) Root hair growth in *Arabidopsis thaliana* is directed by calcium and an endogenous polarity. *Planta*, **203**, 495.

93. Wen, T.-J. and Schnable, P. S. (1994) Analyses of mutants of three genes that influence root hair development in *Zea mays* (Gramineae) suggest that root hairs are dispensable. *Am. J. Bot.*, **81**, 833.

94. Grierson, C. S., Roberts, K., Feldmann, K. A., and Dolan, L. (1997) The *COW1* locus of arabidopsis acts after *RHD2*, and in parallel with *RHD3* and *TIP1*, to determine the shape, rate of elongation, and number of root hairs produced from each site of hair formation. *Plant Physiol.*, **115**, 981.

95. Schiefelbein, J., Galway, M., Masucci, J., and Ford, S. (1993) Pollen tube and root-hair tip growth is disrupted in a mutant of *Arabidopsis thaliana*. *Plant Physiol.*, **103**, 979.

96  Masucci, J. D. and Schiefelbein, J. W. (1996) The *rhd6* mutation of *Arabidopsis thaliana* alters root-hair initiation through an auxin- and ethylene-associated process. *Plant Physiol.*, **106**, 1335.

97. Bibikova, T. N., Jacob, T., Dahse, I., and Gilroy, S. (1998) Localized changes in apoplastic and cytoplasmic pH are associated with root hair development in *Arabidopsis thaliana*. *Development*, **125**, 2925.

98. Wang, H., Lockwood, S. K., Hoeltzel, M. F., and Schiefelbein, J. W. (1997) The *ROOT HAIR DEFECTIVE3* gene encodes an evolutionarily conserved protein with GTP-binding motifs and is required for regulated cell enlargement in *Arabidopsis*. *Genes Dev.*, **11**, 799.

99. TerBush, D. R., Maurice, T., Roth, D., and Novick, P. (1996) The Exocyst is a multiprotein complex required for exocytosis in *Saccharomyces cerevisiae*. *EMBO J.*, **15**, 6483.

100. Gallagher, K. and Smith, L. G. (1997) Asymmetric cell division and cell fate in plants. *Curr. Opin. Cell Biol.*, **9**, 842.

101. Jürgens, G., Grebe, M., and Steinmann, T. (1997) Establishment of cell polarity during early plant development. *Curr. Opin. Cell Biol.*, **9**, 849.

102. Di Laurenzio, L., Wysocka-Diller, J., Malamy, J. E., Pysh, L., Helariutta, Y., Freshour, G., Hahn, M. G., Feldmann, K. A., and Benfey, P. N. (1996) The *SCARECROW* gene regulates an asymmetric division that is essential for generating the radial organization of the *Arabidopsis* root. *Cell*, **86**, 423.

103. Park, S. K., Howden, R., and Twell, D. (1998) The *Arabidopsis thaliana* gametophytic mutation *gemini pollen1* disrupts microspore polarity, division asymmetry and pollen cell fate. *Development*, **125**, 3789.

104. Mayer, U., Buttner, G., and Jurgens, G. (1993) Apical–basal pattern formation in the *Arabidopsis* embryo: studies on the role of the *gnom* gene. *Development*, **117**, 149.

105. Heese, M., Mayer, U., and Jürgens, G. (1998) Cytokinesis in flowering plants: cellular processes and developmental integration. *Curr. Opin. Plant Biol.*, **1**, 486.

106. Cho, S.-O. and Wick, S. M. (1989) Microtubule orientation during stomatal differentiation in grasses. *J. Cell Sci.*, **92**, 581.

107. Cleary, A. L. (1995) F-actin redistributions at the division site in living *Tradescantia* stomatal complexes as revealed by microinjection of rhodamine-phalloidin. *Protoplasma*, **185**, 152.

108. Kennard, J. L. and Cleary, A. L. (1997) Pre-mitotic nuclear migration in subsidiary mother cells of *Tradescantia* occurs in G1 of the cell cycle and requires F-actin. *Cell Motil. Cytoskeleton*, **36**, 55.

109. McCabe, P. F., Valentine, T. A., Forsberg, L. S., and Pennell, R. I. (1997) Soluble signals from cells identified at the cell wall establish a developmental pathway in carrot. *Plant Cell*, **9**, 2225.

110. Pennell, R. I., Janniche, L., Scofield, G. N., Booij, H., de Vries, S. C., and Roberts, K. (1992) Identification of a transitional cell state in the developmental pathway to carrot somatic embryogenesis. *J. Cell Biol.*, **119**, 1371.

111. Shevell, D. E., Leu, W.-M., Gillmor, C. S., Xia, G., Feldmann, K. A., and Chua, N.-H. (1994) *EMB30* is essential for normal cell division, cell expansion, and cell adhesion in *Arabidopsis* and encodes a protein that has similarity to Sec7. *Cell*, **77**, 1051.

112. Vroemen, C. W., Langeveld, S., Mayer, U., Ripper, G., Jurgens, G., Van Kammen, A., and De Vries, S. C. (1996) Pattern formation in the *Arabidopsis* embryo revealed by position-specific lipid transfer protein gene expression. *Plant Cell*, **8**, 783.

113. Busch, M., Mayer, U., and Jurgens, G. (1996) Molecular analysis of the *Arabidopsis* pattern formation of gene *GNOM*: gene structure and intragenic complementation. *Mol. Gen. Genet.*, **250**, 681.

114. Chardin, P. and McCormick, F. (1999) Brefeldin A: the advantage of being uncompetitive. *Cell*, **97**, 153.

115. Peyroche, A., Antonny, B., Robineau, S., Acker, J., Cherfils, J., and Jackson, C. L. (1999) Brefeldin A acts to stabilize an abortive ARF–GDP–Sec7 domain protein complex: involvement of specific residues of the Sec7 domain. *Mol. Cell*, **3**, 275.

116. Chen, R., Hilson, P., Sedbrook, J., Rosen, E., Caspar, T., and Masson, P. H. (1998) The *Arabidopsis thaliana AGRAVITROPIC1* gene encodes a component of the polar-auxin-transport efflux carrier. *Proc. Natl Acad. Sci., USA*, **95**, 15112.

117. Gälweiler, L., Guan, C., Müller, A., Wisman, E., Mendgen, K., Yephremov, A., and Palme, K. (1998) Regulation of polar auxin transport by AtPIN1 in *Arabidopsis* vascular tissue. *Science*, **282**, 2226.

118. Luschnig, C., Gaxiola, R. A., Grisafi, P., and Fink, G. R. (1998) EIR1, a root-specific protein involved in auxin transport, is required for gravitropism in *Arabidopsis thaliana*. *Genes Dev.*, **12**, 2175.

119. Müller, A., Guan, C., Gälweiler, L., Tanzler, P., Huijser, P., Marchant, A., Parry, G., Bennett, M., Wisman, E., and Palme, K. (1998) AtPIN2 defines a locus of Arabidopsis for root gravitropism control. *EMBO J.*, **17**, 6903.

120. Utsuno, K., Shikanai, T., Yamada, Y., and Hashimoto, T. (1998) *Agr*, an *Agravitropic* locus of *Arabidopsis thaliana*, encodes a novel membrane-protein family member. *Plant Cell Physiol.*, **39**, 1111.

121. Goldsmith, M. H. M. (1977) The polar transport of auxin. *Annu. Rev. Plant Physiol.*, **28**, 439.

122. Lomax, T. L., Muday, G. K., and Rubery, P. H. (1995) Auxin transport. In *Plant hormones: physiology, biochemistry and molecular biology*, (ed. P. J. Davis), p.509. Kluwer Academic Publishers, Dordrecht.

123. Sachs, T. (1991) Cell polarity and tissue patterning in plants. *Development (Suppl.)*, **1**, 83.

124. Dolan, L. (1998) Pointing roots in the right direction: the role of auxin transport in response to gravity. *Genes Dev.*, **12**, 2091.

125. Young, L. M., Evans, M., and Hertel, R. (1990) Correlations between gravitropic curvature and auxin movement across gravistimulated roots of *Zea mays*. *Plant Physiol.*, **92**, 792.

126. Evans, M. L. (1991) Gravitropism: Interaction of sensitivity, modulation, and effector redistribution. *Plant Physiol.*, **95**, 1.

127. Rubery, P. H. and Sheldrake, A. R. (1974) Carrier-mediated auxin transport. *Planta*, **188**, 101.

128. Raven, J. A. (1975) Transport of indoleacetic acid in plant cells in relation to pH and electrical potential gradients, and its significance for polar IAA transport. *New Phytol.*, **74**, 163.

129. Bennett, M. J., Marchant, A., Green, H. G., May, S. T., Ward, S. P., Millner, P. A., Walker, A. R., Schulz, B., and Feldmann, K. A. (1996) *Arabidopsis AUX1* gene: a permease-like regulator of root gravitropism. *Science*, **273**, 948.

130. Marchant, A., Kargul, J., May, S. T., Muller, P., Delbarre, A., Perrot-Rechenmann, C., and Bennett, M. J. (1999) AUX1 regulates root gravitropism in *Arabidopsis* by facilitating auxin uptake within root apical tissues. *EMBO J.*, **18**, 2066.

131. Bennett, M. J., Marchant, A., May, S. T., and Swarup, R. (1998) Going the distance with auxin: unravelling the molecular basis of auxin transport. *Phil. Trans. R. Soc. Lond. B Biol. Sci.*, **353**, 1511.

132. Lützelschwab, M., Asard, H., Ingold, U., and Hertel, R. (1989) Heterogeneity of auxin-accumulating membrane vesicles from *Cucurbita* and *Zea*: A possible reflection of cell polarity. *Planta*, **177**, 305.

133. Okada, K., Ueda, J., Komaki, M. K., Bell, C. J., and Shimura, Y. (1991) Requirement of the auxin polar transport system in the early stage of *Arabidopsis* floral bud formation. *Plant Cell*, **3**, 677.

134. Bennett, S. R. M., Alvarez, J. A., Bossinger, G., and Smyth, D. R. (1995) Morphogenesis in *pinoid* mutants of *Arabidopsis thaliana*. *Plant J.*, **8**, 505.

135. Morris, D. A., Rubery, P. H., Jarman, J., and Sabater, M. (1991) Effects of inhibitors of protein synthesis on transmembrane auxin transport in *Cucubita pepo* L. hypocotyl segments. *J. Exp. Bot.*, **42**, 773.

140. Bernasconi, P., Patel, B. C., Reagan, J. D., and Subramanian, M. V. (1996) The N-1-naphthylphthalamic acid-binding protein is an integral membrane protein. *Plant Physiol.*, **111**, 427.

141. Dixon, M. W., Jacobson, J. A., Cady, C. D., and Muday, G. K. (1996) Cytoplasmic orientation of the naphthylphthalamic acid binding protein in zucchini plasma membrane vesicles. *Plant Physiol.*, **112**, 421.

142. Butler, J. H., Hu, S., Brady, S. R., Dixon, M. W., and Muday, G. K. (1998) *In vitro* and *in vivo* evidence for actin association of the naphthylphthalamic acid-binding protein from zucchini hypocotyls. *Plant J.*, **13**, 291.
143. Ruegger, M., Dewey, E., Hobbie, L., Brown, D., Bernasconi, P., Turner, J., Muday, G., and Estelle, M. (1997) Reduced naphthylphthalamic acid binding in the *tir3* mutant of *Arabidopsis* is associated with a reduction in polar auxin transport and diverse morphological defects. *Plant Cell*, **9**, 745.
144. Bellanger, K. and Quatrano, R. S. (2000) Membrane recycling occurs during asymmetric tip growth and cell plate formation in *Fucus* zygotes. *Protoplasma*, In press.
145. Li, H., Lin, Y., Heath, R. M., Zhu, M. X., and Yang, Z. (1999) Control of pollen tube tip growth by a Rop GTPase-dependent pathway that leads to tip-localized calcium influx. *Plant Cell*, **11**, 1731.

# 6 | Cell biology of polarity development in *Xenopus* oocytes and embryos

CAROLYN A. LARABELL

## 1. Introduction

The development of a single cell into a multicellular organism involves a complex series of highly orchestrated events. Establishing polarity is critical for the normal development of all species and, therefore, it is not surprising that it begins very early—even before the egg is fertilized in some species. During development of the vertebrate, *Xenopus laevis*, three distinct axes of polarity are established:

(1) the animal–vegetal axis, established during oogenesis;

(2) the dorsal–ventral axis, established immediately following fertilization; and

(3) the left–right axis, established during embryogenesis.

These axes are established sequentially, with each subsequent axis superimposed upon the preceding axis. In this chapter I will focus on the first two polarization events, since the third has been addressed quite well in recent reviews (1–3).

## 2. The animal–vegetal axis

The animal–vegetal axis is the first axis established during *Xenopus* development and is an important foundation for the specification of the three primary germ layers during embryogenesis. At the end of oogenesis there is a distinct polarity that can be detected externally by the concentration of pigment granules in the animal cortex and reduction of pigment in the vegetal cortex (4) and internally by the positioning of the large nucleus, referred to as the germinal vesicle (GV), towards the animal pole (Fig. 1). More subtle, but equally important, examples of polarity include organelle gradients (such as the positioning of the largest yolk platelets in the vegetal cytoplasm), the asymmetrical distribution of cytoskeletal proteins, and the localization of specific mRNA molecules to either the animal or vegetal pole. The correct

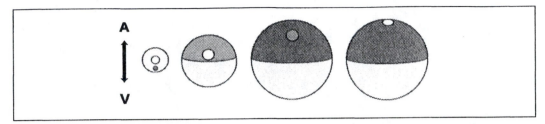

**Fig. 1** Animal–vegetal polarity. The earliest evidence of animal–vegetal polarity is the asymmetrical positioning of the mitochondrial cloud and mRNA molecules at one side of the nucleus, followed by its transport to the oocyte cortex during stage I of oogenesis. Pigment granules begin moving to the animal hemisphere and large yolk platelets move to the vegetal hemisphere during stage IV of oogenesis. Pigment accumulation in the animal hemisphere increases and the nucleus moves toward the animal pole during stage VI. During meiotic maturation, the chromosomes displace the pigment granules at the animal pole, forming the characteristic 'white spot'.

positioning of these components is undeniably important because cells that derive from the animal hemisphere will give rise to the ectodermal structures (including the nervous system and epidermis), cells that derive from the vegetal hemisphere will give rise to endodermal structures (primarily the primitive gut), and cells located in the equatorial (or marginal) zone, the junction of the animal and vegetal hemispheres (5, 6), will give rise to mesoderm (muscle, blood, bone, etc.). The process by which animal–vegetal polarity is established during oogenesis is discussed below.

## 2.1 Organelle asymmetry

The earliest reported evidence of an animal–vegetal asymmetry of organelles in *Xenopus* oocytes is the formation of the mitochondrial cloud that translocates to the vegetal pole during stage I of oogenesis. Oogenesis has been divided into six stages, I–VI (4), that encompass the diplotene stage of meiotic prophase and are characterized by extensive growth and differentiation. Prior to oogenesis, when the cells reach the early diplotene stage (25–50 μm in diameter, often referred to as stage 0), the oocytes are radially symmetrical, transparent, and contain few organelles. The large nucleus of these oocytes, referred to as the germinal vesicle (GV), is centrally located, with the other cellular organelles distributed evenly throughout the surrounding cytoplasm. At this time, several aggregates (typically between two and four) of mitochondria are seen closely apposed to the GV. Since one of these aggregates eventually develops into the mitochondrial cloud during stage I of oogenesis, they are all referred to as pre-mitochondrial clouds. All of these aggregates, including the prospective mitochondrial cloud, are tightly associated with the germinal vesicle at this time, as demonstrated by the fact that they remain attached to a GV that has been isolated from live oocytes (7).

Stage I of oogenesis is distinguished by a significant increase in oocyte size (50–200 μm in diameter) and a marked proliferation of mitochondria in one of the clouds. At the same time, numerous small organelles (0.1 μm diameter), endoplasmic reticulum, and germ plasm are incorporated into this growing cluster, which is now identified as the mitochondrial cloud (4, 8–12). The mechanism by which the one aggregate of mitochondria expands into the mitochondrial cloud is not known, but it has been suggested that it is influenced by the nuclear–centrosomal axis, an alignment of the cytoplasmic bridge, centriole pair, and chromosomes during the mitotic divisions that produced sets of 16 cells (5, 10, 12).

Eventually, in response to some as yet unidentified signal, the mitochondrial cloud begins moving away from its perinuclear position and travels to one specific site in the cell periphery, where it ultimately docks and remains at the cortex. The mechanism by which one cortical region is selected, and by which the mitochondrial cloud is transported to that site at the vegetal pole, is equally as mysterious. It appears unlikely that typical cytoskeletal transport mechanisms are involved since there is no detectable asymmetry in either the microtubule or the filamentous actin network (13, 14). In addition, disassembly of either of these networks using cytoskeletal inhibitors (cytochalasin B or nocodazole) has no effect on the localization of the mitochondrial cloud at the cortex (13–16). Although it has been reported recently that there is an enrichment of spectrin in the mitochondrial cloud, there is no evidence that spectrin is necessary for the transport of the cloud. It is more likely that spectrin provides a structural foundation for, or meshwork in which to trap, cloud contents. Despite the fact that the mechanism of mitochondrial cloud formation and transport to the oocyte periphery is not known, its positioning at the cortex is important for normal embryonic development. There are ample data demonstrating that the movement is spatially and temporally correlated with the transportation and localization of several unique messenger RNA molecules to the oocyte cortex (17). Furthermore, the region of the cell periphery to which the cloud moves ultimately becomes the vegetal pole and, therefore, positioning of the mitochondrial cloud at this site is the earliest reliable indicator of animal–vegetal polarity in the oocyte (13, 17).

Additional manifestations of polarity continue throughout oogenesis with the asymmetrical distribution of other organelles as well as cytoskeletal proteins. Yolk platelets, the large round or oblong organelles (3–14 μm in the long axis) that provide nutrients during embryogenesis, first appear when the oocytes are 300 μm in diameter (early stage III). These platelets, which contain lipovitellin and phosvitin derived by proteolytic cleavage of the precursor protein vitellogenin, are symmetrically taken up by pinocytosis. As the yolk platelets are travelling from the cell surface inward, cortical granules are being formed in the central region of the egg and moving outward, to all regions of the oocyte periphery. When the oocyte is approximately 450 μm in diameter (stage III), pigment granules begin to appear and they, too, move toward the periphery of the oocyte with no detectable asymmetry (4). An elaborate transport and sorting process underlying these steps must be present to generate an inward movement of yolk plates at the same time pigment and cortical granules are moving outward. Although this correlates temporally with the polarization of the microtubule network, a direct connection with these cytoskeletal components remains to be demonstrated.

In early stage IV, when the oocyte is approximately 600 μm in diameter, a non-pigmented patch appears in one region of the oocyte and continues to enlarge until there are two distinct halves: one dark and one light. The animal–vegetal axis is well established by this time and oocyte growth continues with only minor modifications until it reaches the full-grown (1200–1400 μm diameter) stage VI oocyte. During these stages, the larger yolk platelets accumulate in the vegetal hemisphere. The full-grown, stage VI oocyte also demonstrates a distinct polarity of cortical granules based on size and density. There is a single layer of smaller (1–2 μm in diameter) granules in the animal hemisphere and two to three rows of the larger (up to 3 μm in diameter) granules lining the plasma membrane of the vegetal hemisphere. Several plausible hypotheses have been presented (4), but they have not been tested in light of our current understanding of the cytoskeletal polarity that also arises at this time (18, 19).

## 2.2 Cytoskeletal asymmetry

Extensive reviews of the organization of the cytoskeleton in *Xenopus* oocytes have been published recently (18–20). Therefore, I will summarize those findings only briefly in this chapter. During the earliest stages of oogenesis, when the mitochondrial cloud is formed and moved to the cortex and the RNAs are localized via the METRO (see below), there is no detectable asymmetry in cytoskeletal proteins. The development of cytoskeletal polarity is first seen during stages IV–V (15, 21). The smallest stage I oocytes (35–50 μm in diameter) contain actin cables that fill the nucleus, penetrate the mitochondrial cloud, and form a dense cytoplasmic network with no apparent asymmetry by late stage I (14). Because of difficulties visualizing filamentous actin in the dense, yolky, vitellogenic *Xenopus* oocytes, little is known about the inner cytoplasmic actin organization during the later stages of oogenesis. The cortex of oocytes, however, can be viewed with the confocal microscope, and contains a dense meshwork of actin filaments with no obvious asymmetry (18, 22). The intermediate filament protein, cytokeratin, first appears as oocytes reach 75–125 μm in diameter and is seen associated with the surface of the GV, the oocyte cortex, and with the pre-mitochondrial clouds (21). Later in stage I, a network of cytoplasmic cytokeratin filaments links the perinuclear and cortical cytokeratin networks and also penetrates the mitochondrial cloud (21). Cytokeratin organization continues to be radially symmetrical until midway through stage IV, as oocytes reach 600–800 μm diameter, when the keratin network becomes thicker in the vegetal hemisphere than in animal hemisphere (21); however, this polarity is reversed by stage VI.

Microtubules form a radially symmetrical, randomly oriented, dense network throughout the cytoplasm of the smallest stage I oocytes. This radial symmetry prevails until stage III of oogenesis, when they demonstrate a radial symmetry extending from the perinuclear cytoplasm towards the oocyte cortex. During stage IV, as the pigment granules and GV move to the animal hemisphere, the radial organization of microtubules persists in the animal hemisphere, but a less organized pattern of microtubules is seen in the vegetal hemisphere. At this time γ-tubulin is seen in the cortex, suggesting that at least some of the microtubules have their negative ends in the cortex and positive ends extending towards the GV (19). This pattern persists through to stage VI of oogenesis.

## 2.3  RNA localization

Localization of specific RNA molecules to one region of the cell is common to oocytes of vertebrates and invertebrates. The first report of an asymmetrical distribution of specific RNAs in *Xenopus* oocytes was by Rebagliati *et al.* (23) who described that three mRNAs, An1, An2, and An3, were preferentially localized to the animal hemisphere and Vg1 mRNA was restricted to the vegetal hemisphere. Since then additional localized RNAs have been identified. Those restricted to the animal hemisphere include x121 (24), xlan4 (25), Oct-60 (26), An4 (27), and a poly(A) binding protein (28), and those localized to the vegetal hemisphere include the mRNA for Xlsirts (29), Xcat-2 (30), Xcat-3 (31), Xwnt11 (32), VegT (33–36), Xdazl (37), and Xpat (38). Undoubtedly, this list will continue to grow as more localized RNAs are identified. The positioning of specific mRNA molecules at the vegetal pole is among the earliest detectable signs of oocyte polarity, and the mechanism by which they become localized is the subject of enthusiastic investigations.

There is substantial evidence that mRNAs localized at the vegetal pole utilize one of two pathways—an early cytoskeleton-independent pathway, or a late cytoskeleton-dependent pathway—to become positioned at the vegetal cortex (Plate 3). The RNAs localized at the vegetal pole have received a great deal of attention because there is compelling evidence that this region contains important dorsalizing components, as will be discussed later. Many of the vegetally localized RNAs identified to date, including Xlsirts, Xcat2, Xwnt11, Xdazl, and Xpat, utilize the early pathway (37–40). Examining the transport and localization of three of these RNAs has elucidated the spatial and temporal events of this early pathway, which utilizes a specialized region of the mitochondrial cloud known as the messenger transport organizer, or METRO (29, 40). Xlsirts, a family of non-coding RNAs that function in anchoring Vg1 at the vegetal cortex (29, 41), Xcat2, a mRNA with nanos-like zinc fingers but no known function (30), and Xwnt11, which is thought to be involved in axial patterning (32), all become localized during stage I of oogenesis via the METRO. The METRO is a region coincident with the very tip, or leading edge, of the mitochondrial cloud that contains the nuage or germ plasm (40). Localization of these RNAs via the METRO occurs sequentially in a process including differentiation of the mitochondrial cloud, localization of the mRNAs to this cloud, sorting of the RNAs to specific regions of the METRO, translocation of the cloud to the vegetal cortex, and docking at the vegetal cortex.

Differentiation of the mitochondrial cloud occurs early in oogenesis. In the smallest pre-stage I oocytes (25–50 μm in diameter), Xlsirts, Xcat2, and Xwnt11 mRNAs leave the GV and become positioned in all of the mitochondrial aggregates (pre-mitochondrial clouds) adjacent to the GV. By early stage I, however, they have in some way become restricted to the single mitochondrial aggregate that has enlarged and proliferated into the mitochondrial cloud. Exogenous RNAs demonstrate the same behaviour as the endogenous RNAs, suggesting that there is a distinct modification of the pre-mitochondrial aggregates that accompanies formation of the actual mitochondrial cloud. This was demonstrated when fluorescently labelled Xlsirt and Xcat2 mRNAs, which were simultaneously injected into pre-stage I oocytes, segregated from one another within 5 hours and formed discrete cytoplasmic particles with unique morphologies (17). Xcat2 mRNAs formed very small, homogeneous granules, whereas Xlsirts formed large (0.4–1.2 μm in diameter) particles (17). After 16–48 hours, the exogenous RNAs were localized within all of the pre-cloud structures of the pre-stage I oocytes. When, however, the same fluorescent RNAs were injected into stage I oocytes they behaved very differently. When injected into stage I oocytes the RNAs became localized to the METRO region of only the mitochondrial cloud and were not seen in the mitochondrial aggregates (17). It is apparent from these studies that there is a mechanism in stage I oocytes that modifies the properties of the pre-mitochondrial clouds, causing one to become the RNA transport vehicle.

Once the mRNAs have become restricted to the mitochondrial cloud they segregate into different domains within the mitochondrial cloud. Although not yet identified, a mechanism that facilitates sorting within the mitochondrial cloud must exist, since the individual RNAs occupy unique and specific positions within the cloud. When viewed from the vegetal pole, Xlsirts mRNA appears evenly distributed along the surface, while Xcat2 forms a ring around the periphery and Xwnt11 mRNAs are localized within the centre of the cloud (40). An ultrastructural examination using whole mount electron microscopy *in situ* hybridization revealed that Xcat2 mRNA specifically localizes to the germinal granules, whereas Xlsirts and Xwnt11 are associated with a fibrillar network of the germ plasm (42). In separate labelling experiments, the fibrillar network of the germ plasm labels with anti-spectrin antibodies. It is tempting, therefore, to suggest that the fibrillar material with which these RNAs are associated is the spectrin network that fills the mitochondrial cloud, but direct evidence awaits ultrastructural confirmation.

By the end of stage I, the mitochondrial cloud begins moving to the vegetal cortex. The mechanism by which transport occurs, however, is not known. Whatever the mechanism, it does not require typical cytoskeletal transport processes, since disruption of actin (using cytochalasin B) or tubulin (using nocodazole) has no effect on the localization of these mRNAs to the vegetal pole (17). By late stage I or early stage II the METRO-localized RNAs reach the vegetal cortex, where they become anchored. At this time Xlsirt, Xcat2, and Xwnt11 become layered in a disk-like structure, with Xcat2 closest to the cortex and Xlsirts and Xwnt11 following (17). After docking of the cloud and RNAs, the oocyte demonstrates the first unambiguous evidence of the animal–vegetal polarity.

The late pathway for mRNA localization acts during stages III–IV of oogenesis and is responsible for localization of Vg1 RNA, a member of the transforming growth factor-β family that has been implicated in the establishment of both dorsal–ventral (43) and left–right (44) asymmetries of the embryo. The mechanism of action of the late pathway is distinctly different from that of the early METRO pathway, in that it does depend on cytoskeletal proteins. Studies of Vg1 localization revealed that the late pathway occurs as a two-step process; the first step, which begins about late stage II to early stage III, is a microtubule-dependent translocation step, and the second step, which follows immediately, is a microfilament-dependent anchoring step (40, 45). This model has been refined recently by data showing a cytoskeletal-independent step in localization of Vg1 that acts prior to the onset of microtubule-mediated transport (7) (see Chapters 1 and 8 for discussions of other mechanisms for localizing mRNAs).

Prior to Vg1 mRNA transport, while the METRO pathway is operational, Vg1 mRNA is excluded from the mitochondrial cloud and is, instead, homogeneously distributed throughout the cytoplasm of the oocyte. As the mitochondrial cloud and the METRO mRNAs are approaching the vegetal cortex, the distribution of Vg1 mRNA is altered. The RNAs become co-localized, in a microtubule-independent manner, with a structure containing a subdomain of the endoplasmic reticulum (ER) that is located at the trailing edge of the mitochondrial cloud (7). At the same time, a steady depletion of Vg1 mRNA from the animal hemisphere is observed. After the early RNAs have anchored at the vegetal cortex, the ER and Vg1 mRNA form a wedge-like structure around the anchored cloud. The Vg1 mRNA appears to be co-localized with this wedge of ER, as shown by localization patterns of an integral ER protein, TRAPα (46) or an ER luminal protein, GRP78 (7). Transportation of Vg1 into the wedge-like pattern is dependent on microtubules and is disrupted by the use of nocodazole and colchicine (7, 45). Once anchored at the vegetal cortex, however, Vg1 localization is dependent on microfilaments and is disrupted by the use of cytochalasin B (45). The anchoring of Vg1 at the vegetal cortex is also dependent on the prior localization of Xlsirts mRNA to this region, as demonstrated by the fact that elimination of Xlsirts RNA by the injection of antisense Xlsirts oligodeoxynucleotides results in a release of Vg1 mRNA from the cortex (41). This does not affect Xcat2, which remains localized to the vegetal cortex, another example of the differences between the early and late RNA anchoring mechanisms (40). Specific signals that target RNAs to either of the two pathways must exist, as demonstrated by the ability to inappropriately target Vg1 to the METRO pathway by making a chimeric Xlsirt–Vg1 mRNA that links the Xlsirts localization signals to various regions of the full-length Vg1 mRNA (7).

# 3. The dorsal–ventral axis

In *Xenopus*, an elaborate series of cytoplasmic events converts the radially symmetrical egg into a bilaterally symmetrical embryo. This programme is under cytoplasmic control and is biased by the sperm–egg interaction, with the simple restriction that sperm entry occurs only in the animal hemisphere. The site of sperm entry dictates the orientation of the dorsal–ventral axis and is clearly recognized as a dark spot formed as the pigmented cortex contracts around the fertilizing sperm (5) (Fig. 2). Fate-map studies have shown that the sperm entry site ultimately develops into the ventral side of the embryo, whereas the equatorial region opposite the sperm entry

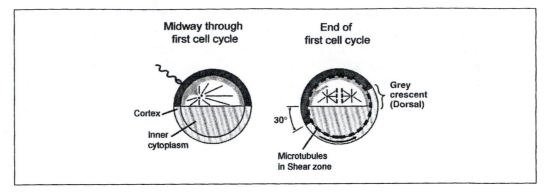

**Fig. 2** Cortical rotation. The cortex moves away from the site of sperm binding and toward the future dorsal side of the egg.

point develops into the dorsoanterior structures. The mechanism by which this occurs has been the subject of intense investigation for years and the molecular details are emerging rapidly. It was shown in the 1980s that a 30° rotation of the egg cortex with respect to the inner cytoplasm occurred during the first cell cycle (47). This event, known as cortical rotation, is dependent on microtubules and is required for normal embryonic development. But for another decade the mechanism by which this 30° cortical–cytoplasmic shift triggered formation of dorsoanterior structures remained quite puzzling. During the past few years, however, it has become quite clear than an elaborate microtubule-mediated transport of organelles and proteins is responsible for this important polarization step that occurs during the first cell cycle.

The first cell-cycle events that are responsible for establishing dorsal–ventral polarity are under cytoplasmic control and can be initiated by pricking the un-fertilized egg with a sharp glass micropipette, as long as the buffer contains calcium ions. The first minutes after fertilization are dedicated to assuring that the egg is protected from polyspermy. The structural block to polyspermy is created as a consequence of a wave of exocytosis that crosses the egg, releasing enzymes and structural proteins into the extracellular space for the assembly of the fertilization envelope (48). This is followed immediately by a wave of endocytosis, required to take up the additional membrane inserted into the cell surface during exocytosis. Both of these events are accompanied by a complete remodelling of the cell surface and cortex, including the cytoskeleton and organelles, with no detectable effects on cell polarity. Once the developing embryo has been protected from secondary sperm binding, the task of dorsal–ventral axis specification begins.

## 3.1  Movements of inner cytoplasm

Confocal microscopy of the vegetal pole of living eggs during early stages of cortical rotation has revealed the most intricate details of the highly orchestrated cortical–cytoplasmic rotation. Under ordinary circumstances, the egg cortex travels along the microtubules attached to the inner cytoplasm, moving toward the future dorsal side and away from the sperm entry point. However, eggs viewed with the microscope are immobilized in agarose wells to reduce movement artefacts. As a consequence, the inner cytoplasm moves along the immobilized cortex in the opposite direction, that is, toward the sperm entry site (5). This has led to some confusing diagrammatic representations and misunderstandings of cortical rotation, which I will try to clarify in this chapter. Movements of yolk platelets, typically labelled with the vital dye Nile Red, reflect the movements of the inner cytoplasm (that region more than 4 μm from the cell surface). Examining eggs labelled with Nile Red throughout rotation has revealed four distinct phases of cortical rotation: initiation, acceleration, translocation, and termination (49). Those eggs examined as soon as possible after fertilization and removal of the jelly coats (at least 10–15 min prior to the onset of rotation) revealed only Brownian movements of organelles. The initiation phase begins between 0.28 and 0.4 normalized time (NT; normalized time of the first cell cycle, in which 0.0 = fertilization and 1.0 = first cleavage), as slight jostling movements of yolk platelets that eventually become directional. Slow, directional movements (<less than 1 μm min$^{-1}$) continue for several more minutes, then a rapid acceleration of yolk platelets is seen between 0.4 and 0.45 NT, the acceleration phase. The yolk platelets continue to accelerate until about 0.5 NT, when the full rotational velocity of 10 μm min$^{-1}$ is achieved, which is the time referred to as the translocation phase. Once rotation is progressing steadily, at about 0.45–0.5 NT, long, winding channels form among the mass of yolk platelets in the region 4–8 μm from the cell surface. These channels, as shown by antibody labelling of fixed cells, contain large bundles of microtubules that are affixed to the inner cytoplasm. The cortical–cytoplasmic rotation continues at this maximal velocity from 0.50 to 0.85 NT, until movements suddenly cease at the end of rotation, known as termination.

## 3.2 Microtubule dynamics

Microtubules and microtubule-mediated transport are critical for establishing dorsal–ventral polarity in normal *Xenopus* development. Eliminating microtubules with microtubule-depolymerizing agents (e.g. nocodazole, colchicine, UV exposure) results in embryos lacking dorsoanterior structures (50). However, the mere presence of microtubules is not enough; they must be aligned in a polarized, parallel array. Exposure of eggs to heavy water (D$_2$O) triggers the precocious production of microtubules in a tangled network, a reduced (<20°) cortical rotation, and hyperdorsalized embryos (51). These enigmatic data forced the question: 'How does a reduced cortical rotation produce excess dorsoanterior structures?' Recent data, discussed below, provide new explanations. First let's examine the formation of the microtubule array. Monitoring the polymerization of microtubules using rhodamine-labelled tubulin revealed that the polymerization and alignment of microtubules accompanies the early stages of rotation rather than precedes rotational movements. Short segments of randomly arranged rhodamine–labelled microtubules are detected in the vegetal pole of eggs at approximately 0.4 NT (49). This time span correlates with the period known as the acceleration phase of rotation, but is after the initial yolk platelet movements detected between 0.28 and 0.4 NT (49). This discrepancy may reflect the inability to detect individual, short, rhodamine-labelled microtubules in living eggs. However, microtubules were not seen in the vegetal subcortical region of fixed eggs, either, until approximately 0.35 NT, when immunolabelling with anti-tubulin antibodies revealed a disordered mesh of microtubules (51). It is also possible, however, that these early rotational movements are caused by movements of the cortex initiated by microtubules at another site in the egg, such as the animal hemisphere, or that they are caused by a non-microtubule-mediated mechanism, such as an actin contraction. This remains to be determined. Whatever the mechanism, the microtubule segments become longer and denser over time, simultaneous with the early movements of cortical rotation, and eventually form the dense parallel array of microtubules that fills the outer 4–8 μm of the egg by 0.5 NT (49).

The microtubules of the parallel array are restricted to that region 4–8 μm from the cell surface referred to as the shear zone, where the cortex and inner cytoplasmic mass shift with respect to one another (49). Once the array is fully formed, rotation achieves a velocity of 10 μm min$^{-1}$, which persists throughout the translocation phase. During the early movements of the translocation phase, the microtubules appear quite wavy, perhaps reflecting cortical cytoplasmic displacements that are slower than the rate of polymerization. As rotation proceeds, however, the microtubules become aligned parallel with each other and in the direction of rotation. The fully formed microtubule array seen in living cells resembles microtubules previously seen using antitubulin antibodies (52, 53). The data obtained from studying microtubule polymerization and array formation in living cells suggest that the microtubules become aligned as a consequence of rotation. It is likely, therefore, that the function of cortical rotation is to align the microtubules in a directional array that facilitates the rapid and efficient transport of organelles and molecules to one specific region of the embryo (49, 54).

During rotation, the microtubules are attached to the inner cytoplasm and the cortex travels along these microtubules towards the future dorsal side of the embryo. Since the negative ends of greater than 90% of the microtubules are at the sperm entry site and the positive ends face the future dorsal side of the egg (55), we presume that these movements are mediated by kinesin-like motors. However, the exact motor protein involved has not yet been identified. Immunocytochemistry has shown that antibodies to a sea-urchin kinesin (SUK2) and a *Xenopus* kinesin (Eg5) bind to the subcortical microtubules (56), but functional studies to test the role of either of these proteins in cortical rotation have not been reported. Other kinesin-like proteins, such as Xklp1 and 2, have been identified in *Xenopus*, but embryos resulting from oligonucleotide knockouts of these proteins do not demonstrate impaired cortical rotation (57).

The biochemical signal(s) responsible for initiation of microtubule polymerization have not been identified. It has been suggested that the subcortical microtubules emanate from, or are continuations of, the sperm aster microtubules. This is based on the fact that immunocytochemistry of fixed eggs reveals a radial arrangement of microtubules extending from the sperm aster to the microtubules in the cell periphery. However, artificially activated eggs lacking a sperm aster are still capable of undergoing cortical rotation and contain microtubule arrays in the egg periphery. So what acts as a microtubule organizing centre (MTOC)? The possibility of γ-tubulin acting as an MTOC was lessened when, although it was seen in the cortex of the meiotically immature, stage VI oocyte, it was absent in the unfertilized egg (13). Our understanding of the proteins involved in the polymerization and alignment of the microtubules, or those proteins involved in destruction of the microtubule array, is also lacking. Three microtubule-associated proteins have been identified in *Xenopus* oocytes and eggs, XMAP215, XMAP230, and XMAP310 (19). One of these, XMAP230, labels cortical microtubules and, when perturbed upon injection of antibodies to XMAP230, blocks normal cortical rotation (58).

## 3.3 Microtubule-mediated transport

During the past decade it became clear that material specifically located at the vegetal pole of the unfertilized egg was required for specification of the dorsal–ventral axis. Cytoplasmic transfer experiments demonstrated that material located at the vegetal pole before rotation had the capacity to induce a dorsal–ventral axis, whereas after rotation the dorsalizing cytoplasm was located at the equator opposite the sperm entry point (59–61). Furthermore, deletion of the vegetal pole before rotation, or the equatorial zone opposite sperm entry after rotation, eliminated dorsal–ventral polarity (62). These data provided strong evidence that dorsalizing material must be moved 90° during cortical rotation—a displacement greater than that accomplished by the 30° shift of the cortex with respect to the cytoplasm. Monitoring cortical rotation in living eggs using confocal microscopy provided the first evidence of such a transport mechanism. Fertilized eggs were double labelled to monitor movements of the inner cytoplasm (using Nile Red to label yolk platelets) and movements of endoplasmic reticulum, mitochondria, and other small membranous organelles (using DiOC6(3), a lipophilic dye), and revealed organelle movements of differing velocities (54). As the cortex rotates with respect to the inner cytoplasm, approximately 10% of those organelles labelled with $DiOC_6(3)$ travel towards the plus ends of microtubules (and the presumptive dorsal region) at 30–50 $\mu$m min$^{-1}$ (54), a velocity significantly faster than the cortical–cytoplasmic rotation of 10 $\mu$m min$^{-1}$. At velocities of 30–50 $\mu$m min$^{-1}$, components located at the vegetal pole could travel from the vegetal pole to the presumptive dorsal side of the embryo (Plate 4). This capability was clearly demonstrated by the ability of fluorescein-filled, carboxylated beads injected into the vegetal pole before rotation to travel that distance, leaving a streak of beads from the vegetal pole to the dorsal side (54). There is thus strong evidence for a transport mechanism capable of delivering dorsalizing components, but what molecules are being transported?

## 3.4 Wnt signalling pathway and polarity

The role of the Wnt signalling pathway in establishment of the dorsal–ventral axis of *Xenopus* embryos is robust. The Wnt pathway was first implicated by data showing that injection of Wnt1 into the ventral region of the early embryo triggers formation of the dorsal–ventral axis (63). This was strengthened by the demonstration that antisense oligonucleotide-mediated depletion of mRNA for β-catenin, a component of the Wnt signalling pathway, blocks formation of the dorsal–ventral axis (64) and, in addition, that overexpression of β-catenin in ventral cells induces formation of a secondary axis and also rescues UV-treated embryos (65, 66). Although it was clear that β-catenin was required for dorsal–ventral axis specification, there was no evidence that this activity occurred prior to the midblastula transition. Attempts to demonstrate an asymmetrical accumulation of β-catenin in early cleavage stage embryos were unsuccessful, likely due to the immunolabelling preparations used in those studies (67). Finally, by utilizing whole-mount labelling approaches that avoid the use of ethanol and other extracting agents, it was revealed that β-catenin accumulates on the prospective dorsal side of the embryo at the end of cortical rotation (54, 68) and continues to accumulate dorsally during the early cleavage stages (68). These studies provided the first evidence of a dorsal–ventral asymmetry of a protein that is both necessary and sufficient for dorsal–ventral axis specification. At the 16–32-cell stage, β-catenin translocates to the dorsal, but not ventral, nuclei (68) and, along with the architectural HMG box transcription factor Xtcf3, activates transcription of the gene *siamois* (69), a dorsal-specific gene required for the activation of the Spemann organizer (70, 71). The dorsal–ventral asymmetry of β-catenin persists through the blastula stage (68, 72), when zygotic transcription begins. The spatial and temporal relationship of the dorsal enrichment of β-catenin at the 16–32-cell stage correlates with the ability of those dorsal animal and vegetal blastomeres to induce a secondary axis when transplanted into the ventral side of embryos (70, 71).

The asymmetrical accumulation of β-catenin is dependent on cortical rotation, as shown by the ability of microtubule depolymerizing agents known to block dorsal–ventral polarity to block its accumulation (68). The dorsal accumulation of β-catenin and dorsal–ventral axis specification are also blocked by expression of glycogen synthase kinase 3 (GSK3), a negative regulator of β-catenin (68, 73). So how does cortical rotation trigger downregulation of GSK3 and the ultimate accumulation of β-catenin? Endogenous β-catenin is not seen at any region of the egg prior to its accumulation on the dorsal side, suggesting that it is phosphorylated, which results in its ubiquitination and consequent degradation. Does β-catenin move to the future dorsal side and away from its negative regulators? Although β-catenin co-localizes (at the level of resolution of the light microscope) with the microtubules on the dorsal side of the egg at the end of the first cell cycle (53), there is no evidence that it actually travels along the microtubules during cortical rotation. Furthermore, studies of exogenous β-catenin–GFP during the first cell cycle failed to detect its movement along microtubules during cortical rotation (Rowning, Moon, and Larabell, unpublished data), suggesting that the exogenous protein is also degraded. If β-catenin is associated with the microtubules, which remains to be determined by ultrastructural analysis, it is likely that it is merely being stabilized by this interaction, perhaps via APC, as has been shown in somatic cells (74).

If β-catenin does not travel along microtubules, what does? Likely candidates include a negative regulator, such as GSK3, and a recently identified protein known as GSK3-binding protein (GBP) that can bind and suppress GSK3 activity *in vivo* (75). Protein localization data for both proteins is lacking, and attempts to monitor movements of GSK3–GFP have been unsuccessful due to negative effects of the exogenous protein on meiotic maturation and activation (Rowning, Moon, and Larabell, unpublished data). Another candidate for regulation of the asymmetrical accumulation of β-catenin is an upstream regulatory protein of the Wnt signalling pathway, known as Dishevelled. Ectopic expression of *Xenopus* Dishevelled (Xdsh) triggers formation of a secondary axis (76), but axis formation was not inhibited by expression of the dominant negative mutant (77). None the less, recent data reveals that Dishevelled–GFP travels along the subcortical microtubules at 30–50 μm min⁻¹ towards the prospective dorsal side of the embryo (78). Furthermore, endogenous Dishevelled accumulates on the dorsal side at the end of the first cell cycle in a microtubule-dependent fashion since it is blocked by UV treatment and perturbed by incubation in $D_2O$ (78). At this site it can trigger downregulation of GSK3 and accumulation of β-catenin (Fig. 3). Discrepancies between these data and observations that the dominant-negative form of Dishevelled was ineffective in blocking axis formation can be explained by the fact that the mutant Dishevelled was injected during the first cell cycle—during the process of cortical rotation. As a result, it is unlikely that enough protein was translated in time to have an effect before the endogenous Dishevelled was transported during rotation, and, therefore, before the downregulation of GSK3. Another possible explanation is that the mutant used was one in which the PDZ domain had been deleted. Such mutants tagged with GFP fail to move along the microtubules during cortical rotation (78). Although it is possible Dishevelled is not the only molecule moved during cortical rotation, it seems clear that it is an important player in the microtubule-mediated dorsal–ventral axis specification.

These data lead to several models for dorsal–ventral axis specification. First,

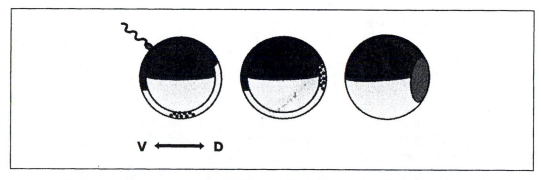

**Fig. 3** Accumulation of β-catenin at the end of cortical rotation. Dishevelled, located at the vegetal pole before cortical rotation, moves along the microtubules during rotation. When it reaches the prospective dorsal side it triggers downregulation of GSK3 and accumulation of β-catenin.

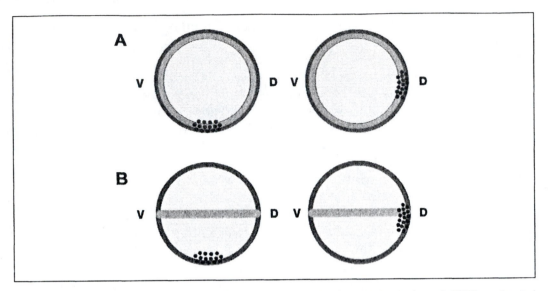

**Fig. 4** Model for cortical rotation and dorsal–ventral axis specification. (A) β-catenin and GSK3 are located around the entire periphery of the egg and Dishevelled is located in vesicles at the vegetal pole. These vesicles move to the future dorsal side during rotation and initiate the accumulation of β-catenin. This model requires an inhibitor of Dishevelled at the vegetal pole or an activator of Dishevelled at the prospective dorsal side. (B) β-catenin and GSK3 are located around the equator of the egg. Dishevelled moves from the vegetal pole to the prospective dorsal side and initiates accumulation of β-catenin.

consider a simplistic process involving only three proteins, β-catenin, GSK3, and Dishevelled (Fig. 4). In this case, there are two possible models:

1. β-catenin and GSK3 are localized in the cortex of the entire egg, whereas Dishevelled is located at the vegetal pole. Transport of Dishevelled along microtubules to the prospective dorsal side would downregulate the phosphorylation of β-catenin by GSK3 and allow accumulation of β-catenin. This, however, would require an inhibitor at the vegetal pole, to prevent accumulation of β-catenin before rotation, or an activator of Dishevelled at the equatorial zone to facilitate accumulation of β-catenin after rotation.

2. The second model would position β-catenin and GSK3 at the equatorial zone, rather than around the entire periphery, of the unfertilized egg. Microtubule-mediated transport of Dishevelled to the equatorial region on the dorsal side would then downregulate the phosphorylation of β-catenin by GSK3 in this region.

Both models are feasible, but the actual distribution of β-catenin and GSK3 prior to cortical rotation is not known. The earliest detectable β-catenin using immuno-cytochemistry is the asymmetrical accumulation at the end of the first cell cycle (53), presumably, due to its rapid phosphorylation and degradation.

Two additional models can be envisioned that build on the first two models but incorporate other proteins known to be components of the Wnt signalling pathway

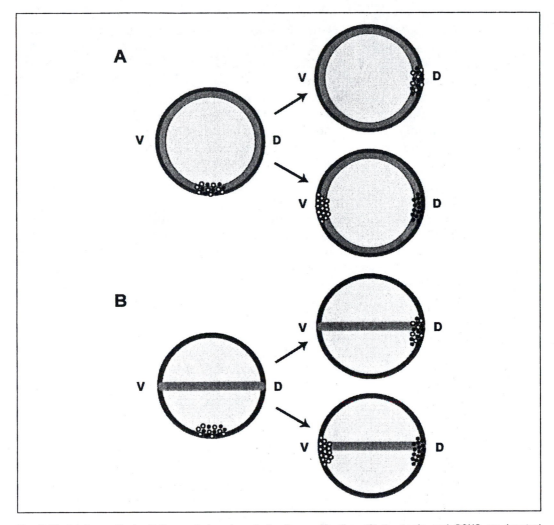

**Fig. 5** Model for cortical rotation and dorsal–ventral axis specification. (A) β-catenin and GSK3 are located around the entire periphery of the egg. Dishevelled and another component(s) of the Wnt signalling pathway are located in vesicles at the vegetal pole. These vesicles follow one of two possible routes during rotation: (1) all vesicles travel to the future dorsal side and initiate the accumulation of β-catenin; or (2) Dishevelled moves to the dorsal side and the other, perhaps inhibitory molecule, moves to the ventral side. (B) β-catenin and GSK3 are located around the equator of the egg. Again, two scenarios are possible for movement of the vesicles: (1) all vesicles travel to the future dorsal side and initiate the accumulation of β-catenin; or (2) Dishevelled moves from the vegetal pole to the prospective dorsal side and the other, perhaps inhibitory, molecule moves away from Dishevelled.

(GBP, Axin, APC, etc.), and variations of these models incorporate the possibility of transport toward the minus ends of the microtubules (Fig. 5). The first of these models, based on β-catenin and GSK3 (as well as other molecules of the Wnt signalling pathway) being localized around the entire egg periphery, places Dishevelled and another protein(s) of the Wnt pathway at the vegetal pole, in separate vesicles.

The vesicles could either move to the same region and effect accumulation of β-catenin, perhaps by forming a complex at this site, or they could move in opposite directions, moving an inhibitor away from dorsally directed Dishevelled. The final model, which is based on the localization of β-catenin and GSK3 (as well as other molecules of the Wnt pathway) at the equatorial zone, moves Dishevelled dorsally and a potential 'suppressor of dorsal' protein ventrally. All of these models will require extensive testing in the near future. To complicate this task, recent data indicate that there are an increasing number of proteins in the complex that can regulate the accumulation of β-catenin.

## 4. Conclusions

In order for a single cell to develop into a multicellular organism, an elaborate sequence of programmed events must occur. An important step in this process is the establishment of polarity. The earliest polarization events in *Xenopus* development occur during oogenesis when the animal–vegetal axis is established, an important foundation for the specification of the three primary germ layers during embryogenesis. In the smallest oocytes, a process of mRNA localization begins positioning specific molecules at the vegetal pole, utilizing both cytoskeletal-independent and microtubule-dependent mechanisms, where they will be used during embryogenesis. The distribution of cytoskeletal components and organelles also becomes polarized as oogenesis proceeds. At fertilization, another programme is initiated which will trigger a microtubule-mediated cortical rotation that results in the establishment of dorsal–ventral polarity and, ultimately, left–right asymmetry. The role of cortical rotation in setting up the dorsal–ventral axis is well established. We now know that a specific molecule of the Wnt signalling pathway, Dishevelled, is transported along microtubules during cortical rotation to the region that will develop into dorsoanterior structures. At this site, it triggers downregulation of GSK3 and, ultimately, the accumulation of β-catenin. The role of the other components of the Wnt signalling pathway in dorsal–ventral axis specification are not as well known at this time. This is the first molecular mechanism linking cortical rotation with activation of a maternal Wnt pathway, asymmetrical accumulation of β-catenin, and dorsal–ventral axis specification.

## References

1. Wood, W. B. (1997) Left-right asymmetry in animal development. *Annual Review of Cell and Developmental Biology*, **13**, 53–82.
2. Ramsdell, A. F. and Yost, H. J. (1998) Molecular mechanisms of vertebrate left-right development. *Trends in Genetics*, **14**, 459–65.
3. Yost, H. J. (1998) Left-right development in Xenopus and zebrafish. *Seminars in Cell and Developmental Biology*, **9**, 61–6.
4. Dumont, J. N. (1972) Oogenesis in *Xenopus laevis* (Daudin). I. Stages of oocyte development in laboratory maintained animals. *Journal of Morphology*, **136**, 153–79.

5. Gerhart, J. C. (1980) Mechanisms regulating pattern formation in the amphibian egg and early embryo. In *Molecular organization and cell function*, (ed. R. Goldberger), Vol. 2, pp. 133–316. Plenum Press, New York.

6. Heasman, J. (1997) Patterning the Xenopus blastula. *Development*, **124**, 4179–91.

7. Kloc, M. and Etkin, L. D. (1998) Apparent continuity between the messenger transport organizer and late RNA localization pathways during oogenesis in Xenopus. *Mechanisms of Development*, **73**, 95–106.

8. al-Mukhtar, K. A. and Webb, A. C. (1971) An ultrastructural study of primordial germ cells, oogonia and early oocytes in *Xenopus laevis*. *Journal of Embryology and Experimental Morphology*, **26**, 195–217.

9. Billett, F. S. and Adam, E. (1976) The structure of the mitochondrial cloud of *Xenopus laevis* oocytes. *Journal of Embryology and Experimental Morphology*, **36**, 697–710.

10. Coggins, L. W. (1973) An ultrastructural and radioautographic study of early oogenesis in the toad *Xenopus laevis*. *Journal of Cell Science*, **12**, 71–93.

11. Czolowska, R. (1969) Observations on the origin of the 'germinal cytoplasm' in *Xenopus laevis*. *Journal of Embryology and Experimental Morphology*, **22**, 229–51.

12. Heasman, J., Quarmby, J., and Wylie, C. C. (1984) The mitochondrial cloud of Xenopus oocytes: the source of germinal granule material. *Developmental Biology*, **105**, 458–69.

13. Gard, D. L., Affleck, D., and Error, B. M. (1995) Microtubule organization, acetylation, and nucleation in *Xenopus laevis* oocytes: II. A developmental transition in microtubule organization during early diplotene. *Developmental Biology*, **168**, 189–201.

14. Roeder, A. D. and Gard, D. L. (1994) Confocal microscopy of F-actin distribution in Xenopus oocytes. *Zygote*, **2**, 111–24.

15. Gard, D. L. (1991) Organization, nucleation, and acetylation of microtubules in *Xenopus laevis* oocytes: a study by confocal immunofluorescence microscopy. *Developmental Biology*, **143**, 346–62.

16. Wylie, C. C., Heasman, J., Parke, J. M., Anderton, B., and Tang, P. (1986) Cytoskeletal changes during oogenesis and early development of *Xenopus laevis*. *Journal of Cell Science Supplement*, **5**, 329–41.

17. Kloc, M., Larabell, C., and Etkin, L. D. (1996) Elaboration of the messenger transport organizer pathway for localization of RNA to the vegetal cortex of Xenopus oocytes. *Developmental Biology*, **180**, 119–130.

18. Gard, D. L. (1995) Axis formation during amphibian oogenesis: reevaluating the role of the cytoskeleton. *Current Topics in Developmental Biology*, **30**, 215–52.

19. Gard, D. L. (1999) Confocal microscopy and 3-D reconstruction of the cytoskeleton of Xenopus oocytes. *Microscopy Research and Technique*, **44**, 388–414.

20. Chang, P., Pérez-Mongiovi, D., and Houliston, E. (1999) Organisation of Xenopus oocyte and egg cortices. *Microscopy Research and Technique*, **44**, 415–29.

21. Gard, D. L., Cha, B. J., and King, E. (1997) The organization and animal–vegetal asymmetry of cytokeratin filaments in stage VI Xenopus oocytes is dependent upon F-actin and microtubules. *Developmental Biology*, **184**, 95–114.

22. Larabell, C. A. (1995) Cortical cytoskeleton of the Xenopus oocyte, egg, and early embryo. *Current Topics in Developmental Biology*, **31**, 433–53.

23. Rebagliati, M. R., Weeks, D. L., Harvey, R. P., and Melton, D. A. (1985) Identification and cloning of localized maternal RNAs from Xenopus eggs. *Cell*, **42**, 769–77.

24. Kloc, M., Reddy, B., Crawford, S., and Etkin, L. D. (1991) A novel 110-kDa maternal CAAX box-containing protein from Xenopus is palmitoylated and isoprenylated when expressed in baculovirus. *Journal of Biological Chemistry*, **266**, 8206–12.

25. Reddy, B. A., Kloc, M., and Etkin, L. D. (1992) The cloning and characterization of a localized maternal transcript in *Xenopus laevis* whose zygotic counterpart is detected in the CNS. *Mechanisms of Development*, **39**, 143–50.

26. Hinkley, C. S., Martin, J. F., Leibham, D., and Perry, M. (1992) Sequential expression of multiple POU proteins during amphibian early development. *Molecular and Cellular Biology*, **12**, 638–49.

27. Hudson, J. W., Alarcon, V. B., and Elinson, R. P. (1996) Identification of new localized RNAs in the Xenopus oocyte by differential display PCR. *Developmental Genetics*, **19**, 190–8.

28. Schroeder, K. E. and Yost, H. J. (1996) Xenopus poly(A) binding protein maternal RNA is localized during oogenesis and associated with large complexes in blastula. *Developmental Genetics*, **19**, 268–76.

29. Kloc, M., Spohr, G., and Etkin, L. D. (1993) Translocation of repetitive RNA sequences with the germ plasm in Xenopus oocytes. *Science*, **262**, 1712–14.

30. Mosquera, L., Forristall, C., Zhou, Y., and King, M. L. (1993) A messenger RNA localized to the vegetal cortex of Xenopus oocytes encodes a protein with a nanos-like zinc finger domain. *Development*, **117**, 377–86.

31. Elinson, R. P., King, M. L., and Forristall, C. (1993) Isolated vegetal cortex from Xenopus oocytes selectively retains localized mRNAs. *Developmental Biology*, **160**, 554–62.

32. Ku, M. and Melton, D. A. (1993) Xwnt-11—a maternally expressed Xenopus Wnt gene. *Development*, **119**, 1161–73.

33. Zhang, J. and King, M. L. (1996) Xenopus VegT RNA is localized to the vegetal cortex during oogenesis and encodes a novel T-box transcription factor involved in mesodermal patterning. *Development*, **122**, 4119–29.

34. Lustig, K. D., Kroll, K. L., Sun, E. E., and Kirschner, M. W. (1996) Expression cloning of a Xenopus T-related gene (Xombi) involved in mesodermal patterning and blastopore lip formation. *Development*, **122**, 4001–12.

35. Stennard, F., Carnac, G., and Gurdon, J. B. (1996) The Xenopus T-box gene, Antipodean, encodes a vegetally localised maternal mRNA and can trigger mesoderm formation. *Development*, **122**, 4179–88.

36. Horb, M. E. and Thomsen, G. H. (1997) A vegetally localized T-box transcription factor in Xenopus eggs specifies mesoderm and endoderm and is essential for embryonic mesoderm formation. *Development*, **124**, 1689–98.

37. Houston, D. W., Zhang, J., Maines, J. Z., Wasserman, S. A., and King, M. L. (1998) A *Xenopus* DAZ-like gene encodes an RNA component of germ plasm and is a functional homologue of *Drosophila* boule. *Development*, **125**, 171–80.

38. Hudson, C. and Woodland, H. R. (1998) Xpat, a gene expressed specifically in germ plasm and primordial germ cells of *Xenopus laevis*. *Mechanisms of Development*, **73**, 159–68.

39. Forristall, C., Pondel, M., Chen, L. H., and King, M. L. (1995) Patterns of localization and cytoskeletal association of two vegetally localized RNAs, Vg1 and Xcat-2. *Development*, **121**, 201–8.

40. Kloc, M. and Etkin, L. D. (1995) Two distinct pathways for the localization of RNAs at the vegetal cortex in Xenopus oocytes. *Development*, **121**, 287–97.

41. Kloc, M. and Etkin, L. D. (1994) Delocalization of Vg1 mRNA from the vegetal cortex in Xenopus oocytes after destruction of Xlsirt RNA. *Science*, **265**, 1101–3.

42. Kloc, M., Larabell, C., Chan, A. P. Y., and Etkin, L. D. (1998) Contribution of METRO pathway localized molecules to the organization of the germ cell lineage. *Mechanisms of Development*, **75**, 81–93.

43. Thomsen, G. H. and Melton, D. A. (1993) Processed Vg1 protein is an axial mesoderm inducer in Xenopus. *Cell*, **74**, 433–41.

44. Hyatt, B. A., Lohr, J. L., and Yost, H. J. (1996) Initiation of vertebrate left-right axis formation by maternal Vg1. *Nature*, **384**, 62–5.

45. Yisraeli, J. K., Sokol, S., and Melton, D. A. (1990) A 2-step model for the localization of maternal messenger RNA in Xenopus oocytes—involvement of microtubules and microfilaments in the translocation and anchoring of Vg1 messenger RNA. *Development*, **108**, 289–98.

46. Deshler, J. O., Highett, M. I., and Schnapp, B. J. (1997) Localization of Xenopus Vg1 mRNA by vera protein and the endoplasmic reticulum. *Science*, **276**, 1128–31.

47. Vincent, J. P. and Gerhart, J. C. (1987) Subcortical rotation in Xenopus eggs: an early step in embryonic axis specification. *Developmental Biology*, **123**, 526–39.

48. Larabell, C. A. and Chandler, D. E. (1988) The extracellular matrix of Xenopus laevis eggs: a quick-freeze, deep-etch analysis of its modification at fertilization. *Journal of Cell Biolog*, **107**, 731–41.

49. Larabell, C. A., Rowning, B. A., Wells, J., Wu, M., and Gerhart, J. C. (1996) Confocal microscopy analysis of living Xenopus eggs and the mechanism of cortical rotation. *Development*, **122**, 1281–9.

50. Gerhart, J., Danilchik, M., Doniach, T., Roberts, S., Rowning, B., and Stewart, R. (1989) Cortical rotation of the Xenopus egg: consequences for the anteroposterior pattern of embryonic dorsal development. *Development*, **107**, (Suppl.), 37–51.

51. Scharf, S. R., Rowning, B., Wu, M., and Gerhart, J. C. (1989) Hyperdorsoanterior embryos from Xenopus eggs treated with $D_2O$. *Developmental Biology*, **134**, 175–88.

52. Schroeder, M. M. and Gard, D. L. (1992) Organization and regulation of cortical microtubules during the first cell cycle of Xenopus eggs. *Development*, **114**, 699–709.

53. Elinson, R. P. and Rowning, B. (1988) A transient array of parallel microtubules in frog eggs: potential tracks for a cytoplasmic rotation that specifies the dorso-ventral axis. *Developmental Biology*, **128**, 185–97.

54. Rowning, B. A., Wells, J., Wu, M., Gerhart, J. C., Moon, R. T., and Larabell, C. A. (1997) Microtubule-mediated transport of organelles and localization of beta-catenin to the future dorsal side of Xenopus eggs. *Proceedings of the National Academy of Sciences of the United States of America*, **94**, 1224–9.

55. Houliston, E. and Elinson, R. P. (1991) Evidence for the involvement of microtubules, ER, and kinesin in the cortical rotation of fertilized frog eggs. *Journal of Cell Biology*, **114**, 1017–28.

56. Houliston, E., Leguellec, R., Kress, M., Philippe, M., and Leguellec, K. (1994) The kinesin-related protein Eg5 associates with both interphase and spindle microtubules during Xenopus early development. *Developmental Biology*, **164**, 147–59.

57. Robb, D. L., Heasman, J., Raats, J., and Wylie, C. (1996) A kinesin-like protein is required for germ plasm aggregation in Xenopus. *Cell*, **87**, 823–31.

58. Cha, B. J. and Gard, D. L. (1999) XMAP230 is required for the organization of cortical microtubules and patterning of the dorsoventral axis in fertilized Xenopus eggs. *Developmental Biology*, **205**, 275–86.

59. Yuge, M., Kobayakawa, Y., Fujisue, M., and Yamana, K. (1990) A cytoplasmic determinant for dorsal axis formation in an early embryo of *Xenopus laevis*. *Development*, **110**, 1051–6.

60. Fujisue, M., Kobayakawa, Y., and Yamana, K. (1993) Occurrence of dorsal axis-inducing activity around the vegetal pole of an uncleaved Xenopus egg and displacement to the equatorial region by cortical rotation. *Development*, **118**, 163–70.

61. Holowacz, T. and Elinson, R. P. (1993) Cortical cytoplasm, which induces dorsal axis formation in Xenopus, is inactivated by UV irradiation of the oocyte. *Development*, **119**, 277–85.

62. Sakai, M. (1996) The vegetal determinants required for the Spemann organizer move equatorially during the first cell cycle. *Development*, **122**, 2207–14.

63. McMahon, A. P. and Moon, R. T. (1989) Ectopic expression of the proto-oncogene int-1 in Xenopus embryos leads to duplication of the embryonic axis. *Cell*, **58**, 1075–84.

64. Heasman, J., Crawford, A., Goldstone, K., Garner-Hamrick, P., Gumbiner, B., McCrea, P., Kintner, C., Noro, C. Y., and Wylie, C. (1994) Overexpression of cadherins and under-expression of beta-catenin inhibit dorsal mesoderm induction in early Xenopus embryos. *Cell*, **79**, 791–803.

65. Funayama, N., Fagotto, F., McCrea, P., and Gumbiner, B. M. (1995) Embryonic axis induction by the armadillo repeat domain of beta-catenin: evidence for intracellular signaling. *Journal of Cell Biology*, **128**, 959–68.

66. Guger, K. A. and Gumbiner, B. M. (1995) beta-Catenin has Wnt-like activity and mimics the Nieuwkoop signaling center in Xenopus dorsal-ventral patterning. *Developmental Biology*, **172**, 115–25.

67. Fagotto, F. and Gumbiner, B. M. (1994) Beta-catenin localization during Xenopus embryogenesis: accumulation at tissue and somite boundaries. *Development*, **120**, 3667–79.

68. Larabell, C. A., Torres, M., Rowning, B. A., Yost, C., Miller, J. R., Wu, M., Kimelman, D., and Moon, R. T. (1997) Establishment of the dorso–ventral axis in Xenopus embryos is presaged by early asymmetries in beta-catenin that are modulated by the Wnt signaling pathway. *Journal of Cell Biology*, **136**, 1123–36.

69. Molenaar, M., van de Wetering, M., Oosterwegel, M., Peterson-Maduro, J., Godsave, S., Korinek, V., Roose, J., Destrée, O., and Clevers, H. (1996) XTcf-3 transcription factor mediates beta-catenin-induced axis formation in Xenopus embryos. *Cell*, **86**, 391–9.

70. Miller, J. R. and Moon, R. T. (1996) Signal transduction through beta-catenin and specification of cell fate during embryogenesis. *Genes and Development*, **10**, 2527–39.

71. Harland, R. and Gerhart, J. (1997) Formation and function of Spemann's organizer. *Annual Review of Cell and Developmental Biology*, **13**, 611–67.

72. Schneider, S., Steinbeisser, H., Warga, R. M., and Hausen, P. (1996) Beta-catenin translocation into nuclei demarcates the dorsalizing centers in frog and fish embryos. *Mechanisms of Development*, **57**, 191–8.

73. Yost, C., Torres, M., Miller, J. R., Huang, E., Kimelman, D., and Moon, R. T. (1996) The axis-inducing activity, stability, and subcellular distribution of beta-catenin is regulated in Xenopus embryos by glycogen synthase kinase 3. *Genes and Development*, **10**, 1443–54.

74. Polakis, P. (1997) The adenomatous polyposis coli (APC) tumor suppressor. *Biochimica et Biophysica Acta-Reviews On Cancer*, **1332**, F127–F147.

75. Yost, C., Farr, G. H., 3rd, Pierce, S. B., Ferkey, D. M., Chen, M. M., and Kimelman, D. (1998) GBP, an inhibitor of GSK-3, is implicated in Xenopus development and oncogenesis. *Cell*, **93**, 1031–41.

76. Sokol, S. Y., Klingensmith, J., Perrimon, N., and Itoh, K. (1995) Dorsalizing and neuralizing properties of Xdsh, a maternally expressed Xenopus homolog of dishevelled. *Development*, **121**, 3487.

77. Sokol, S. Y. (1996) Analysis of Dishevelled signalling pathways during Xenopus development. *Current Biology*, **6**, 1456–67.

78. Miller, J. R., Rowning, B. A., Larabell, C. A., Yang-Snyder, J. A., Bates, R. L., and Moon, R. T. (1999) Establishment of the dorsal–ventral axis in Xenopus embryos coincides with the dorsal enrichment of Dishevelled that is dependent on cortical rotation. *Journal of Cell Biology*, **146**, 427–38.

# 7 | Cell polarity in response to chemoattractants

ORION D. WEINER, GUY SERVANT, CAROLE A. PARENT,
PETER N. DEVREOTES, and HENRY R. BOURNE

## 1. Introduction

To carry out their biological responsibilities, many eukaryotic cells depend on their ability to polarize and migrate toward a source of chemoattractant ligand. This crucial ability allows single-cell organisms to hunt and mate, axons to find their way in the developing nervous system, and cells in the innate immune system to find and kill invading pathogens. How do eukaryotic cells interpret a chemotactic gradient? Which signalling molecules carry information from the external world to internal cellular responses? What are the final effectors for cell polarity and migration? How are polarity responses co-ordinated in space and time? To address these and other questions, we will focus on two especially useful systems for the study of eukaryotic chemotaxis: neutrophils and *Dictyostelium discoideum*.

Neutrophils are cells of the innate immune system. All animals from sponges to humans have some version of these amoeboid cells programmed to find and kill invading pathogens. Neutrophils find bacteria by following gradients of formylated peptides released by the bacteria. When isolated in an unpolarized state and presented with a gradient of chemoattractant, neutrophils polarize and migrate towards the highest concentration of chemoattractant (Fig. 1). Constantly interpreting the gradient, they unerringly follow a moving micropipette containing chemoattractant.

*Dictyostelium discoideum* is a free-living soil amoeba that feeds on bacteria. Under starvation conditions, *Dictyostelium* aggregates to form a multicellular mound that undergoes a complex developmental programme, culminating in the production of hardy spores. Chemotaxis is necessary for *Dictyostelium* to find bacteria during the vegetative phase and to form multicellular aggregates in response to starvation. Both neutrophils and *Dictyostelium* detect and respond to shallow chemical gradients, as small as 2% across the cells' diameter; none the less, the large dynamic range of their responses allows them to respond to concentrations of chemoattractant varying over several orders of magnitude (1).

Chemotaxis, or the directed movement of cells in response to chemotactic gradients, was first discovered more than a century ago in bacteria. In contrast to the

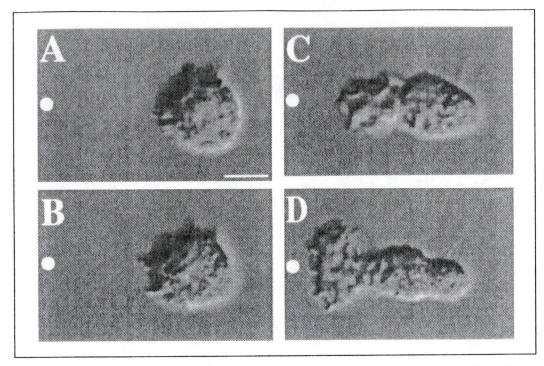

**Fig. 1** Polarization of neutrophil in response to gradient of chemoattractant. Nomarski images of unpolarized neutrophil responding to a micropipette containing the chemoattractant FMLP (white circle) at (A) 5 s, (B) 30 s, (C) 81 s, and (D) 129 s. Bar = 5 μm. Figure reprinted from ref. 96 with permission from ***Nature Cell Biology***.

mechanism(s) used by eukaryotes, the *temporal* mechanism used by prokaryotes to interpret a chemotactic gradient and translate it into directed movement is well understood: a bacterium senses the local concentration of chemoattractant as a function of time; an increase in chemoattractant concentration over time makes the bacterium more likely to persist in forward movement and less likely to tumble and travel in a random new direction. The resulting longer duration of runs directed toward the chemoattractant produces a biased random walk that eventually delivers the bacterium close to the highest concentration of chemoattractant. For bacterial chemotaxis, we understand the basic input of the system (the chemoattractant concentration at successive points in time), the basic output (turning the flagellar motor in the direction that generates smooth movement or random tumbling), and much of the molecular machinery in between (2).

What are the basic requirements for eukaryotic chemotaxis? First, a neutrophil or an amoeba needs an external gradient of a *chemotactic ligand* and *a receptor* that transmits a signal into the cell upon binding the ligand. The receptor (or the signal it generates) needs an *adaptation* (or background subtraction) mechanism to allow responses to shallow gradients over a large range of ligand concentrations. Each cell must *interpret* the gradient—that is, identify the portion of its surface that receives the most intense external signal. The cells need *second messengers* to transmit information

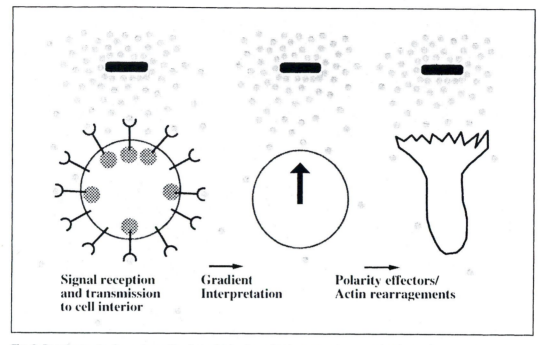

**Fig. 2** Requirements for eukaryotic chemotaxis. In order to respond appropriately to chemotactic gradients, eukaryotic cells must contain receptors that transmit a signal to the cell interior upon binding chemoattractant. Each cell must manipulate this information to determine which region of its surface is exposed to maximal chemoattractant. Finally, the cell must transmit this information to the final effectors responsible for spatial regulation of actin rearrangements and cell motility.

about the chemoattractant gradient from receptors to the final effectors that determine cell polarity and mediate cell movement. Here we shall focus on the effects of spatial gradients of chemoattractant on the actin cytoskeleton and the regulation of actin polymerization, which is necessary for morphological polarization and migration of neutrophils, *Dictyostelium*, and almost all motile eukaryotic cells (Fig. 2).

## 2. Chemotactic receptors and G proteins

To interpret external gradients, eukaryotic cells require receptors to relay to the cell interior information about ligand concentration outside the cell. Both neutrophils and *Dicytostelium* cells use ligand-sensing transmembrane proteins known as G-protein-coupled receptors (GPCRs). Upon binding specific extracellular ligands, GPCRs undergo conformational changes that lead to activation of trimeric G-proteins, which are located on the cytoplasmic side of the plasma membrane. In its inactive form, the G-protein trimer includes a GDP-bound $\alpha$ subunit, associated with a stable $\beta\gamma$ heterodimer. Interaction of the G protein with ligand-bound receptor induces the $\alpha$ subunit to exchange its bound GDP for GTP, with the result that $\alpha$–GTP dissociates from $\beta\gamma$. In their dissociated states, both $\alpha$–GTP and $\beta\gamma$ can interact with down-

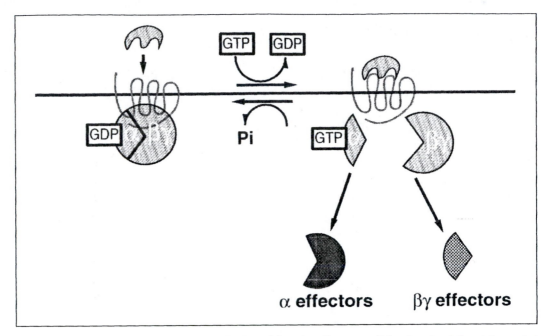

**Fig. 3** Overview of trimeric G-protein cycle. Binding of ligand to the extracellular domain of G-protein-coupled receptors induces GTP-charging of α and dissociation of α from βγ. In their dissociated states, both α and βγ can interact with downstream effectors. Hydrolysis of GTP by α regenerates the inactive G protein and terminates the signal. For details, see text.

stream effectors, which recognize surfaces of the two proteins that are inaccessible in the heterotrimer. Hydrolysis of bound GTP by α and reassociation of α–GDP with βγ regenerate an inactive G protein and terminate the signal (Fig. 3). Structurally and functionally heterogeneous, mammalian trimeric G-proteins are made up of polypeptides from three large families, encoded by at least 16 α, 5 β, and 11 γ genes. Individual G proteins are usually denoted by their distinctive α subunits, each of which regulates a different subset of downstream effectors. In addition to chemotactic ligands, GPCRs in mammalian cells detect and relay signals mediated by a host of hormones and neurotransmitters, as well as sensory stimuli, including light, sound, odorants, and tastants.

While any GPCR can relay information about the concentration of an extracellular ligand, only $G_i$-coupled receptors trigger chemotaxis of mammalian cells. Neutrophils respond to a very large number of chemotactic signals. In addition to the formylated peptides produced by bacteria, these include interleukin-8 (IL-8), a component of the complement cascade (C5a), and several other chemokines that are produced by endothelial cells, immunocytes, and other inflammatory cells at sites of tissue injury. All the GPCRs stimulated by these and other chemoattractants in mammals couple to the $G_i$ class of trimeric G-proteins, as indicated by the sensitivity of chemotactic responses to inhibition by a bacterial toxin, pertussis toxin (PTX), which specifically attaches ADP-ribose to a cysteine in the C terminus of $\alpha_i$, thereby uncoupling $G_i$ from GPCR stimulation. Several other $G_i$-coupled receptors whose ligands are not classically considered chemoattractants also mediate chemotaxis in cultured cells, but GPCRs coupled to other G proteins do not (3, 4).

What is special about $G_i$-coupled receptors? One obvious distinction is that only $G_i$-coupled receptors activate $\alpha_i$. Is $\alpha_i$–GTP then a necessary mediator of chemotaxis? This $\alpha$ subunit is probably not a necessary mediator, as indicated by experiments in which HEK293 cells were tricked into using a $G_i$-coupled receptor to activate a G-protein trimer that does not contain $\alpha_i$ (5). The cells were made to express an $\alpha_q/\alpha_z$ chimera, in which the C-terminal four residues of $\alpha_q$ were replaced by the corresponding residues of $\alpha_z$, a member of the $\alpha_i$ family that can be activated by $G_i$-coupled receptors but whose C terminus lacks the cysteine that confers sensitivity to inhibition by PTX. In PTX-treated cells, CXCR1, a $G_i$-coupled receptor for IL-8, could mediate a chemotactic response (migration across a filter toward a chamber containing chemoattractant, in a device called a Boyden chamber) if the cells expressed the $\alpha_q/\alpha_z$ chimera, but not in untransfected cells or in cells expressing recombinant $\alpha_q$. As expected, in PTX-treated $\alpha_q/\alpha_z$-expressing cells, IL-8 stimulated an $\alpha_q$ effector, phospholipase C, but had no effect on an $\alpha_i$ effector, adenylyl cyclase. Thus, although a $G_i$-coupled GPCR is required for chemotaxis, specific $\alpha_i$-dependent signals are not. Until this experiment is repeated in a professional chemotactic cell, like the neutrophil, this result does not rule out the possibility that $\alpha_i$–GTP transmits messages that contribute to efficient chemotaxis of cells that move faster than HEK293 cells; the result suggests, none the less, that $\alpha_i$–GTP is probably not *required* for neutrophil chemotaxis.

If $\alpha_i$ is dispensable for chemotaxis, what components downstream of the GPCR are necessary? In mammalian cells, one necessary component is the $\beta\gamma$ subunit released by G-protein activation. Chemotaxis in HEK293 or lymphocyte cells is blocked by expression of either of two proteins that bind and sequester free $\beta\gamma$ (3, 4). One of these proteins, the $\alpha$ subunit ($\alpha_t$) of $G_t$, can be activated by rhodopsin, but not by $G_i$-coupled receptors; consequently, recombinant $\alpha_t$ expressed in HEK293 cells or lymphocytes remains in its GDP-bound form and sequesters free $\beta\gamma$, thereby inhibiting chemotaxis. In HEK293 cells, expression of a C-terminal fragment of the $\beta$-adrenergic receptor kinase ($\beta$ARK) similarly sequesters $\beta\gamma$ and prevents chemotaxis (the C-terminal fragment binds $\beta\gamma$ but lacks the kinase catalytic domain). Although $\alpha_t$ and $\beta$ARK sequester free $\beta\gamma$, they do not prevent CXCR1 from mediating an $\alpha_i$–GTP-dependent response to IL-8-inhibition of adenylyl cyclase.

What roles do $\alpha$ and $\beta\gamma$ subunits play in the chemotactic responses of *Dictyostelium*? Homologous recombination, or site-directed insertion of foreign DNA at specific locations in the *Dictyostelium* genome, makes it possible to inactivate specific genes in this haploid organism—to create, for example, cells with null alleles for any one of the eight $\alpha$ genes or for the single $\beta$ or $\gamma$ gene. This approach has been used to study chemotaxis toward two ligands that are detected by different GPCRs: folate, a bacterial product that tells amoebae where to find their prey, and cAMP, the ligand that mediates aggregation of starved amoebae to form a slug that will later produce spores. Chemotaxis toward folate is impaired in *Dicytostelium* null for one $\alpha$ subunit,

$\alpha_4$, while a different null mutation (in the $\alpha_2$ gene) impairs chemotaxis toward cAMP; in neither case does the null $\alpha$ gene affect chemotaxis toward the other ligand (6, 7). Thus the two receptors mediate chemotaxis by activating G-protein trimers containing different $\alpha$ subunits, but the results do not tell us whether the $\alpha$ subunit specificity reflects a requirement for interaction with specific receptors or with specific downstream effectors. The latter interpretation, implying that the two chemotactic responses depend on different downstream effectors, seems possible but unlikely. The former interpretation suggests that specific $\alpha$ subunits in *Dicytostelium* act primarily as tools for coupling release of free $\beta\gamma$ to ligand stimulation of a specific subset of GPCRs, which discriminate among $\alpha$ subunits of the trimers they activate—that is, a role similar to that of $\alpha_i$ in mammalian cells.

What about $\beta\gamma$? Cells null for the gene encoding the single $\beta$ subunit of *Dictyostelium* do not migrate toward any chemoattractant (8). This result indicates that $\beta\gamma$ is essential for chemotaxis but by itself does not tell us whether $\beta\gamma$ signals directly to downstream effectors for chemotaxis or whether the primary function of $\beta\gamma$ is to mediate receptor activation of $\alpha$ subunits. The amenability of *Dicytostelium* to homologous recombination made it possible to address this question, by rescuing $\beta$ null cells with mutated versions of the $\beta$ gene (9). One $\beta$ mutant supported chemotaxis very poorly, but allowed proper coupling of G protein to receptor (assayed by a GTP-induced loss in ligand affinity) and proper activation of G protein by ligand-bound receptor (assayed by ligand-induced actin polymerization). This result strongly suggests that in *Dicytostelium* $\beta\gamma$ directly regulates downstream effectors for chemotaxis. Indeed, it seems likely that both *Dicytostelium* and neutrophils use $\beta\gamma$ as the principal mediator of chemotaxis, and specific $\alpha$ subunits to couple the process to specific GPCRs.

## 3. Adaptation

In both neutrophils and *Dictyostelium*, exposure to chemoattractant elicits a number of transient responses, including actin polymerization, cell-shape changes, activation of adenylyl cyclase, and phosphorylation of myosin heavy and light chains (1). After the immediate transient response to a given concentration of chemoattractant, the cells become refractory to stimulation with that concentration, but can respond to a chemotactic stimulus of greater intensity. This process is called 'desensitization' or 'adaptation'. The ability to adapt to a given concentration of chemoattractant probably contributes to the ability of neutrophils and *Dictyostelium* to undergo chemotaxis over ranges of ligand concentration that span several orders of magnitude. Although adaptation could occur at many different levels in the signalling cascade, we will focus on adaptation at the levels of GPCR and G protein, which is best understood.

A common mechanism for adaptation of G-protein-mediated signals is to phosphorylate the GPCR, thereby marking it for physical uncoupling from G protein. This marking process is crucial for proper visual transduction in the rod photoreceptors of the vertebrate retina. Photoexcited rhodopsin itself initiates adaptation

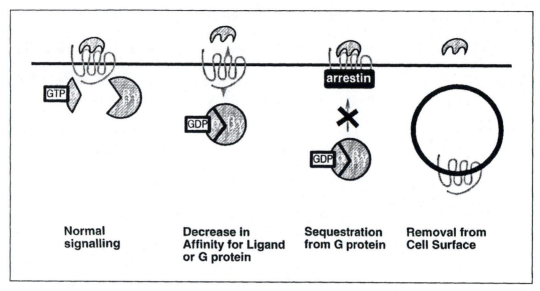

**Normal signalling**     **Decrease in Affinity for Ligand or G protein**     **Sequestration from G protein**     **Removal from Cell Surface**

**Fig. 4** Adaptation at the level of the G-protein-coupled receptor. G-protein-coupled receptor signalling can be downregulated through a decrease in affinity for ligand or G protein (some types of receptor phosphorylation produce these effects), sequestration of receptor from G protein (arrestin is thought to operate in this fashion), or removal of the receptor from the cell surface (for example, clathrin-mediated receptor internalization). For comparison, normal GPCR signalling is shown at far left.

by activating rhodopsin kinase, a member of a large class of G-protein receptor kinases (GRKs). The kinase, in turn, phosphorylates rhodopsin on multiple serine and threonine residues. This phosphorylation marks the receptor for binding by another protein, arrestin, which sterically prevents rhodopsin from coupling to and activating $G_t$. For other GPCRs, such as the β-adrenergic receptor, βγ helps to recruit the GRK (e.g. βARK) to the receptor (10). In these cases, members of the arrestin family act not only to turn off the GPCR but also as adapters between the GPCR and clathrin, leading to internalization of the GPCR (but not of the G protein) (11). Thus the GRK/arrestin mechanism, activated by the GPCR in combination with βγ, uncouples the GPCR from the G protein not only by sterically blocking their association but also by physically separating the GPCR from the G protein (Fig. 4).

Does eukaryotic chemotaxis require a GRK/arrestin mechanism for adaptation? Although GPCRs induce phosphorylation of chemotactic receptors in both neutrophils and *Dicytostelium*, available evidence indicates that neither receptor phosphorylation nor receptor internalization is necessary for adaptation during chemotaxis. For instance, chemotaxis and adaptation of multiple cellular responses, including actin polymerization and activation of adenylyl and guanylyl cyclases, are unaffected in *Dictyostelium* cells whose wild-type cAMP receptor is replaced by a cAMP receptor lacking sites for phosphorylation (12). Similarly, mutation of all the carboxy-terminal serine and threonine residues in the CCR2B chemokine receptor markedly impairs ligand-dependent internalization of this GPCR but has no effect on its ability to mediate chemotaxis in pre-B lymphocytes (13), and phosphorylation of the *N*-formyl peptide receptor is required for receptor internalization but not chemotaxis of myeloid cells (14).

It is likely that cells can use alternative mechanisms to uncouple GPCRs from G proteins. One such mechanism would reverse the order of regulation so that the uncoupler and not the receptor is marked or regulated (in contrast to GRK-mediated desensitization where the receptor to be uncoupled is marked by phosphorylation). Imagine, for instance, that GPCR activation leads to localized activation of an uncoupler, making it able to bind the GPCR and prevent activation of the G protein. Alternatively, GPCR activation could induce signals that alter the GPCR's micro-environment (physical location or interaction with the actin cytoskeleton) and thereby alter signal transduction from GPCR to signalling components downstream. Indeed, GPCRs and G proteins are reported to interact differently with the actin cytoskeleton of stimulated neutrophils, and certain signalling cascades are potentiated by depolymerization of the actin cytoskeleton.

Other adaptation mechanisms could act directly on the G protein. Strong candidates for roles in such a mechanism belong to the growing family of RGS (regulators of G-protein signalling) proteins, which increase the rate of GTP hydrolysis by $\alpha_i$ and $\alpha_q$, thereby decreasing the duration of G-protein activation and the intensity of the transmitted signal. The activity of an RGS is essential for the GPCR-mediated mating response of the budding yeast, *Saccharomyces cerevisiae*. During mating this organism responds to a pheromone gradient by orienting its polarity toward the source of pheromone, the mating partner (described in Chapter 2). Mutational inactivation of the yeast RGS protein renders the mating response pathway hypersensitive, resulting in saturation of the response at very low concentrations of pheromone and consequently preventing the organism from mating with partners that express a normal amount of pheromone. However, such a yeast cell *can* mate with partners expressing very low concentrations of pheromone and can correctly orient its polarity in response to a sufficiently low pheromone gradient (15, 16).

While these results suggest that the RGS protein sets the *perceived* intensity of the pheromone signal within a range that the yeast cell can interpret, they do not tell us whether the RGS machinery participates in actual adaptation, defined as *a stimulus-triggered change* in responsiveness. It is quite likely that extracellular signals do in fact regulate the activities of RGS proteins, including that of *S. cerevisiae*. For instance, most mammalian RGS proteins contain large N-terminal domains, separate from the GTPase-stimulating RGS domain itself; these domains appear to control the sub-cellular location of RGS activity and may, in some cases, make the RGS protein more effective in damping signals mediated by specific receptors. The N-terminal domain of the yeast RGS protein probably contains a site of regulation, as suggested by a point mutation in that domain which produces a dominant increase in the protein's RGS activity and a decrease in pheromone signalling (17). In summary, RGS proteins furnish an ideal mechanism for damping the signal intensity perceived by a cell, a mechanism that should allow a cell to expand its dynamic range of responsiveness, perhaps in a stimulus-dependent fashion. Very little is known about the importance of RGS signalling for chemotaxis. The observation (18) that certain overexpressed

RGS proteins impair chemotaxis of leukocytes identifies RGS proteins that can inhibit signals, but does not address the actual role of any RGS protein in regulating chemotaxis. Given the large number of different RGS proteins, specifying their individual roles by studying loss-of-function mutations appears a daunting task. As an alternative, a more general role of RGS proteins could be defined in experiments requiring that chemotaxis be mediated by a mutant $G\alpha_i$ that cannot interact with known RGS proteins (19) (for instance, by making the mutant resistant to PTX and using PTX to inactivate endogenous $\alpha_i$).

Proteins that sequester $\alpha$ or (more likely) $\beta\gamma$ would also decrease transmission of the chemotactic signal to downstream effectors. Such proteins certainly exist, as shown by identification (20–23) of $\beta\gamma$-sequestering proteins in *S. cerevisiae* (where their role in regulating signals is poorly understood). Other evidence raises the possibility that chemoattractants can act at a level downstream of the G protein to induce adaptation. For instance, preincubation of *Dicytostelium* with cAMP decreases the ability of the hydrolysis-resistant GTP analogue GTP$\gamma$S (which directly activates G protein in a receptor-independent fashion) to activate adenylyl cyclase. This effect could reflect either adaptation downstream of the G protein or a decrease in the pool of G protein available for stimulation. Similarly, stimulation of neutrophils with IL-8 or C5a decreases FMLP-mediated IP$_3$ production without affecting the ability of the FMLP receptor to activate G protein, as determined by a GTP$\gamma$S-binding assay (24).

# 4. Interpreting the chemotactic gradient

To mount an appropriately graded response, most cells need only decide *how much signal* they are receiving. In addition, chemotaxis requires a cell to decide *where the signal is coming from*, by comparing signal intensity over the cell's entire surface, in which some areas are exposed to greater concentrations of chemoattractant than others. Consider, for instance, two cells at different distances from a point source of chemoattractant (Fig. 5). Both cells move toward the source and exhibit greater actin polymerization and accumulation of filamentous actin on their up-gradient edges (as discussed in a later section). The down-gradient edge of cell 1 shows morphology characteristic of a trailing edge, despite the fact that it is exposed to a higher concentration of chemoattractant than the up-gradient edge of cell 2, in which the characteristic asymmetry of the actin cytoskeleton directly parallels that seen in cell 1. In other words, behaviour of any portion of a cell depends not only on the absolute intensity of the signal it receives, but also on that intensity *relative* to the intensity of signals received by other portions of the cell. Such comparisons require communication between different regions of the cell surface. What is the nature of this communication, and what is the basis of the comparison? At which level(s) of the signalling cascade is this comparison performed?

## 4.1 Asymmetrical intracellular signals

To understand how cells interpret the chemotactic gradient, it would be helpful to pinpoint the level at which asymmetry first appears in the signalling cascade between

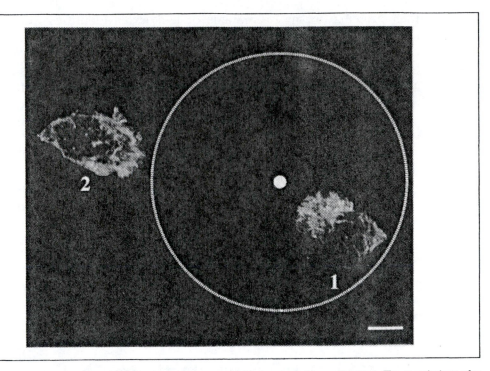

**Fig. 5** Actin staining of two neutrophils converging on a point source of chemoattractant. The morphology of a region of a cell depends on signal intensity elsewhere in the cell rather than on the absolute level of chemo-attractant. Thus the down-gradient region of cell 1 exhibits less actin ruffling than the up-gradient region of cell 2, despite the fact that the down-gradient region of cell 1 is exposed to a higher concentration of chemoattractant (for relative distances, note the position of the circle centred on the point source of chemoattractant). Bar = 5 μm.

ligand detection and actin polymerization. Are the primary sensors, the GPCRs, asymmetrically localized? Despite conflicting inferences drawn from observations of fixed mammalian cells, chemotactic GPCRs tagged with green fluorescent protein (GFP) are distributed uniformly on the surfaces of living *Dictyostelium* (25) and neutrophils (26) during chemotaxis. In contrast, actin accumulation, actin polymerization, and a variety of actin-associated proteins are asymmetrically distributed during chemotaxis (discussed in the final section). Thus the decision for directional polarization must occur at a level between localization of the GPCR and the actin cytoskeleton.

If receptor localization is uniform during chemotaxis, what about receptor *activity*? Does activity of the receptor and/or the G protein constitute a direct readout of the concentration of extracellular ligand, or does it exhibit signs of local amplification at the leading edge? To address this question, we need a spatial readout of receptor activity, distinct from the actin cytoskeleton and proximal to it in the signalling cascade. One approach is to determine the subcellular distribution of a GFP-tagged protein that is recruited to the plasma membrane upon activation of the GPCR. If the site for docking of the GFP-tagged probe is sufficiently discrete, it can be used as an indirect spatial indicator of GPCR activity.

Several properties of a *Dicytostelium* signalling protein, the cytosolic regulator of adenylyl cyclase (CRAC) made it a useful probe for GPCR activation and for the location of free βγ generated in response to a chemotactic stimulus. *Dicytostelium* cells genetically lacking CRAC are capable of chemotaxis toward cAMP, but cannot activate adenylyl cyclase, increase cAMP secretion, or aggregate normally (27). G-protein activation in wild-type *Dicytostelium*, in response either to extracellular cAMP or to GTPγS, induces CRAC to translocate from the cytosol to the plasma membrane (28). The pleckstrin homology (PH) domain of CRAC is sufficient for stimulus-mediated recruitment to the plasma membrane. It is likely that free βγ or a signal generated in response to βγ recruits CRAC to the plasma membrane, because GTPγS induces CRAC translocation to the plasma membrane in *Dicytostelium* cells genetically lacking every component of the pathway *except* Gβ.

On this basis, GFP-tagged CRAC was used as a probe—direct or indirect—for the location of free βγ generated by a chemotactic stimulus (29). In cells stimulated with a uniform concentration of cAMP, CRAC–GFP transiently translocates to the plasma membrane in a symmetrical fashion and returns to the cytosol within 1–3 minutes. Increases in chemoattractant concentration elicit repeated cycles of symmetrical recruitment and release of CRAC–GFP from the plasma membrane. In contrast, gradients of chemoattractant recruit CRAC–GFP to the up-gradient surface of the cells; the asymmetry of this recruitment substantially exceeds the asymmetry of receptor occupancy, inferred from the chemoattractant's extracellular concentration (Fig. 6). The apparent lack of GPCR-generated signalling at the back of the cell depends not on an absolute inability to respond, but instead on the relative intensities of signals in different parts of the cell, as inferred from a simple observation (29): replacement of the chemotactic gradient by a high and *uniform* concentration of chemoattractant causes CRAC–GFP to be recruited symmetrically to the entire cell surface.

Thus GPCRs are uniformly distributed through the plasma membrane during chemotaxis, but their activity is NOT. Asymmetry of the CRAC–GFP signal in excess of the asymmetry of external ligand concentration strongly suggests that the cells' perception of the gradient is amplified at the level of the GPCR or the G protein. Moreover, signalling on the down-gradient surface must somehow be inhibited, because this surface is exposed to chemoattractant but does not show a response to the GPCR. A second probe constructed from a PI3 kinase effector, the PH domain of AKT tagged with GFP, shows similarly asymmetrical patterns of apparent GPCR activity in both *Dicytostelium* (30) and neutrophils (105).

## 4.2 Initiation and maintenance of the asymmetrical signal: models

How might signals communicated between different regions of the plasma membrane induce apparent inhibition of the chemotactic signal at the down-gradient

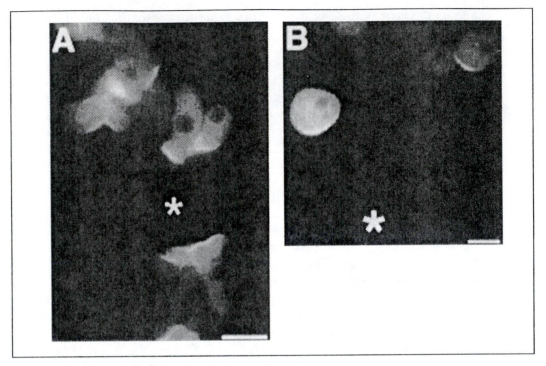

**Fig. 6** (A) CRAC–GFP is exclusively found at the leading edge of chemotaxing *Dictyostelium* amoebae. CRAC-null cells expressing CRAC–GFP were exposed to a gradient of chemoattractant generated by a micropipette containing 1 μM cAMP (asterisk). The image represents cells 90 seconds after exposure to the micropipette. Bar = 7 μm. (B) *Dictyostelium* amoebae can sense chemoattractant gradients in the absence of actin polymerization. CRAC-null cells expressing CRAC–GFP were treated with latrunculin-A (final concentration 0.5 μM) for 15 min and then exposed to a micropipette containing 1 μM cAMP (asterisk) for 65 seconds. Bar = 12 μm.

surface? One potential mechanism would rely on exquisitely well-tuned *global inhibition* (31). Suppose that the sum of GPCR activities produces a rapidly diffusing global inhibitor, which reduces, by an absolute amount, the signal transmitted by each GPCR. If this inhibition were sufficiently well-tuned, only the GPCRs on the up-gradient surface would exhibit net activation (Fig. 7). This conceptually straight-forward model requires relatively simple machinery—a diffusible regulator that controls responsiveness to the signal. A disadvantage is that regulation by inhibition alone may not amplify small differences effectively enough.

A potentially more effective mechanism combines *global inhibition* with *local enhancement* of the signal. This more complex model is thought to explain how developing organisms solve a similar problem in using gradients of morphogens to create polarity and sculpt the shapes of organs and tissues. In order to produce *exactly one* signalling organizer at the site of the maximal concentration of a morphogen in a gradient (or even in the presence of an initially uniform concentration of morpho-gen), each responding cell in the tissue produces both a *long-range inhibitor* of activity and a *short-range enhancer* of activity. The combination of positive and negative regulation results in positive feedback that amplifies small initial asymmetries in morphogen concentration (32–34).

Applying this model to chemotaxis, we focus our hypothesizing lens on a single cell, rather than on a developing tissue (Fig. 7). GPCRs at the plasma membrane of a single cell generate both a global inhibitor of signalling AND a local activator of signalling (which might, for example, *inhibit inhibition* locally). It is important to note that local enhancement is not simply the local activation of G proteins by GPCRs but is a layer of regulation superimposed on normal signalling whereby each activated

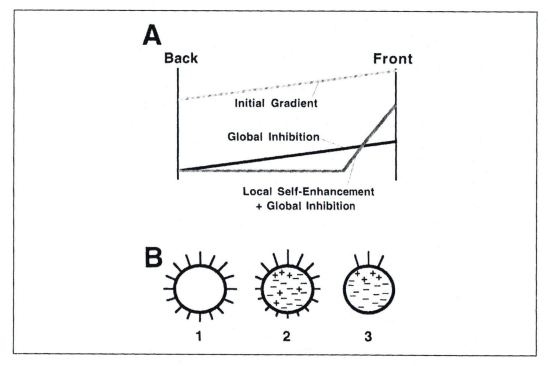

**Fig. 7** (A) Comparison of types of signal processing for gradient interpretation. The horizontal axis represents the position along a cell exposed to a chemotactic gradient to the right. The vertical axis represents the intensity of an intracellular signal generated by the chemoattractant. The light grey line represents the signal in response to the initial gradient in the absence of processing. The black line represents the signal following global inhibition. Note that the percentage difference in signal from front to back of the cell increases, but the absolute difference does not. The dark grey line represents the signal following global inhibition and local self-enhancement. Note that both the percentage and absolute difference from front to back of the cell increase. (B) Spatial example of global inhibition and local self-enhancement amplification of a gradient. Each activated receptor generates a global inhibitor of signalling (–) and a local self-enhancer of signalling (+), with the net result that each receptor favours signalling in its own region and inhibits signalling elsewhere. It is important to note that local enhancement is not simply the local activation of G proteins by GPCRs but is a layer of regulation superimposed on normal signalling, whereby each activated GPCR increases its own activity in a short-range fashion. Lines outside of the circles represent the level of intracellular signal generated in response to the chemoattractant. (1) Cell initially exposed to a chemotactic gradient experiences slight asymmetry in intracellular signalling. (2) Global inhibition and local self-activation generated by each receptor dramatically amplify the difference in intracellular signalling across the cell when iterated over time (3).

GPCR increases its own activity in a short-range fashion. To make the model more concrete, imagine possible biochemical mechanisms in which the activated GPCR stimulates synthesis of a rapidly diffusible global inhibitor that damps signalling throughout the cell—by stimulating an appropriate RGS protein, by sequestering free G-protein βγ subunits, or by otherwise uncoupling the GPCR from its G-protein target. Localized signal enhancers would act in the opposite direction—for instance, by inducing phosphorylation of an RGS protein to decrease its ability to accelerate GTP hydrolysis by the $\alpha_i$ subunit.

Available evidence restricts the number of potential mechanisms for either global inhibition or local enhancement of the signal. Neither type of regulation is likely to be mediated solely by altering localization, phosphorylation, ligand affinity, or internalization of GPCRs, for reasons outlined above; similarly, neither is likely to be exerted solely at the level of actin polymerization (see below). Note, however, that our emphasis on regulation at the level of receptor and G protein depends on an assumption that may not be correct. This assumption is that free βγ in the plasma membrane of cells exposed to a chemotactic gradient *directly* mediates recruitment of the cytoplasmic proteins used as markers for receptor activity. Several observations suggest that the asymmetrical signals studied so far may instead reflect regulation of signalling events downstream of GPCR and G protein. CRAC–GFP is recruited normally to macropinosomes, even in *Dictyostelium* cells lacking Gβ (C. Parent and P. Devreotes, unpublished). AKT–GFP, a marker recruited by chemoattractants to the up-gradient edges of both *Dictyostelium* (30) and neutrophils (105), is considered a specific probe for PI(3,4)P$_2$ and PI(3,4,5)P$_3$, rather than for βγ.

Wherever regulation takes place, the simple global inhibition model exhibits two disadvantages in comparison to the combined inhibition/enhancement model. First, it can generate a robustly asymmetrical signal only by exquisite fine tuning and thresholding; to allow signalling solely on the cell's up-gradient edge, the inhibitor must damp the signal by *precisely* the right amount to drop the down-gradient (but not up-gradient) edge below the threshold for response. Quantitatively precise inhibition is not as crucial in a model that incorporates positive feedback as well. Secondly, it is difficult to imagine how a model based on global inhibition alone could account for the ability of neutrophils in a *uniform concentration of ligand* to establish unequivocal polarity of their actin cytoskeletons. Indeed, a uniform concentration of chemoattractant can also recruit the GFP-tagged PH domain of AKT asymmetrically to the leading edge of a neutrophil (105). In contrast, as in a developing tissue responding to a morphogen, the combination of positive and negative signals may not merely generate appropriate polarity more efficiently in response to a small gradient, but may also generate random polarity in response to small stochastic fluctuations in a uniform concentration of ligand.

This apparent advantage of the combined inhibition/enhancement model also implies a potential disadvantage—that is, it might generate an internal gradient of signal intensity so strong that the cell finds it difficult to change polarity in response to changing external conditions. Slight variations in the model could make it easier to change polarity (106). For instance, a cell could 'clean the slate' by cyclically destroying the strong amplified peak of internal activity to allow a different interpretation of the external gradient during the next activation cycle; this could account for the cyclical pattern of receptor activation in *Dicytostelium* during chemotaxis, as assayed by cycles of recruitment and dissociation of CRAC–GFP during chemotaxis (29). Alternatively, it may be easier to move an amplified internal peak of activity than to destroy it. In *Dictyostelium* cells unable to polymerize actin, CRAC–GFP redistributes smoothly in response to a moving external point source of chemoattractant, without exhibiting cyclical turn-off and turn-on of the signal (29). Similarly, neutrophils tend to reorient an existing front rather than to generate a new front in response to a moving point source of chemoattractant (35).

Neutrophils and *Dicytostelium* cells are unlikely to interpret gradients using a simple temporal mechanism, like that of prokaryotes, because they pursue a moving point source of chemoattractant smoothly (35, 36), rather than approach it in a biased random walk. In contrast to the simple temporal mechanism of gradient interpretation in which cells compare levels of global signalling before and after movement of the entire cell, for the 'pilot pseudopodia model' cells compare levels of signalling at each point of their surface during extension of projections—the region of the surface experiencing the largest *temporal* increase in signalling (i.e. extending maximally up the chemotactic gradient) exhibits continued extension. The pilot pseudopodia model is difficult to distinguish behaviourally from spatial gradient interpretation in which the region of the cell surface experiencing maximal signalling (in an absolute sense) becomes the leading edge. None the less, one straightforward prediction is that temporal—but not spatial—interpretation of a stable gradient should require movement of the cell, or part of it. In contrast to this prediction, a stationary neutrophil extends its first projection towards a point source of chemoattractant (36). Moreover, CRAC–GFP is recruited preferentially to the up-gradient edge of *Dicytostelium* cells that are paralysed by a toxin, latrunculin, that causes depolymerization of the dynamic actin cytoskeleton, inhibits actin polymerization, and prevents actin-dependent morphological changes; the GFP-tagged PH domain of AKT shows similarly asymmetrical recruitment to the plasma membrane of latrunculin-treated neutrophils (105). Thus interpretation of the gradient takes place upstream of actin polymerization and does not require motility, suggesting that the primary interpretation of a gradient is spatial, although an additional temporal mechanism cannot be ruled out.

Interpretation of a pheromone gradient by *S. cerevisiae* may differ from interpretation of chemoattractant gradients by neutrophils and *Dictyostelium*. Whereas GPCRs in the latter cells are distributed uniformly throughout the cell surface during chemotaxis (25, 26), exposure of yeast cells to pheromone induces a massive internalization of mating factor receptors, followed by synthesis of new receptors which later reappear at the tip of the mating projection. Moreover, depolymerization of the yeast actin cytoskeleton inhibits proper induction of morphological polarity and asymmetrical accumulation of mating factor receptors and other polarity markers in response to stimulation with uniform pheromone (37). In contrast, as we have seen, loss of actin polymerization does not prevent asymmetrical signalling polarity (i.e.

distribution of markers for receptor activation) in either neutrophils or *Dicytostelium*. Thus development of polarity in yeast cells appears to require actin rearrangements, perhaps as a way of reinforcing asymmetrical distribution of receptors, while *Dicytostelium* and neutrophils polarize without rearranging the actin cytoskeleton or redistributing GPCRs. An important caveat: the yeast experiments were performed in a uniform concentration of pheromone, rather than (like neutrophils and *Dicytostelium*) in a gradient, raising the possibility that a pheromone *gradient* could induce yeast cell polarity even in the absence of the actin cytoskeleton.

# 5. Polarity effectors: Rho-GTPases

Once a cell has interpreted the gradient, how does it point itself in the right direction and move? Rapidly accumulating evidence indicates that members of the family of Rho-GTPases transmit spatial interpretation of the chemoattractant gradient from GPCRs and trimeric G-proteins at the cell surface to the ultimate effectors of polarity and motility. In *Dictyostelium*, neutrophils, and virtually every motile eukaryotic cell, these effectors regulate polymerization of actin and rearrangements of the actin cytoskeleton. Deferring more detailed description of actin polymerization to the next section of this chapter, here we describe the roles of three Rho-GTPases—Cdc42, Rac, and Rho—in chemotaxis. Rapidly accumulating evidence is beginning to identify parts of the machinery responsible for regulating each GTPase and to define the mechanisms underlying their intertwined but distinct functions in rearranging the actin cytoskeleton. We begin by introducing the GTPases and their effects on actin polymerization in cells and cell extracts, move to genetic analysis of their roles in intact cells responding to a gradient of chemoattractant, and finally focus in greater detail on each individual member of the family.

## 5.1 Rho-GTPases in cell extracts

Early evidence that GTPases regulate actin polymerization in chemotactic cells came from experiments in which a hydrolysis-resistant GTP analogue, GTPγS, was introduced into neutrophils (by electroporation or permeabilization of the plasma membrane): the analogue, which directly activates G proteins, induced polymerization of actin. Development of a cell-free system, in which GTPγS stimulates actin polymerization of cytosolic extracts from neutrophils or *Dictyostelium* cells (38), opened the way to identifying the specific G proteins involved. The effect of GTPγS was not impaired in cytosol from *Dictyostelium* mutants lacking Gβ, indicating that one or more G proteins distinct from the heterotrimers suffice to support actin polymerization. These G proteins belong to the Rho family of small GTPases, as indicated by experiments in which GTPγS-induced actin polymerization is blocked by three classes of inhibitors (38–40): *Clostridium difficile* toxin (which glucosylates Rho-GTPases and prevents their interaction with effectors), Rho-GDI (which prevents nucleotide exchange and insertion of Rho-GTPases into membranes), or dominant-negative constructs (which are thought to sequester proteins that activate the Rho-GTPases).

First identified by analysing mutations in *S. cerevisiae* (41), the Rho family of GTPases is implicated in a vast array of cell functions, including modulation of the actin and tubulin cytoskeletons, adhesion, secretion, transcription, cell proliferation, and neoplastic transformation. Like other GTPases in the Ras superfamily, Rho-GTPases cycle between GTP- and GDP-bound forms with the assistance of a variety of proteins which enhance GTP-loading (guanine nucleotide exchange factors, or GEFs) or GTP-hydrolysis (GTPase activating proteins, or GAPS), and their GTP- and GDP-bound forms interact with different subsets of effectors and regulatory proteins. Specific point mutations produce Rho-family proteins that are constitutively active (GTP-bound) or inactive (GDP-bound).

The corresponding dominant-positive and dominant-negative effects of such Rho mutants helped to identify their effects on the actin cytoskeleton. Injection of different constitutively activated Rho-GTPases into tissue-culture cells produces distinctive rearrangements of the actin cytoskeleton: activated Cdc42 induces thin finger-like projections, called filopodia; activated Rac induces sheet-like ruffles, called lamellipodia; and activated Rho induces formation of actin bundles, called stress fibres (42). Dominant-negative versions of these proteins inhibit the corresponding actin rearrangements produced by extracellular stimuli (42). (As described below and in Chapter 2, yeast and mice genetically lacking specific members of the Rho family are beginning to extend our understanding of the roles played by these proteins in cell polarity and chemotaxis.)

So far, cell-free extracts have proved useful for studying effects of Cdc42 on actin polymerization, but not for understanding biochemical roles of other Rho family members or for analysing the links that connect activation of Rho proteins to ligand-stimulation of GPCRs in the plasma membrane. In cytosolic extracts from neutrophils and *Dictyostelium*, activated Cdc42 induces actin polymerization and dominant-negative Cdc42 blocks GTPγS-induced actin polymerization, but the corresponding Rac mutants have no effect—despite the fact that activated Rac induces actin polymerization in whole or permeabilized cells (see below).

## 5.2 Rho-GTPases in whole cells

Chemotaxis of macrophages, which are larger than neutrophils and more accessible to microinjection of foreign proteins, provides a useful experimental model for studying the effects of mutant Rho-GTPases. Videomicroscopy allows quantitative observation of individual microinjected cells migrating up a gradient of a chemoattractant (43). Dominant-activated versions of Cdc42, Rac, and Rho all inhibit cell motility and chemotaxis, but exert different effects on cell morphology: activated Cdc42 induces filopodia around the entire cell periphery, activated Rac induces ruffling throughout the cell periphery, and activated Rho induces cell rounding and inhibits cell adhesion (43) (Fig. 8). Similarly, dominant-negative versions of the same GTPases inhibit chemotaxis but exert different effects on cell motility and morphology: dominant-negative Cdc42 randomizes the direction of motility of cells exposed to a chemoattractant, but the cells show polarized morphology and increased

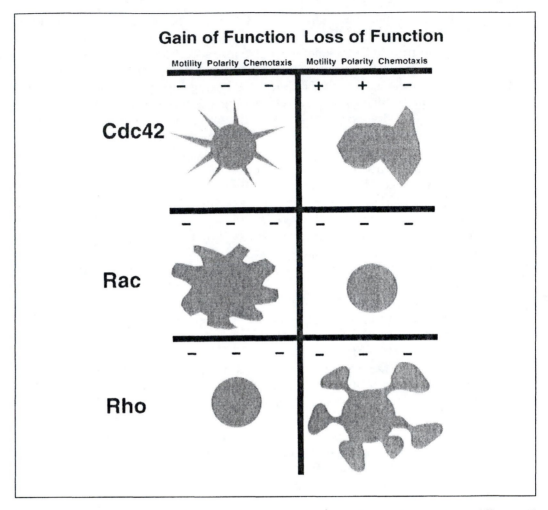

**Fig. 8** Phenotypes of macrophages for manipulations that effect gain or loss of function for Rho-GTPases. All phenotypes represent macrophages injected with dominant-activated or dominant-negative versions of Rho-GTPases, except for the Rho loss-of-function studies, which were performed with the botulinum C3 toxin, which inactivates endogenous Rho.

motility (43). Dominant-negative Rac inhibits both migration and morphological cell polarity. Dominant-negative Rho produces a dendritic cell morphology, with many extensions in all directions. A consistent interpretation of these observations is that Cdc42 matches cell polarity to the gradient of chemoattractant (perhaps in part by regulating Rac), while Rac plays a key role in chemoattractant-induced actin polymerization and generation of lamellipodia, and Rho is involved in cell adhesion and retracting aberrant cell projections (perhaps by modulating myosin, the actin motor protein; see below).

How does activation of a GPCR regulate activities of the Rho-GTPases? Here the most comprehensive answer comes from genetic analysis of the role of Cdc42 in the pheromone response of *S. cerevisiae*. Pheromone activation of the yeast GPCR induces release of $\beta\gamma$, which binds to an adaptor protein, Far1. Far1 also interacts with Cdc24, a GEF that activates Cdc42. Genetic manipulations that prevent interaction of $\beta\gamma$, Far1, and Cdc24 prevent activation of Cdc42, recruitment of this GTPase to the cell surface, and oriented polarization in response to gradients of pheromone (44–46) (for more details, see Chapter 2). In larger eukaryotes, a number of specific and nonspecific GEFs are candidates for transmitting signals from GPCRs to Rho-family GTPases. A direct biochemical mechanism is understood in only one case, however: a GEF for Rho, called p115 RhoGEF, also serves as an RGS for $\alpha_{12}$ and $\alpha_{13}$ (i.e. it increases the rate at which these $\alpha$ subunits hydrolyse GTP). Moreover, association of $\alpha_{13}$ with p115 RhoGEF activates its GEF activity (47). The specific GEFs responsible for activating Rac and Cdc42 in response to specific stimuli are not yet identified.

Knockout mice lacking a Rac protein, Rac2, reveal the crucial importance of this Rho-GTPase in chemotaxis (48). Neutrophils from these mice show marked defects in chemoattractant-induced actin polymerization and chemotaxis. (Embryonic lethality, the quite different phenotype of mice genetically lacking another Rac, Rac1, probably reflects the more ubiquitous expression of Rac1; Rac2 is restricted to the immune system.)

## 5.3 Cdc42 and actin polymerization

How do Rho-GTPases regulate rearrangements of the actin cytoskeleton? Here we understand the biochemical machinery best for Cdc42, principally because of this protein's ability to stimulate actin polymerization in cytosolic extracts (Fig. 9). Cdc42 does not induce pure G-actin to polymerize, but can do so in the presence of protein fractions prepared from cytosol of frog eggs; one such fraction contains a heptameric protein complex, called Arp2/3 (49). This complex, first identified as the profilin-binding complex of *Acanthamoeba* (50) and conserved from yeast to humans (51, 52), contains two actin-related proteins, Arp2 and Arp3 (50). Structural models position amino acids required to interact with actin at the barbed ends of Arp2 and Arp3, suggesting that the complex acts as a nucleus for barbed-end actin polymerization (53) (the biochemistry of actin polymerization is described more fully below). Indeed, the human Arp2/3 complex is necessary and sufficient to mediate ActA-dependent actin polymerization at the surface of an intracellular bacterial pathogen, *Listeria monocytogenes* (54, 55), where the complex stimulates nucleation of actin filaments that elongate only from their barbed ends (56, 57). Defective actin-dependent functions in conditional Arp2 and Arp3 mutants (52, 58, 59) indicate a critical role for the Arp2/3 complex in controlling the yeast actin cytoskeleton. Consistent with the notion that Cdc42 and the Arp2/3 complex act in a common pathway, function-blocking antibodies to the complex inhibit Cdc42- or GTP$\gamma$S-mediated actin polymerization in *Acanthamoeba* extracts (60). Because cytosol extracts can serve as models only for some types of actin polymerization, this observation does not indicate whether the Arp2/3

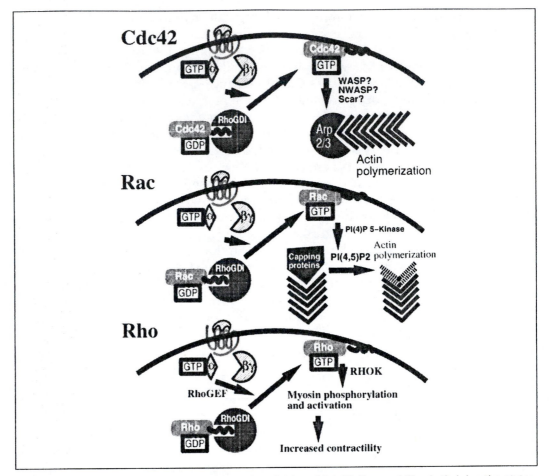

**Fig. 9** Modulation of the actin cytoskeleton through activation of Cdc42, Rac, and Rho. In their inactive states, Cdc42, Rac, and Rho are in their GDP-bound forms and reside in the cytoplasm with their lipid tails embedded in Rho-GDIs, proteins that prevent both GTP-charging and membrane association of the Rho-GTPases. Upon receptor activation, the Rho-GTPases are charged with GTP and driven into a membrane-bound pool. In general, it is not known whether dissociation from Rho-GDI or activation of exchange factors is the regulated step for Rho-GTPase activation. It is not known what product of receptor activation leads to activation of Cdc42 and Rac, but $\alpha_{13}$–GTP is known to directly activate the exchange factor for Rho. Activated Cdc42 induces *de novo* actin polymerization by causing activation of the Arp2/3 complex. Activated Rac induces liberation of barbed end of actin filaments by stimulating formation of the phosphoinositide PI(4,5)$P_2$ which removes capping proteins from the barbed end of actin filaments. Rac is also thought to stimulate Arp2/3 complex activation and inhibit myosin-based contractility (see text). Activated Rho induces cell contractility by promoting activation and cortical association of the actin motor protein, myosin.

complex is necessary for actin polymerization induced by other signals, including Rac activation. In other cells, overexpression of protein fragments postulated to interfere with recruitment of the Arp2/3 complex does prevent actin rearrangements in response to activation of Rac and Rho, as well as Cdc42 (61), but the specific target of inhibition in these experiments is difficult to assess.

By itself the Arp2/3 complex nucleates actin polymerization rather weakly in the absence of activating proteins, such as the ActA protein of *L. monocytogenes* (56, 57). Three eukaryotic proteins that mimic the potentiating effect of ActA are candidates for roles as important links between Cdc42 activation and regulation of the actin cytoskeleton: the mammalian Wiskott–Aldrich syndrome protein (WASP, which is expressed predominantly in haematopoietic cells) (55), N-WASP (a close relative expressed more widely) (62), and a WASP-related protein of *Dictyostelium*, Scar-1 (63). Scar-1 mutants in *Dicytostelium* show defects in actin polymerization and chemotaxis (64). WASP mutations cause the human immunodeficiency disorder of patients with the Wiskott–Aldrich syndrome, whose haematopoetic cells exhibit a variety of defects related to actin function, including defects in monocyte chemo-taxis, platelet function, T-cell signalling, and actin polymerization (65 and references therein). Injection of WASP into cultured cells induces actin polymerization (66), and N-WASP potentiates Cdc42-induced formation of filopodia in neuronal cells (67).

The combined efforts of several laboratories are just beginning to explore *how* these proteins link Rho-GTPases to actin polymerization of actin. A G-protein-binding domain (GBD) enables WASP and N-WASP to bind specifically to GTP-bound forms of Rac or Cdc42 (66). WASP, N-WASP, and Scar-1 contain a conserved C-terminal domain, which suffices for binding and activation of the Arp2/3 complex (61, 63). In cell-injection experiments, WASP lacking the GBD induces actin polymer-ization more potently than does full-length WASP (A. Abo, personal communica-tion), raising the possibility that Cdc42 or Rac activate normal WASP by abrogating an autoinhibitory effect of the GBD and exposing the C-terminal Arp2/3 complex activation domain. Consistent with this idea, Cdc42–GTP (in the presence of PI(4,5)P$_2$) increases the ability of purified N-WASP to stimulate the Arp2/3 complex *in vitro* (62). Furthermore, N-WASP depletion prevents Cdc42-mediated actin poly-merization in cell extracts (62). However, the complete cascade is not likely to be as simple as Cdc42–GTP activates N-WASP which in turn activates the Arp2/3 com-plex. First, the ability of N-WASP to activate the Arp2/3 complex *in vitro* requires activated Cdc42 *and* PI(4,5)P$_2$ (62), but activated Cdc42-mediated actin polymer-ization in extracts is insensitive to the sequestration of PI(4,5)P$_2$ (38). Thus, the role for PI(4,5)P$_2$ in normal N-WASP activation is uncertain. Secondly, N-WASP exists in a large protein complex in extracts and only partially purifies with the biochemical fraction necessary to mediate Cdc42-dependent activation of the Arp2/3 complex (L. Ma and R. Rohatgi, personal communication). Biochemical fractionation of cell-free systems (38, 49) should eventually make it possible to fully reconstitute Cdc42-dependent activation of the Arp2/3 complex.

## 5.4 Rac and actin polymerization

A primary effect of Rac activation is to increase the *availability* of barbed ends by promoting uncapping of previously formed filaments (Fig. 9). Here the best evidence comes from studies of permeabilized platelets (68), blood cells that promote form-ation of thrombi in response to thrombin and other agents and that are essential for normal haemostasis. In the resting state, most barbed ends of actin filaments are capped by barbed-end capping proteins, such as gelsolin (see below for general description of actin polymerization and capping proteins). Activation of the thrombin receptor increases the proportion of *uncapped* barbed ends from around 4% to 20–25%, resulting in massive polymerization of actin and striking changes in morphology. The number of exposed barbed ends correlates strongly with the amount of actin polymerization, making their exposure a convenient assay for signalling events that trigger actin polymerization. Indeed, these signals can act in detergent-permeabilized platelets: the proportion of barbed ends increases in response to activation of the GPCR for thrombin, addition of constitutively active Rac, or GTPγS.

In this system, ligand activation of the thrombin receptor triggers transient accumulation of a polyphosphoinositide, $PI(4,5)P_2$, which closely parallels the kinetics of barbed-end exposure. The following series of observations in permeabilized platelets (68) suggests that $PI(4,5)P_2$ is necessary and sufficient to induce actin polymerization, that it acts downstream of one or more G proteins, including Rac, and that it promotes polymerization by removing gelsolin or some other capping protein from barbed ends:

1. Direct addition of Rac-GTP increases accumulation of $PI(4,5)P_2$.
2. Addition of $PI(4,5)P_2$ induces exposure of barbed ends.
3. Barbed-end exposure triggered by activation of the thrombin receptor is inhibited by a GDP analogue that blocks G-protein activation, and this inhibition can be overcome by addition of $PI(4,5)P_2$.
4. Because $PI(4,5)P_2$ induces gelsolin and other capping proteins to dissociate from the barbed ends of actin filaments in purified systems (69), the $PI(4,5)P_2$-binding domain of gelsolin was used to sequester endogenous $PI(4,5)P_2$; this sequestration inhibits exposure of barbed ends triggered by activation of the thrombin receptor, by activated Rac, and by GTPγS.
5. Finally, Rac is thought to increase production of $PIP_2$ by binding to and activating PI(4)P 5-kinase, which converts PI(4)P to $PI(4,5)P_2$; PI(4)P 5-kinase lacking its Rac-interaction domain acts as a dominant negative for Rac-induced actin polymerization in platelets (J. Hartwig, personal communication).

It is not known whether the product of the kinase reaction, $PI(4,5)P_2$, always regulates the actin cytoskeleton directly, as it does in platelets. In some systems $PI(4,5)P_2$ appears to act upstream of Rho-GTPases; probably by facilitating nucleotide exchange on Cdc42. The polyphosphoinositide suffices to induce Cdc42-mediated actin polymerization in *Xenopus* extracts (40) and potentiates GTPγS-mediated actin polymerization in neutrophil extracts (38).

More recently a putative protein effector of Rac, known as WAVE, has been identified. Dominant-negative versions of WAVE prevent Rac-mediated actin polymerization, and overexpression of WAVE induces actin polymerization even in the presence of dominant-negative Rac, suggesting that WAVE is a downstream effector of Rac (70). WAVE shares sequence similarity to WASP and N-WASP and, importantly, contains a conserved C terminus, which has been shown in Scar-1 and WASP to be sufficient to activate the Arp2/3 complex (63). These data suggest that Rac may mediate its effects on actin polymerization through both PI(4,5)P$_2$-mediated uncapping of actin filaments *and* WAVE-mediated activation of the nucleation ability of the Arp2/3 complex.

## 5.5 Regulation of the actin cytoskeleton by Rho

In contrast to Cdc42 and Rac, Rho seems to control actin rearrangements indirectly, by regulating activity of myosin (Fig. 9) rather than polymerization of actin. Indeed, botulinum C3 toxin—which ADP-ribosylates and inactivates Rho—impairs neutrophil chemotaxis without preventing chemoattractant-induced actin polymerization (71). If Rho is not required for polymerizing actin, what role might it play in chemotaxis?

The most likely function of Rho in chemotaxis is to regulate cell contractility, allowing a cell to retract or inhibit surface projections that otherwise prevent the cell from moving. This idea is consistent with the appearance of macrophages, described above, in which Rho is inactivated by C3 toxin (43): these macrophages, their surfaces covered with dendritic projections, are unable to move. If loss of Rho allows unbridled formation of projections, it is not surprising that expression of activated Rho inhibits formation of actin projections triggered by Rac or Cdc42 in several systems (42). Rho is thought to modulate cell contractility by activating Rho-kinase (RHOK), which phosphorylates the light chain of the actin motor protein, myosin. This phosphorylation leads to bundling of actin filaments into contractile fibres (72, 73). Neutrophils treated with a RHOK inhibitor spread spontaneously in the absence of chemotactic stimulation (O. Weiner, unpublished observation). Also consistent with the idea that contractility and process formation are competing processes, Rac opposes the effect of Rho on cell contractility. Rac activates the kinase Pak1, which in turn phosphorylates and inhibits myosin light chain kinase (MLCK), reducing myosin phosphorylation and blocking the contractile effect of myosin (74).

## 6. Other regulators: Ca$^{2+}$ and cGMP

One ubiquitous second-messenger molecule, intracellular Ca$^{2+}$ ion, is *not* absolutely required for chemotaxis. Neutrophils do exhibit transient elevations of intracellular Ca$^{2+}$ during random motility and in response to stimulation by chemoattractant; motility and chemotaxis can proceed normally, however, when Ca$^{2+}$ elevations are inhibited by calcium chelators and calcium ionophores (75, 76). It is likely that intracellular Ca$^{2+}$ does play an adjuvant role, however, in facilitating ability of the trailing edge of a motile cell to detach from the underlying substrate; detachment *is* inhibited when calcium-depleted neutrophils are plated on certain substrates, such as fibronectin, but directional polarity is not (76). This Ca$^{2+}$-dependent detachment from fibronectin is thought to reflect severing by calcium-activated proteases of integrin contacts with the extracellular matrix.

Analysis of several *Dicytostelium* mutants has suggested that another second messenger, cGMP, does play a role in chemotaxis (77). These mutants, found by screening for defective chemotaxis toward both cAMP and folic acid, are defective in cGMP regulation; chemoattractant-induced accumulation of cAMP and IP$_3$, however, are normal in most of the mutants. Thus in comparison to two other second messengers, cGMP appears to play a much more important role in one or more signal cascades leading from GCPR activation to chemotactic effectors.

One such cascade may regulate cyclical association of myosin with the cell cortex. Thus, elevated cGMP leads to activation of myosin heavy chain kinase (MHCK), which, along with myosin II, associates with the cortex in response to chemoattractant stimulation. Next, MHCK phosphorylates myosin II, causing it to dissociate from the cortex (78, 79). In keeping with this idea, both defects in cGMP accumulation and MHCK mutations result in persistent association of myosin with the cortex (78–80), while overexpression of MHCK causes myosin to distribute primarily to the cytosol and blocks formation of cell polarity in response to chemoattractant (80). All of these perturbations severely impair chemotaxis (78–80). It is unlikely that myosin II phosphorylation represents the only important response to cGMP, however, because mutations of the myosin II heavy chain show almost normal chemotaxis (81). So far, a role for cGMP in leucocyte chemotaxis has not been reported; perhaps myosin regulation by cGMP in *Dicytostelium* plays a role similar to Rho-mediated regulation of myosin in mammalian cells.

# 7. Co-ordination of actin polymerization, polarization, and directed movement

Both neutrophils and *Dicytostelium* undergo massive bursts of actin polymerization in response to chemotactic gradients, and this polymerization is necessary for chemo-taxis: inhibition of actin polymerization by cell-permeable toxins (cytochalasin, which caps the barbed ends of actin filaments, or latrunculin, which sequesters actin monomers) prevents chemoattractants from inducing polar morphology and directed migration. How then is actin polymerization coupled to these responses? Where does chemoattractant-induced actin polymerization take place? This section will describe co-ordination of the spatial distribution of polymerizing actin filaments with cell polarity and directed migration in response to gradients of chemoattractant. We begin with a brief description of actin biochemistry and explain how addition of actin monomers to polymers probably produces protrusions of the plasma membrane. We then describe a highly informative model of actin polymerization, intracellular motility of *L. monocytogenes*, a pathogenic bacterium, and conclude by describing chemoattractant-induced foci of actin polymerization at the up-gradient edge of migrating cells.

## 7.1 Growing actin filaments and Brownian ratchets

Actin, one of the most highly conserved proteins known, is found in all eukaryotes. In the presence of physiological salt concentrations and ATP, monomeric actin assembles into actin filaments. An actin filament is a polar structure, the ends of which are designated as 'pointed' and 'barbed' (based on their appearance in electron micrographs after decoration by a myosin fragment). Monomers are added to the barbed end 10 times faster than to the pointed end (82). Because the cytosolic concentration of actin monomers far exceeds the critical concentration for their addition to either end of a pre-existing filament, cells use two kinds of proteins to prevent unregulated polymerization: monomeric actin-binding proteins, including thymosin-β4 and profilin, and capping proteins, including gelsolin and capping protein, which associate with barbed ends of actin filaments and prevent addition of monomers (83). Thus, cells can increase the formation of actin polymers in two ways—by generating nuclei for polymerization *de novo* (a process that can be accelerated by the Arp2/3 complex, as described above), or by generating an increased number of barbed (fast-growing) ends by uncapping or severing pre-existing actin filaments (Fig. 10).

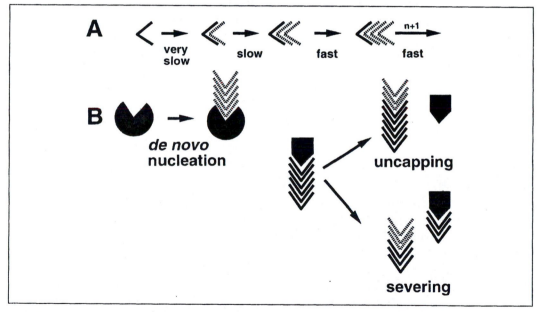

**Fig. 10** Review of actin dynamics. (A) Polymerization of actin (black Vs) is limited by formation of actin trimers. Formation of actin dimers is a very slow event, formation of an actin trimers is also slow, but the addition of actin subunits to trimers or greater polymers proceeds much more rapidly. Actin polymerization (hatched Vs) proceeds in the barbed end (as opposed to pointed end) direction. (B) Means of generating nuclei for actin polymerization. Actin polymerization is limited in cells through actin sequestering proteins and proteins that cap the fast-growing barbed end of actin filaments. Actin polymerization can be induced in three basic ways. *De novo* nucleation of actin polymerization is accomplished through molecules that simulate actin dimers or trimers; the Arp2/3 complex is thought to act in this fashion. Alternatively, barbed ends can be generated from pre-existing capped actin filaments either by uncapping (such as PI(4,5)P$_2$-mediated uncapping of filaments) or severing (gelsolin and cofilin are thought to act in this fashion).

How might growth of an actin filament extend the plasma membrane toward a chemoattractant? Even in the absence of accessory proteins or signalling molecules, actin polymerization suffices to induce surface protrusions of a vesicle loaded with a high concentration of actin monomers; such protrusions extend at rates similar to those observed at leading edges of neutrophils and *Dicytostelium* (84). How does a growing actin polymer exert force on the membrane? According to one explanation, using a thermal Brownian ratchet (85), growth of actin polymers does not actively push the membrane, but instead makes Brownian motion of the membrane unidirectional: Brownian motion randomly pushes the membrane back and forth; movement in the forward direction allows a monomer to be added to the barbed end of a filament apposed to the membrane, and addition of a monomer to the filament opposes movement in the backward direction. A slightly more complex model, the 'elastic Brownian ratchet' (86), agrees somewhat better with experimental observations. In the latter model, elasticity of the actin filament increases the likelihood that thermal fluctuations will create gaps between the membrane and the filament large enough for addition of an actin monomer. The simple Brownian ratchet allows only the membrane to undergo thermal fluctuations because the filament itself is not elastic. In the elastic ratchet model, the filament's flexibility allows it to vibrate away from the membrane, providing additional opportunities for addition of monomer, and also allowing the filament to exert an active force on the membrane (in contrast to the simple Brownian ratchet) when the elongated filament bends back into its original position. As a result, the ratchet in this model can actively push the membrane forward. The elastic ratchet would generate force optimally when filaments are oriented at an angle of 48° relative to the plane of the membrane (86)—in close agreement with the angle (45°–55°) observed experimentally at leading edges of lamellipodia (87, 88).

## 7.2 Motility of *L. monocytogenes*

This bacterium evades the immune system by living in the cytoplasm of host cells and moves from one cell to another by virtue of its ability to move rapidly, exert force on the plasma membrane, and push itself into the cytosol of a second cell. Its rapid movement leaves in its wake a long tail of polymerized actin (89). Actin polymerization appears to provide the driving force for *Listeria* motility (Fig. 11), as suggested by the observation that these organisms move at the same rate at which an actin filament polymerizes, and by experiments in which prevention of actin polymerization also prevents *Listeria* motility (90, 91). This motility requires only one bacterial protein, ActA, but requires host factors in addition to G-actin. Testing biochemical fractions of cell extracts for proteins required for *Listeria*-induced actin polymerization led to isolation of the Arp2/3 complex; the ability of this complex to nucleate actin is greatly potentiated by the bacterial ActA protein (as described above) (57).

In the absence of other host factors, *Listeria* do not move in the presence of ActA and the Arp2/3 complex, although they form actin 'clouds' (rather than tails). Thus

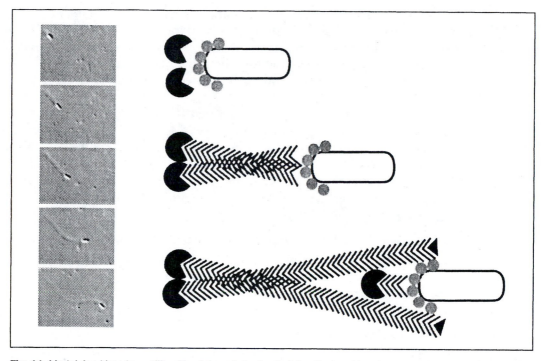

**Fig. 11** Model for *Listeria* motility. The intracellular bacterial pathogen *Listeria monocytogenes* exhibits actin-based motility in the cytosol of infected cells (Nomarski image to the left of the figure shows moving *Listeria* leaving an comet-like actin tail in its wake). The bacterium expresses a protein, ActA (circles), on its surface. This recruits and activates the Arp2/3 complex (black 'Pacman' shapes), which stimulates *de novo* nucleation of actin polymerization (black Vs). The force of actin polymerization propels the bacterium by an elastic Brownian ratchet (see text), and actin filaments are capped (black triangles) after they leave the posterior region of the bacterium. For the sake of simplicity, other proteins also required for *Listeria* motility, such as VASP (104), are not represented.

motility probably requires additional proteins that act, in part, by controlling the orientation of filaments that push the bacterium forward. How might this occur? Normally, actin polymerizes only at the portion of the actin tail adjacent to that bacterial surface. Actin tails are composed of long, axial and short, randomly oriented filaments (92). Thus it is likely (92, 93) that actin filaments nucleated at the surface of *Listeria* stop polymerizing when they leave the zone of actin polymerization near the bacterial surface, and become capped at their barbed ends. Thus, normal motility requires other host factors, possibly including proteins that induce cross-linking of filaments in the actin tail, which may be required for action of the elastic Brownian ratchet.

## 7.3  Specifying *where* actin polymerizes during chemotaxis

Like motility of *Listeria* in the cytoplasm, directed movement during chemotaxis probably requires precise positioning of the sites of actin polymerization. Recently it has become possible to analyse the spatial distribution of filament nucleation and polymerization by assessing incorporation into the actin cytoskeleton of tagged (fluorescent) actin monomers introduced into cells by microinjection or permeabilization. This approach has shown that actin polymerizes preferentially at the leading edges of lamellipodia in fibroblasts and at the tips of filopodia in neuronal growth cones (94, 95).

Similar experiments in permeabilized neutrophils (96) suggest that exposure to chemoattractants causes polymerization foci to form at or just under the plasma membrane in multiple discrete sites located predominantly at the tips of ruffles that protrude from a cell's leading edge or pseudopodium. A gradient of chemoattractant biases spatial distribution of these foci to the up-gradient edge of the cell. Two observations (96) suggest that the Arp2/3 complex plays a crucial role in determining sites of polymerization:

(1) the complex dynamically redistributes to the up-gradient surface of neutrophils;
(2) a pool of the complex that preferentially resists methanol extraction co-localizes with sites of actin polymerization at the tips of finger-like actin bundles that project into the pseudopodium.

These data led to a model (96) in which stimulation of chemoattractant receptors leads to organization of polymerization foci, at or just under the plasma membrane, which are functionally equivalent to the zone of actin polymerization generated by the ActA protein at the posterior surface of *L. monocytogenes*. In this model (Fig. 12), actin polymerization at the surface of the polymerization focus propels it and the cell membrane forward, forming an actin 'finger', analogous to a *Listeria* tail. Polymerization takes place only in the zone of actin polymerization at the tip of the growing finger, and filaments are capped at their barbed ends after they leave this zone. Asymmetrical establishment and/or maintenance of polymerization foci produce the cytoskeletal and morphological rearrangements responsible for moving the cell up a chemotactic gradient. The model predicts that inhibiting the activity of the Arp2/3 complex will block chemotaxis and formation of actin fingers. A comprehensive test of the model, of course, will require reconstitution of the signal cascade from chemotactic GPCRs to the final effectors for actin polymerization.

## 7.4 Autonomous cell polarization and motility

The effects of chemoattractants on cell polarity, actin rearrangements, and motility are superimposed upon an *intrinsic ability of motile eukaryotic cells to establish and maintain polarity*, even in the absence of extracellular stimuli. Indeed, as described below, the actin cytoskeleton itself can exhibit an intrinsic capacity to establish and maintain polarity. It will be an important challenge to determine to how the chemotactic response harnesses these intrinsic abilities to generate directed polarity and movement.

Small cell fragments of keratocytes provide an instructive example of the cytoskeleton's ability to regulate itself (97). Such fragments are non-motile and lack

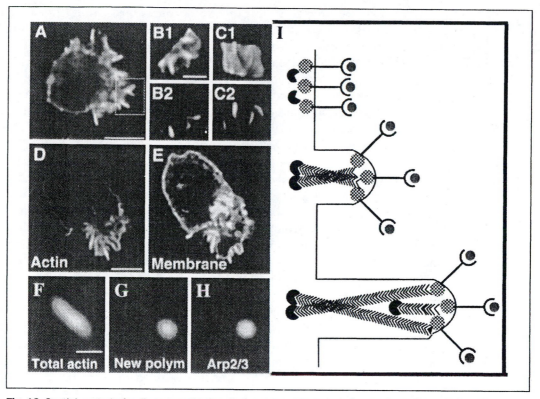

**Fig. 12** Spatial control of actin polymerization during neutrophil chemotaxis. (A) Phalloidin staining of a neutrophil stimulated with uniform chemoattractant, representing total actin. Bar = 5 μm. (B, C) Three-dimensional reconstruction of the boxed region of pseudopodium from (A): (B) represents a top view of the pseudopodium and (C) represents a side view. (B1, C1) Phalloidin staining, representing total actin. Bar = 5 μm. (B2, C2) TMR-actin stain, representing newly polymerized actin. (D) Actin staining of a neutrophil stimulated with uniform chemoattractant. (E) The cell surface of the same neutrophil as (D), detected by virtue of GFP attached to a chemotactic receptor. Note that the cell surface perfectly corresponds to the morphology of the tips of the finger-like actin bundles. (F–H) Finger-like projection of a neutrophil stimulated with uniform chemoattractant, permeabilized in the presence of TMR-actin, and then fixed and stained for antibodies to the Arp2/3 complex. (F) Phalloidin stain, representing total actin. Bar = 0.5 μm. (G) Newly incorporated actin. (H) Arp2/3 complex. (I) Model for chemoattractant-stimulated actin polymerization during neutrophil chemotaxis (similar to the *Listeria* model from Fig. 11). Binding of ligand to GPCR chemoattractant receptor (grey circle in wineglass shape) leads to generation of internal signals (checkerboard circles) which locally recruit and activate the Arp2/3 complex (black 'Pacman' shapes). The Arp2/3 complex stimulates *de novo* nucleation of actin polymerization (black Vs). The force of actin polymerization propels the cell membrane by an elastic Brownian ratchet (see text), and actin filaments are capped (black triangle) after they leave the region of activated G proteins. A good candidate for recruitment of the Arp2/3 complex is activated Cdc42. Panels A–H reprinted from ref. 96 with permission from *Nature Cell Biology*.

polarity of actin and myosin until they are subjected to mechanical stimulation; the fragments then establish anisotropic arrays of actin and myosin and begin to move. Polarity and motility are maintained long after the external mechanical stimulation is removed, suggesting that an internal mechanism maintains polarity and motility. It is thought to do so by generating a positive feedback loop that reinforces polarity and motility (97). How might it do so? An unperturbed, non-polar cell or cell fragment represents a balance among multiple actin protrusions, driven by growing actin polymers and distributed over the entire cell surface but coupled to one another through membrane tension. Myosin-based contraction at the region of mechanical perturbation is thought to unbalance the system—that is, compression of the actin–myosin system inhibits protrusion in the perturbed region, and unbalanced protrusion at the other edge of the cell fragment generates net motility. Translocation of the fragment further reinforces accumulation of myosin and contraction at the back of the cell, and protrusions reduce the likelihood of myosin-based contraction at the front; both effects reinforce polarity and motility.

Note that this cytoskeletal-based mechanism for establishing persistent polarity somewhat resembles the model, described above, in which global inhibition and local activation allow the cell to interpret a gradient of chemoattractant. In the cell fragment one signal, myosin-based contraction, is autocatalytic and acts at short range, while subsequent development of polarity and motility further reinforces this signal and acts over a longer range to prevent propagation of the myosin-based contraction to other parts of the moving cell.

Just as induction of polarity and motility does not always require an extracellular stimulus, interpretation of the chemotactic gradient does not absolutely require rearrangement of the cytoskeleton, as shown by inability of latrunculin to prevent GPCR-induced recruitment of GFP-tagged proteins to the leading edge of *Dicytostelium* and neutrophils (described above). It is likely that these potentially independent functions co-operate with one another in a cell moving toward a source of chemoattractant. A challenge for the future is to unravel the connections that mediate such co-operation.

# 8. Perspectives and future directions

Recent advances in studies of eukaryotic chemotaxis furnish tantalizing glimpses into specific signalling pathways and cytoskeletal rearrangements. We still do not know, however, how the cell choreographs the vast congeries of regulatory events to produce the dance of chemotaxis. Closer approaches to this central mystery will follow several roads; here we sketch approaches to questions that we find especially provocative.

## 8.1 $G_i$-coupled receptors

To mediate chemotaxis, must the GPCR do more than simply link binding of a chemoattractant to activation of a trimeric G-protein? $G_i$-linked GPCRs are required for chemotaxis of mammalian cells, while $\alpha_i$ itself is not (5); these observations suggest that we should search for a unique property of $G_i$-coupled receptors, distinct from their selectivity for $G_i$. Such properties might include the following:

1. $G_i$-coupled receptors serve as docking sites for RGS or other proteins that enhance a cell's ability to interpret the chemotactic gradient.

2. $G_i$-coupled receptors signal directly to chemotactic effectors distinct from the G-trimer, as reported (98) for certain GPCRs in *Dicytostelium*.

3. $G_i$-coupled receptors activate G-trimers containing a structurally specific subset of βγ subunits.

4. By associating with one another or with other membrane proteins, $G_i$-coupled receptors are targeted specifically to membrane subdomains specialized for chemotactic signalling.

Some of these possibilities could be tested by analysing phenotypes of GPCR chimeras that combine amino-acid sequences of a chemotactic $G_i$-coupled receptor with complementary sequences derived from a GPCR that does not mediate chemotaxis.

## 8.2   G-protein βγ versus α subunits

Since Gβγ appears to couple directly to chemotactic effectors, what are the functions of the α subunit? Does it merely provide a 'handle' for the GPCR to trigger release of βγ, or does it transmit a separate signal? Experiments in HEK293 cells (5), described above, suggest that the GTP-bound $\alpha_i$ subunit may not play a necessary role in chemotaxis. It is likely that chemotactic GPCRs activate G-trimers in addition to $G_i$, and other α subunits may carry messages crucial for chemotaxis. For instance, activation of $G_{13}$ (or $G_{12}$) can activate Rho by interacting with p115 RhoGEF (47), and loss of $\alpha_{13}$ produces knockout mice whose embryonic fibroblasts show severely impaired migratory responses to thrombin (99), which stimulates a GPCR coupled to both $G_i$ and $G_{13}$ (100). In addition, as noted above, G-protein α subunits could perform a key role by serving as substrates for regulation by RGS and other proteins, thereby regulating the half-life of βγ release upon receptor activation. These issues may be most effectively addressed by genetic approaches in *Dicytostelium* and mammalian cells—that is, by knocking out specific α subunits and/or expressing α subunits with point mutations analogous to Gβ mutations used to dissect the signalling role of βγ (9).

## 8.3   Inhibitory signals that may help to interpret the gradient

Exposure of a neutrophil or *Dicytostelium* cell to a chemotactic gradient rapidly induces inhibition of the GPCR-mediated signal, manifested in two ways, asymmetry of the signal and generalized adaptation (also termed densitization): the first produces a specific decrease in the intensity of signals transmitted at the back of the cell, relative to the front, while the second reversibly diminishes the cell's overall responsiveness to the stimulus. At present we do not know how—or even whether—the two kinds of inhibition relate to one another, although both are likely to play essential roles in chemotaxis. As noted above, we do know that chemotaxis is *not* likely to depend on certain well-established adaptation/desensitization mechanisms (including phosphorylation or internalization of the GPCR, or a decrease of its ability to bind ligand).

None the less, identifying the inhibitor(s) required for chemotaxis represents a crucial challenge. One difficulty is that inhibitors may act by distinct mechanisms at multiple stages of the signal cascade. When sensitive genetic screens, *in vitro* assays, or biochemical fractionation do identify an inhibitor, it will be necessary to identify both the substrate(s) for inhibition and the signalling protein(s) that generate the inhibitor; candidates for all these roles include ligand-occupied GPCRs, G-protein α subunits, released βγ subunits, RGS proteins, and many others.

## 8.4 Spatial readouts for GPCR-dependent signals

We can usually distinguish between mutations or pharmacological manipulations that impair a cell's general ability to move from those that prevent it from moving in the right direction. Within the latter class, however, we must learn to better distinguish between mutations and drugs that prevent primary interpretation of the chemotactic gradient and those that transmit polarized signals to downstream effectors. Recent observations (29; 105) that latrunculin fails to prevent asymmetrical recruitment of GFP-tagged receptor activity markers to the plasma membrane are promising first efforts in this direction. These results have not yet specified precisely where the amplification occurs in the signalling cascade, for reasons noted above. In this regard, it will be important to design assays that can determine the spatial distribution of other GPCR-dependent signals, including ligand-induced conformational change of the GPCR itself, free βγ, α–GTP, and others.

## 8.5 Effectors for chemoattractant-mediated actin polymerization

Many of the effectors that link chemotactic receptor activation with actin polymerization remain unknown. Extract systems have proven powerful for elucidating signalling events downstream of the Rho-GTPases or immediately upstream of the Arp2/3 complex, but permeabilized cell systems or genetic analyses will probably be required to link these processes with receptor activation. Actin polymerization mutants have generally been poorly represented in most *Dictyostelium* chemotaxis screens because of redundancy of the genes involved, necessity of these genes for cell viability, or a selection bias for the screens. Recent evidence for the final possibility comes from the observation that general chemotaxis mutants, such as Gβ null mutants, also exhibit a defect in bacterial phagocytosis. Mutants defective in phagocytosis would be overlooked in traditional screens because selection is typically performed by seeding individual clones on bacterial lawns. A recent screen designed to isolate mutants defective in chemotaxis and phagocytosis has yielded 10 mutants defective in chemoattractant-induced actin polymerization, suggesting that this approach will prove valuable in elucidating the signal transduction cascade from the chemotactic receptor to the actin polymerization machinery (107).

## 8.6 Signal cascades mediated by Rho-GTPases

Despite the key roles of Rho-GTPases as polarity effectors for chemotaxis, we know little of the specific biochemical mechanisms that regulate their activation or mediate their effects on polarity. Assessing changes in morphology represents an indirect and potentially incomplete strategy for identifying such mechanisms. For instance, the ability of Cdc42 to induce formation of filopodia, and the loss of both filopodia and directed motility in macrophages lacking Cdc42 function, are proposed (43) to indicate a role of filopodia in sensing the chemotactic gradient. Alternatively, Cdc42 could play morphologically 'invisible' roles in chemotaxis (e.g. by directly regulating other Rho-GTPases). Similarly, the successive appearance of filopodia, lamellipodia, and stress fibres after activation of Cdc42 is proposed (101) to represent a hierarchy of GTPase regulation, but the order of events in time does not constitute direct, compelling evidence. More precise dissection of the chemotactic signalling cascade will require assays for other activities, including robust methods for measuring the fraction of a cell's complement of a particular Rho-GTPase that is bound to GTP. One such strategy, recently developed as a means for measuring activated Cdc42 (102), Rac (102), and Rho (103) is to co-immunoprecipitate the GTPase with an appropriate downstream effector. Applied to other second messengers whose effectors preferentially recognize the second messenger's activated state, this approach should prove valuable in ordering signalling events.

## 8.7 Spatial readouts for signals mediated by Rho-GTPases

Determining the roles of Rho-GTPases will also require new tools for assaying and manipulating their activities in space. For instance, dominant-negative and dominant-active mutants of several Rho-GTPases produce similar defects in chemotaxis. Do these G proteins play a rather permissive role, fulfilled by cycling between active and inactive forms (and prevented by mutations that abrogate their abilities to bind or hydrolyse GTP)? Alternatively, and perhaps more likely, do these mutations prevent a specific spatial distribution, required for normal chemotaxis, of the Rho-family protein's effector-stimulating activity? To distinguish between permissive and spatially instructive roles of a Rho-GTPase (or any other signalling protein) in rearranging the cytoskeleton, two kinds of assays will be useful:

1. assays that determine the distribution of the protein's activity in space (e.g. a GFP-tagged effector protein that binds preferentially to the GTP-bound form of a Rho-GTPase);

2. methods that restrict the spatial distribution of the activated Rho-GTPase (e.g. localizing a protein that recruits or otherwise targets an activated version of the Rho-GTPase to a specific region of the plasma membrane).

In sum, because chemotaxis is an inherently spatial function, we can fully understand it only by resolving and reconstituting its components and their activities in the three-dimensional space of the cell.

## Acknowledgements

We thank H. Meinhardt, R. D. Mullins, D. Kalman, and members of the Bourne, Devreotes, and Sedat labs for helpful discussion and J. Hartwig, L. Ma, R. Rohatgi, and S. van Es for communicating data prior to publication. O. Weiner is a Howard Hughes Medical Institute Predoctoral Fellow.

## References

1. Devreotes, P. N. and Zigmond, S. H. (1988) Chemotaxis in eukaryotic cells: a focus on leukocytes and Dictyostelium. *Annu. Rev. Cell Biol.*, **4**, 649–86.
2. Berg, H. C. (1988) A physicist looks at bacterial chemotaxis. *Cold Spring Harbor Symp. Quant. Biol.*, **53**, (Pt 1), 1–9.
3. Neptune, E. R. and Bourne, H. R. (1997) Receptors induce chemotaxis by releasing the betagamma subunit of Gi, not by activating Gq or Gs. *Proc. Natl Acad. Sci., USA*, **94**, 14489–94.
4. Arai, H., Tsou, C. L., and Charo, I. F. (1997) Chemotaxis in a lymphocyte cell line transfected with C-C chemokine receptor 2B: evidence that directed migration is mediated by betagamma dimers released by activation of Gαi-coupled receptors. *Proc. Natl Acad. Sci., USA*, **94**, 14495–9.
5. Neptune, E. R., Iiri, T., and Bourne, H. R. (1999) Gαi is not required for chemotaxis mediated by Gi-coupled receptors. *J. Biol. Chem.*, **274**, 2824–8.
6. Kumagai, A., Hadwiger, J. A., Pupillo, M., and Firtel, R. A. (1991) Molecular genetic analysis of two G alpha protein subunits in *Dictyostelium. J. Biol. Chem.*, **266**, 1220–8.
7. Hadwiger, J. A., Lee, S., and Firtel, R. A. (1994) The G alpha subunit G alpha 4 couples to pterin receptors and identifies a signaling pathway that is essential for multicellular development in *Dictyostelium. Proc. Natl Acad. Sci., USA*, **91**, 10566–70.
8. Wu, L., Valkema, R., Van Haastert, P. J., and Devreotes, P. N. (1995) The G protein beta subunit is essential for multiple responses to chemoattractants in *Dictyostelium. J. Cell. Biol.*, **129**, 1667–75.
9. Jin, T., Amzel, M., Devreotes, P. N., and Wu, L. (1998) Selection of Gβ subunits with point mutations that fail to activate specific signaling pathways *in vivo*: dissecting cellular responses mediated by a heterotrimeric G protein in *Dictyostelium discoideum. Mol. Biol. Cell*, **9**, 2949–61.
10. Lefkowitz, R. J., Inglese, J., Koch, W. J., Pitcher, J., Attramadal, H., and Caron, M. G. (1992) G-protein-coupled receptors: regulatory role of receptor kinases and arrestin proteins. *Cold Spring Harbor Symp. Quant. Biol.*, **57**, 127–33.
11. Goodman, O. B. Jr, Krupnick, J. G., Santini, F., Gurevich, V. V., Penn, R. B., Gagnon, A. W., Keen, J. H., and Benovic, J. L. (1996) Beta-arrestin acts as a clathrin adaptor in endocytosis of the $\beta_2$-adrenergic receptor. *Nature*, **383**, 447–50.
12. Kim, J. Y., Soede, R. D., Schaap, P., Valkema, R., Borleis, J. A., Van Haastert, P. J., Devreotes, P. N., and Hereld, D. (1997) Phosphorylation of chemoattractant receptors is not essential for chemotaxis or termination of G protein-mediated responses. *J. Biol. Chem.*, **272**, 27313–18.
13. Arai, H., Monteclaro, F. S., Tsou, C. L., Franci, C., and Charo, I. F. (1997) Dissociation of chemotaxis from agonist-induced receptor internalization in a lymphocyte cell line transfected with CCR2B. Evidence that directed migration does not require rapid modulation of signaling at the receptor level. *J. Biol. Chem.*, **272**, 25037–42.
14. Hsu, M. H., Chiang, S. C., Ye, R. D., and Prossnitz, E. R. (1997) Phosphorylation of the N-formyl peptide receptor is required for receptor internalization but not chemotaxis. *J. Biol. Chem.*, **272**, 29426–9.

15. Segall, J. E. (1993) Polarization of yeast cells in spatial gradients of alpha mating factor. *Proc. Natl Acad. Sci., USA*, **90**, 8332–6.

16. Dorer, R., Pryciak, P. M., and Hartwell, L. H. (1995) *Saccharomyces cerevisiae* cells execute a default pathway to select a mate in the absence of pheromone gradients. *J. Cell Biol.*, **131**, 845–61.

17. Dohlman, H. G., Apaniesk, D., Chen, Y., Song, J., and Nusskern, D. (1995) Inhibition of G-protein signaling by dominant gain-of-function mutations in Sst2p, a pheromone desensitization factor in *Saccharomyces cerevisiae*. *Mol. Cell Biol.*, **15**, 3635–43.

18. Bowman, E. P., Campbell, J. J., Druey, K. M., Scheschonka, A., Kehrl, J. H., and Butcher, E. C. (1998) Regulation of chemotactic and proadhesive responses to chemoattractant receptors by RGS (regulator of G-protein signaling) family members. *J. Biol. Chem.*, **273**, 28040–8.

19. Lan, K. L., Sarvazyan, N. A., Taussig, R., Mackenzie, R. G., DiBello, P. R., Dohlman, H. G., and Neubig, R. R. (1998) A point mutation in Gαo and Gαi1 blocks interaction with regulator of G protein signaling proteins. *J. Biol. Chem.*, **273**, 12794–7.

20. Spain, B. H., Koo, D., Ramakrishnan, M., Dzudzor, B., and Colicelli, J. (1995) Truncated forms of a novel yeast protein suppress the lethality of a G protein alpha subunit deficiency by interacting with the beta subunit. *J. Biol. Chem.*, **270**, 25435–44.

21. Whiteway, M. S., Wu, C., Leeuw, T., Clark, K., Fourest-Lieuvin, A. ,Thomas, D. Y., and Leberer, E. (1995) Association of the yeast pheromone response G protein beta gamma subunits with the MAP kinase scaffold Ste5p. *Science*, **269**, 1572–5.

22. Kao, L. R., Peterson, J., Ji, R., Bender, L., and Bender, A. (1996) Interactions between the ankyrin repeat-containing protein Akr1p and the pheromone response pathway in *Saccharomyces cerevisiae*. *Mol. Cell. Biol.*, **16**, 168–78.

23. Pryciak, P. M. and Hartwell, L. H. (1996) AKR1 encodes a candidate effector of the G beta gamma complex in the *Saccharomyces cerevisiae* pheromone response pathway and contributes to control of both cell shape and signal transduction. *Mol. Cell. Biol.*, **16**, 2614–26.

24. Richardson, R. M., Ali, H., Tomhave, E. D., Haribabu, B., and Snyderman, R. (1995) Cross-desensitization of chemoattractant receptors occurs at multiple levels. Evidence for a role for inhibition of phospholipase C activity. *J. Biol. Chem.*, **270**, 27829–33.

25. Xiao, Z., N. Zhang, D. B. Murphy, and P. N. Devreotes. (1997) Dynamic distribution of chemoattractant receptors in living cells during chemotaxis and persistent stimulation. *J Cell Biol*, **139**, p. 365–74.

26. Servant, G., Weiner, O. D., Neptune, E. R., Sedat, J. W., and Bourne, H. R. (1999) Dynamics of a chemoattractant receptor in living neutrophils during chemotaxis. *Molec. Biol. Cell*, **10**, 1163–78.

27. Insall, R., Kuspa, A., Lilly, P. J., Shaulsky, G., Levin, L. R., Loomis, W. F., and Devreotes, P. (1994) CRAC, a cytosolic protein containing a pleckstrin homology domain, is required for receptor and G protein-mediated activation of adenylyl cyclase in *Dictyostelium*. *J. Cell Biol.*, **126**, 1537–45.

28. Lilly, P. J. and Devreotes, P. N. (1995) Chemoattractant and GTP gamma S-mediated stimulation of adenylyl cyclase in *Dictyostelium* requires translocation of CRAC to membranes. *J. Cell Biol.*, **129**, 1659–65.

29. Parent, C. A., Blacklock, B. J., Froehlich, W. M., Murphy, D. B., and Devreotes, P. N. (1998) G protein signaling events are activated at the leading edge of chemotactic cells. *Cell*, **95**, 81–91.

30. Ruedi Meili, C. E., Lee, S., Reddy, T. B. K., Hui Ma, and Firtel, R. A. (1999) Chemoattractant-mediated transient activation and membrane localization of Akt/PKB is required for efficient chemotaxis to cAMP in *Dictyostelium*. *EMBO J.*, **18**, (8), 2092–105.

31. Fisher, P. R. (1990) Pseudopodium activation and inhibition signals in chemotaxis by *Dictyostelium discoideum* amoebae. *Semin. Cell Biol.*, **1**, 87–97.

32. Turing, A. M. (1990) The chemical basis of morphogenesis. 1953 [classical article]. *Bull. Math. Biol.*, **52**, 153–97; discussion 119–52.

33. Gierer, A. and Meinhardt, H. (1972) A theory of biological pattern formation. *Kybernetik*, **12**, 30–9.

34. Meinhardt, H. and Gierer, A. (1974) Applications of a theory of biological pattern formation based on lateral inhibition. *J. Cell Sci.*, **15**, 321–46.

35. Gerisch, G. and Keller, H. U. (1981) Chemotactic reorientation of granulocytes stimulated with micropipettes containing fMet–Leu–Phe. *J. Cell Sci.*, **52**, 1–10.

36. Zigmond, S. H. (1974) Mechanisms of sensing chemical gradients by polymorphonuclear leukocytes. *Nature*, **249**, 450–2.

37. Ayscough, K. R. and Drubin, D. G. (1998) A role for the yeast actin cytoskeleton in pheromone receptor clustering and signalling. *Curr. Biol.*, **8**, 927–30.

38. Zigmond, S. H., Joyce, M.,Borleis, J., Bokoch, G. M., and Devreotes, P. N. (1997) Regulation of actin polymerization in cell-free systems by GTPγS and Cdc42. *J. Cell Biol.*, **138**, 363–74.

39. Katanaev, V. L. and Wymann, M. P. (1998) GTPγS-induced actin polymerisation *in vitro*: ATP- and phosphoinositide-independent signalling via Rho-family proteins and a plasma membrane-associated guanine nucleotide exchange factor. *J. Cell Sci.*, **111**, 1583–94.

40. Ma, L., Cantley, L. C., Janmey, P. A., and Kirschner, M. W. (1998) Corequirement of specific phosphoinositides and small GTP-binding protein Cdc42 in inducing actin assembly in *Xenopus* egg extracts. *J. Cell Biol.*, **140**, 1125–36.

41. Johnson, D. I. and Pringle, J. R. (1990) Molecular characterization of CDC42, a *Saccharomyces cerevisiae* gene involved in the development of cell polarity. *J. Cell Biol.*, **111**, 143–52.

42. Hall, A. (1998) Rho GTPases and the actin cytoskeleton. *Science*, **279**, 509–14.

43. Allen, W. E., Zicha, D., Ridley, A. J., and Jones, G. E. (1998) A role for Cdc42 in macrophage chemotaxis. *J. Cell Biol.*, **141**, 1147–57.

44. Butty, A. C., Pryciak, P. M., Huang, L. S., Herskowitz, I., and Peter, M. (1998) The role of Far1p in linking the heterotrimeric G protein to polarity establishment proteins during yeast mating. *Science*, **282**, 1511–16.

45. Nern, A. and Arkowitz, R. A. (1998) A GTP-exchange factor required for cell orientation. *Nature*, **391**, 195–8.

46. Nern, A. and Arkowitz, R. A. (1999) A Cdc24p–Far1p–Gβγ protein complex required for yeast orientation during mating. *J. Cell Biol.*, **144**, 1187–202.

47. Kozasa, T., Jiang, X., Hart, M. J., Sternweis, P. M., Singer, W. D., Gilman, A. G., Bollag, G., and Sternweis, P. C. (1998) p115 RhoGEF, a GTPase activating protein for Gα12 and Gα13. *Science*, **280**, 2109–11.

48. Roberts, A. W., *et al.* (1999) Deficiency of the hematopoietic cell-specific Rho family GTPase Rac2 is characterized by abnormalities in neutrophil function and host defense. *Immunity*, **10**, 183–96.

49. Ma, L., Rohatgi, R., and Kirschner, M. W. (1998) The Arp2/3 complex mediates actin polymerization induced by the small GTP-binding protein Cdc42. *Proc. Natl Acad. Sci., USA*, **95**, 15362–7.

50. Machesky, L. M., Atkinson, S. J., Ampe, C., Vandekerckhove, J., and Pollard, T. D. (1994) Purification of a cortical complex containing two unconventional actins from *Acanthamoeba* by affinity chromatography on profilin-agarose. *J. Cell Biol.*, **127**, 107–15.

51. Welch, M. D., DePace, A. H., Verma, S., Iwamatsu, A., and Mitchison, T. J. (1997) The human Arp2/3 complex is composed of evolutionarily conserved subunits and is localized to cellular regions of dynamic actin filament assembly. *J. Cell Biol.*, **138**, 375–84.

52. Winter, D., Podtelejnikov, A. V., Mann, M., and Li, R. (1997) The complex containing actin-related proteins Arp2 and Arp3 is required for the motility and integrity of yeast actin patches. *Curr. Biol.*, **7**, 519–29.

53. Kelleher, J. F., Atkinson, S. J., and Pollard, T. D. (1995) Sequences, structural models, and cellular localization of the actin-related proteins Arp2 and Arp3 from *Acanthamoeba*. *J. Cell Biol.*, **131**, 385–97.

54. Welch, M. D., Iwamatsu, A., and Mitchison, T. J. (1997) Actin polymerization is induced by Arp2/3 protein complex at the surface of *Listeria monocytogenes*. *Nature*, **385**, 265–9.

55. Yarar, D., To, W., Abo, A., and Welch, M. D. (1999) The Wiskott–Aldrich syndrome protein directs actin-based motility by stimulating actin nucleation with the Arp2/3 complex. *Curr. Biol.*, **9**, 555–8.

56. Mullins, R. D., Heuser, J. A., and Pollard, T. D. (1998) The interaction of Arp2/3 complex with actin: nucleation, high affinity pointed end capping, and formation of branching networks of filaments. *Proc. Natl Acad. Sci., USA*, **95**, 6181–6.

57. Welch, M. D., Rosenblatt, J., Skoble, J., Portnoy, D. A., and Mitchison, T. J. (1998) Interaction of human Arp2/3 complex and the *Listeria monocytogenes* ActA protein in actin filament nucleation. *Science*, **281**, 105–8.

58. McCollum, D., Feoktistova, A., Morphew, M., Balasubramanian, M., and Gould, K. L. (1996) The *Schizosaccharomyces pombe* actin-related protein, Arp3, is a component of the cortical actin cytoskeleton and interacts with profilin. *EMBO J.*, **15**, 6438–46.

59. Moreau, V., Madania, A., Martin, R. P., and Winsor, B. (1996) The *Saccharomyces cerevisiae* actin-related protein Arp2 is involved in the actin cytoskeleton. *J. Cell Biol.*, **134**, 117–32.

60. Mullins, R. D. and Pollard, T. D. (1999) Rho-family GTPases require the Arp2/3 complex to stimulate actin polymerization in *Acanthamoeba* extracts. *Curr. Biol.*, **9**, 405–15.

61. Machesky, L. M. and Insall, R. H. (1998) Scar1 and the related Wiskott–Aldrich syndrome protein, WASP, regulate the actin cytoskeleton through the Arp2/3 complex. *Curr. Biol.*, **8**, 1347–56.

62. Rohatgi, R., Ma, L., Miki, H., Lopez, M., Kirchhausen, T., Takenawa, T., and Kirschner, M. W. (1999) The interaction between N-WASP and the Arp2/3 complex links Cdc42-dependent signals to actin assembly. *Cell*, **97**, 221–31.

63. Machesky, L. M., Mullins, R. D., Higgs, H. N., Kaiser, D. A., Blanchoin, L., May, R. C., Hall, M. E., and Pollard, T. D. (1999) Scar, a WASP-related protein, activates nucleation of actin filaments by the Arp2/3 complex. *Proc. Natl Acad. Sci., USA*, **96**, 3739–44.

64. Bear, J. E., Rawls, J. F., and Saxe, C. L. R. (1998) SCAR, a WASP-related protein, isolated as a suppressor of receptor defects in late *Dictyostelium* development. *J. Cell Biol.*, **142**, 1325–35.

65. Zicha, D., Allen, W. E., Brickell, P. M., Kinnon, C., Dunn, G. A., Jones, G. E., and Thrasher, A. J. (1998) Chemotaxis of macrophages is abolished in the Wiskott–Aldrich syndrome. *Br. J. Haematol.*, **101**, 659–65.

66. Symons, M., Derry, J. M., Karlak, B., Jiang, S., Lemahieu, V., McCormick, F., Francke, U., and Abo, A. (1996) Wiskott–Aldrich syndrome protein, a novel effector for the GTPase CDC42Hs, is implicated in actin polymerization. *Cell*, **84**, 723–34.

67. Miki, H., Sasaki, T., Takai, Y., and Takenawa, T. (1998) Induction of filopodium formation by a WASP-related actin-depolymerizing protein N-WASP. *Nature*, **391**, 93–6.

68. Hartwig, J. H., Bokoch, G. M., Carpenter, C. L. Janmey, P. A., Taylor, L. A., Toker, A., and Stossel, T. P. (1995) Thrombin receptor ligation and activated Rac uncap actin filament barbed ends through phosphoinositide synthesis in permeabilized human platelets. *Cell*, **82**, 643–53.

69. Janmey, P. A., Iida, K., Yin, H. L., and Stossel, T. P. (1987) Polyphosphoinositide micelles and polyphosphoinositide-containing vesicles dissociate endogenous gelsolin–actin complexes and promote actin assembly from the fast-growing end of actin filaments blocked by gelsolin. *J. Biol. Chem.*, **262**, 12228–36.

70. Miki, H., Suetsugu, S., and Takenawa, T. (1998) WAVE, a novel WASP-family protein involved in actin reorganization induced by Rac. *EMBO J.*, **17**, 6932–41.

71. Stasia, M. J., Jouan, A., Bourmeyster, N., Boquet, P., and Vignais, P. V. (1991) ADP-ribosylation of a small size GTP-binding protein in bovine neutrophils by the C3 exoenzyme of *Clostridium botulinum* and effect on the cell motility. *Biochem. Biophys. Res. Commun.*, **180**, 615–22.

72. Kimura, K. *et al.* (1996) Regulation of myosin phosphatase by Rho and Rho-associated kinase (Rho-kinase). *Science*, **273**, 245–8.

73. Chrzanowska-Wodnicka, M. and Burridge, K. (1996) Rho-stimulated contractility drives the formation of stress fibers and focal adhesions. *J. Cell Biol.*, **133**, 1403–15.

74. Sanders, L. C., Matsumura, F., Bokoch, G. M., and de Lanerolle, P. (1999) Inhibition of myosin light chain kinase by p21-activated kinase. *Science*, **283**, 2083–5.

75. Zigmond, S. H., Slonczewski, J. L., Wilde, M. W., and Carson, M. (1988) Polymorpho-nuclear leukocyte locomotion is insensitive to lowered cytoplasmic calcium levels. *Cell Motil. Cytoskeleton*, **9**, 184–9.

76. Marks, P. W. and Maxfield, F. R. (1990) Transient increases in cytosolic free calcium appear to be required for the migration of adherent human neutrophils. *J. Cell Biol.*, **110**, 43–52.

77. Kuwayama, H., Ishida, S., and Van Haastert, P. J. (1993) Non-chemotactic *Dictyostelium discoideum* mutants with altered cGMP signal transduction. *J. Cell Biol.*, **123**, 1453–62.

78. Liu, G., Kuwayama, H., Ishida, S., and Newell, P. C. (1993) The role of cyclic GMP in regulating myosin during chemotaxis of *Dictyostelium*: evidence from a mutant lacking the normal cyclic GMP response to cyclic AMP. *J. Cell Sci.*, **106**, 591–5.

79. Dembinsky, A., Rubin, H., and Ravid, S. (1996) Chemoattractant-mediated increases in cGMP induce changes in *Dictyostelium* myosin II heavy chain-specific protein kinase C activities. *J. Cell Biol.*, **134**, 911–21.

80. Abu-Elneel, K., Karchi, M., and Ravid, S. (1996) *Dictyostelium* myosin II is regulated during chemotaxis by a novel protein kinase C. *J. Biol. Chem.*, **271**, 977–84.

81. De Lozanne, A. and Spudich, J. A. (1987) Disruption of the *Dictyostelium* myosin heavy chain gene by homologous recombination. *Science*, **236**, 1086–91.

82. Pollard, T. D. (1986) Rate constants for the reactions of ATP- and ADP-actin with the ends of actin filaments. *J. Cell Biol.*, **103**, 2747–54.

83. Pantaloni, D. and Carlier, M. F. (1993) How profilin promotes actin filament assembly in the presence of thymosin beta 4. *Cell*, **75**, 1007–14.

84. Miyata, H., Nishiyama, S., Akashi, K., and Kinosita, K. Jr (1999) Protrusive growth from giant liposomes driven by actin polymerization. *Proc. Natl Acad. Sci., USA*, **96**, 2048–53.

85. Peskin, C. S., Odell, G. M., and Oster, G. F. (1993) Cellular motions and thermal fluctuations: the Brownian ratchet. *Biophys. J.*, **65**, 316–24.

86. Mogilner, A. and Oster, G. (1996) Cell motility driven by actin polymerization. *Biophys. J.*, **71**, 3030–45.

87. Small, J. V., Herzog, M., and Anderson, K. (1995) Actin filament organization in the fish keratocyte lamellipodium. *J. Cell Biol.*, **129**, 1275–86.

88. Svitkina, T. M., Verkhovsky, A. B., McQuade, K. M., and Borisy, G. G. (1997) Analysis of the actin–myosin II system in fish epidermal keratocytes: mechanism of cell body translocation. *J. Cell Biol.*, **139**, 397–415.

89. Dabiri, G. A., Sanger, J. M., Portnoy, D. A., and Southwick, F. S. (1990) *Listeria monocytogenes* moves rapidly through the host-cell cytoplasm by inducing directional actin assembly. *Proc. Natl Acad. Sci., USA*, **87**, 6068–72.

90. Sanger, J. M., Sanger, J. W., and Southwick, F. S. (1992) Host cell actin assembly is necessary and likely to provide the propulsive force for intracellular movement of *Listeria monocytogenes*. *Infect. Immun.*, **60**, 3609–19.

91. Theriot, J. A., Mitchison, T. J., Tilney, L. G., and Portnoy, D. A. (1992) The rate of actin-based motility of intracellular *Listeria monocytogenes* equals the rate of actin polymerization. *Nature*, **357**, 257–60.

92. Sechi, A. S., Wehland, J., and Small, J. V. (1997) The isolated comet tail pseudopodium of *Listeria monocytogenes*: a tail of two actin filament populations, long and axial and short and random. *J. Cell Biol.*, **137**, 155–67.

93. Marchand, J. B., Moreau,P., Paoletti, A., Cossart, P.,Carlier, M. F., and Pantaloni, D. (1995) Actin-based movement of *Listeria monocytogenes*: actin assembly results from the local maintenance of uncapped filament barbed ends at the bacterium surface. *J. Cell Biol.*, **130**, 331–43.

94. Symons, M. H. and Mitchison, T. J. (1991) Control of actin polymerization in live and permeabilized fibroblasts. *J. Cell Biol.*, **114**, 503–13.

95. Okabe, S. and Hirokawa, N. (1991) Actin dynamics in growth cones. *J. Neurosci.*, **11**, 1918–29.

96. Weiner, O. D., Servant, G., Welch, M. D., Mitchision, T. J., Sedat, J. W., and Bourne, H. R. (1999) Spatial control of actin polymerization during neutrophil chemotaxis. *Nature Cell Biol.*, **1**, 75–81.

97. Verkhovsky, A. B., Svitkina, T.M., and Borisy, G. G. (1999) Self-polarization and directional motility of cytoplasm. *Curr. Biol.*, **9**, 11–20.

98. Chen, M. Y., Insall, R. H., and Devreotes, P. N. (1996) Signaling through chemoattractant receptors in *Dictyostelium*. *Trends Genet.*, **12**, 52–7.

99. Offermanns, S., Mancino, V., Revel, J.P., and Simon, M. I. (1997) Vascular system defects and impaired cell chemokinesis as a result of Gα13 deficiency. *Science*, **275**, 533–6.

100. Offermanns, S., Laugwitz, K. L., Spicher, K., and Schultz, G. (1994) G proteins of the G12 family are activated via thromboxane A2 and thrombin receptors in human platelets. *Proc. Natl Acad. Sci., USA*, **91**, 504–8.

101. Nobes, C. D. and Hall, A. (1995) Rho, rac, and cdc42 GTPases regulate the assembly of multimolecular focal complexes associated with actin stress fibers, lamellipodia, and filopodia. *Cell*, **81**, 53–62.

102. Benard, V., Bohl, B. P., and Bokoch, G. M. (1999) Characterization of Rac and Cdc42 activation in chemoattractant-simulated human neutrophils using a novel assay for active GTPases. *J. Biol. Chem.*, **274**, 13198–204.

103. Ren, X. D., Kiosses, W. B., and Schwartz, M. A. (1999) Regulation of the small GTP-binding protein Rho by cell adhesion and the cytoskeleton. *EMBO J.*, **18**, 578–85.

104. Laurent, V., Loisel, T. P., Harbeck, B., Wehman, A., Grobe, L., Jockusch, B. M., Wehland, J., Gertler, F. B., and Carlier M. F. (1999) Role of proteins of the Ena/VASP family in actin-based motility of *Listeria monocytogenes*. *J. Cell Biol.*, **144**, 1245–58.

105. Servant, G., Weiner, O. D., Herzmark, P., Bhalla, T., Sedat, J. W., and Bourne, H. R. (2000) Neutrophil chemotaxis: Rho GTPases are required for polarized signals. *Science*, in press.

106. Meinhardt, H. (1999) Orientation of chemotactic cells and growth cones: models and mechanisms. *J. Cell Sci.*, **112**, 2867–2874.

107. van Es S, and Devreotes, P. N. (1999) Molecular basis of localized responses during chemotaxis in amoebae and leukocytes. *Cell Mol. Life Sci.*, **55**, 1341–51.

# 8 | Genetic analysis of intrinsically asymmetrical cell division

FABIO PIANO and KENNETH KEMPHUES

## 1. Introduction

Intrinsically asymmetrical cell division is a major mechanism for generating cell diversity. In this process, referred to simply as asymmetrical division in this chapter, cytoplasmic and cortical asymmetries within a mother cell combine with an oriented cell division to produce daughter cells with different cytoplasmic components and consequently different cell fates (1). As illustrated in Fig. 1, there are three processes that contribute to asymmetrical cell divisions:

(1) establishment of polarity;

(2) localization of determinants;

(3) oriented cell division.

Over the past decade, genetic analyses in a number of organisms have been successful in beginning to unravel the molecular mechanisms that underlie each of these processes. In this chapter we will review progress in three of these systems: mating-type switching in *Saccharomyces cerevisiae*, specification of cell fate in the nervous system of *Drosophila melanogaster*, and early embryogenesis in *Caenorhabditis elegans*.

Our discussion will begin by introducing the experimental systems, and subsequently will focus on what is known about molecular mechanisms in each. A question we address is to what extent different organisms, and even different asymmetrical divisions within an organism, rely on conserved mechanisms. Overall, with the exception of a heavy reliance on the actin cytoskeleton, the trend is more toward diversity than toward conservation.

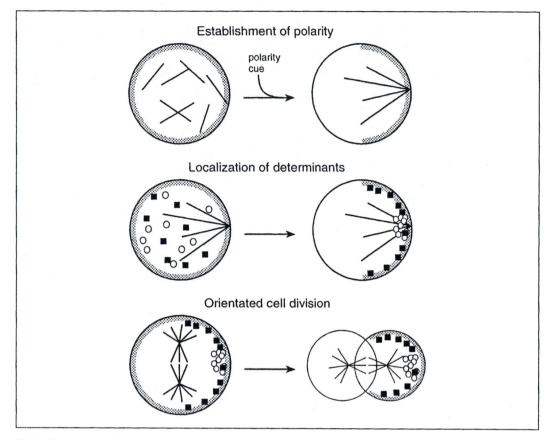

**Fig. 1** Components of asymmetrical cell divisions. Top: intrinsic or extrinsic cues induce cell polarity in the form of organized cytoplasmic filament system (lines) or restriction of cortical components (shading). Middle: cell fate determining molecules (circles and squares) localize to one pole. Bottom: cell division occurs across the axis of polarity, producing daughters with different amounts of the determinants. In animal cells, the site of cytokinesis is determined by the placement of the mitotic apparatus.

## 2. Genetic models for the study of asymmetrical inheritance of cell-fate determinants

### 2.1 Mating-type switching in *S. cerevisiae*

Yeast is the simplest and best understood of the three systems. The daughters of cell division in haploid yeast are known to differ in only one character, the ability to switch between two alternative mating types **a** and α. The larger daughter can switch mating type at high frequency while the smaller daughter cannot (2, 3) (Fig. 2A). (Yeast divides by budding; the larger daughter that produced the bud is called the 'mother cell' and the cell produced from the bud is called the 'daughter cell'.) This pattern of switching most likely evolved to facilitate rapid formation of diploid zygotes among haploid progeny of single spores.

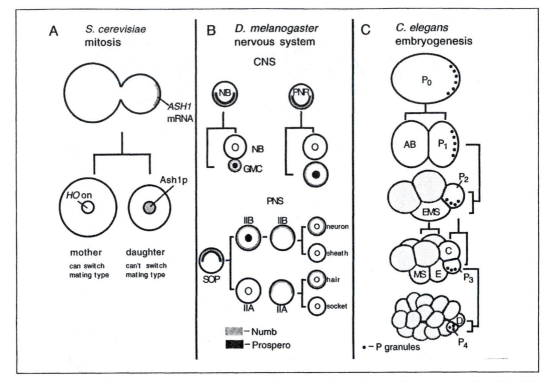

**Fig. 2** Asymmetrical divisions in yeast (A), flies (B), and worms (C). Localized molecules are indicated. In (C), asymmetrically dividing cells and their daughters are indicated; other blastomeres are shaded. Lineage relationships are indicated.

The switch between mating types **a** and α is accomplished through gene conversion that is initiated by mother-cell-specific expression of the *HO* endonuclease (3, 4). When *HO* is ectopically expressed in daughter cells it allows them to switch mating type (5). Thus a key step in switching mating type hinges on the differential regulation of *HO* in mother versus daughter cells.

Genetic studies have identified Ash1p as a protein that fulfils all the criteria for being a cell-fate determinant that is asymmetrically distributed during cell division (6, 7). Ash1p is required in the daughter cell to specifically turn off *HO* transcription. If mislocalized to the mother cell, it will inhibit mating-type switching by blocking expression of *HO*. Conversely, if Ash1p is removed from the daughter cell, it will enable it to switch mating type as if it were a mother cell. We will review what is known about the mechanism by which Ash1p is inherited asymmetrically during division.

## 2.2  Neuronal fate determination in *D. melanogaster*

Studies of asymmetrical cell divisions in *D. melanogaster* focus on multiple cell lineages in the nervous system. Two fate-specification processes have been studied in detail: the specification of neuronal precursors in the central nervous system (CNS), and specification of cells that produce external sense organs (ES) in the peripheral nervous system (PNS) (Fig. 2B; and see refs 8–11 for recent reviews). Cells of the CNS are derived from the neuroectoderm while cells of the PNS are derived from precursor cells scattered throughout the ectoderm in embryos and the epithelium in imaginal discs. Asymmetrical divisions in the CNS occur in the epithelium that gives rise to the larval brain, the procephalic neurogenic region (PNR), and in neuroblasts (NB). Neuroblasts delaminate from the neuroectoderm and divide along their apical–basal axes. Most neuroblasts act like stem cells, producing a small daughter, called a ganglion mother cell (GMC), and a large daughter which remains a neuroblast and divides several more times to produce additional GMCs. The GMC also divides asymmetrically to produce either neuronal or glial daughters with distinct fates. In the PNR, the neurogenic cells also divide along their apical–basal axes, but they do not delaminate. The fates of the daughters are not known. In the PNS, focus has been on the simplest ES, consisting of four cells that are derived from a single precursor called the SOP (sensory organ precursor). Like NBs, SOPs delaminate from the epithelium before dividing. But unlike the NBs, SOPs divide parallel to the plane of the epithelium, to produce daughters IIA and IIB which each undergo another asymmetrical cell division. IIA divides to produce a socket and a hair cell and IIB divides to produce a sheath cell and a neuron.

Genetic analysis identified the transcription factor Prospero as an asymmetrically localized protein that appears to act by directly influencing transcription patterns in GMC daughters, and in a parallel genetic analysis, Numb, a membrane-associated protein with a PTB domain, has been shown to be a localized determinant that acts by inhibiting the response to extracellular signals in one of the SOP daughters. Surprisingly, both proteins are asymmetrically localized in most of the cells of both PNS and CNS, but Numb has no apparent role in most of the NBs and Prospero has only a minor, and perhaps indirect role, in most of the PNS. We will review the mechanism by which these proteins act and the means by which they are localized.

## 2.3 Blastomere fate determination in *C. elegans* early embryogenesis

Asymmetrical divisions play an essential role in establishing the fates of the six founder cells of the early *C. elegans* embryo (Fig. 2C; for recent reviews, see refs 12, 13). The founders each initially produce a clone of cells with specific cell cycle rates and developmental fates. For example, the AB clone cycles fastest and gives rise primarily to neurons and skin, and the $P_4$ clone divides slowest and gives rise only to germ cells. However, in contrast to asymmetrical divisions in yeast and in fly neuronal precursors, in the early asymmetrical divisions of *C. elegans* it has not been possible to identify single proteins, such as Ash1p or Numb, whose presence or absence determines the fate of the daughter cells. Although such molecules may be identified in the future, current results indicate that the fates of at least four founders are specified by combinatorial interactions among a set of maternally encoded proteins that are translated or maintained in particular cells as a consequence of the five asymmetrical cell divisions (12, 14).

Although no single localized cell-fate determinant has been identified, asymmetrical divisions have been easy to follow using P granules as markers. P granules are ribonucleoprotein particles that segregate with the germ line and play a role in its development. They are uniformly distributed in the cytoplasm of the newly fertilized egg but become localized to one pole of the zygote late in the first cell cycle and are partitioned into $P_1$ by an oriented cleavage (15–17). The same processes of localization and partitioning of the granules occur in the divisions of $P_1$, $P_2$, and $P_3$, resulting in the exclusive segregation of P granules into $P_4$, the founder of the germ line. The role of P granules is not fully understood, but one role consistent with their localization is in development of the germ line. Loss of PGL-1, a protein component of P granules, does not eliminate the granules but results in a temperature-sensitive germ-line proliferation defect (18). Mislocalization of P granules correlates with loss of founder-cell identities (as described below), indicating that if they are not themselves determinants of early blastomere identity, they are good markers for the process that localizes such determinants. We will review what is known about the mechanisms that localize these granules and other factors during the five asymmetrical divisions, considering similarities and differences in mechanisms.

# 3. Asymmetrical division in *S. cerevisiae*

## 3.1 Identification of Ash1p and proteins required for its localization

As mentioned above, the question of how mother cells and not daughter cells in *S. cerevisiae* are capable of mating-type switching can be reduced to how the *HO* gene is differentially regulated in the two cells. By focusing on *HO* expression, several proteins have been discovered that are required for the asymmetrical division. One of these, Ash1p, functions as a localized determinant; the others act to localize Ash1p.

### 3.1.1 Identification of factors regulating mother-cell-specific *HO* expression

Two *cis*-regulatory elements, URS1 (upstream regulatory sequence 1) and URS2, are required for the correct expression of *HO* (19, 20). URS1 is required to restrict expression of *HO* to the mother cell and URS2 is required to restrict *HO* expression to the late $G_1$ phase of the cell cycle (21–23). Initial screens to identify regulators of *HO* relied on a phenotype of failure to express *HO*. These screens identified 10 genes required for *HO* expression (the switch genes, *SWI1–SWI10*); however, all but one of them are general transcriptional regulators that, although required to express *HO*, do not influence mother-cell-specific expression (24, 25). One, *SWI5*, encodes a protein that binds to URS1 and plays a role in directing mother-cell-specific *HO* transcription (23, 26). Swi5p, however, shows no asymmetry in its expression or turnover, so it cannot be the most immediate determinant of mother- and daughter-cell differences

(23, 27). Another gene required for mother-specific *HO* expression, *SIN3/SDI1* (28, 29), also shows no asymmetry (30). Five additional genes with roles in mother-cell-specific regulation of *HO* were identified in a screen for mutants specifically defective in URS1 function (31). These genes, called *SHE* genes for Swi5p-dependent *HO* expression, will be discussed below in more detail. Finally, three different screens, two designed to find mutations that allow daughter cells to switch, and one designed to identify suppressors of *she* mutation-induced failure to express *HO*, converged on a single gene, *ASH1* (asymmetric synthesis of *HO*) (6, 7).

### 3.1.2  Ash1p is a determinant that is localized to daughter cells

Ash1p is a 558-amino-acid protein that contains a GATA-1-like putative zinc-finger motif (6, 7). Evidence that Ash1p is a localized determinant that distinguishes mother from daughter is compelling. Ash1p accumulates to a much greater extent in the nuclei of daughter cells than mother cells and disruption of *ASH1* can cause daughters and mothers to switch with frequencies greater than 90%. In addition, over-expression of Ash1p in a wild-type background results in significant reduction of switching in mothers. Finally, Ash1p accumulates to an equal extent in mother cells and daughter cells of *she* mutants (which are unable to switch in either cell). Thus, Ash1p acts as a repressor of *HO* transcription. It appears to do this by blocking Swi5p-mediated recruitment of the Swi/Snf chromatin-remodelling complex (32).

## 3.2  Polarity establishment

Establishment of polarity in *S. cerevisiae* is very well understood and is reviewed in Chapter 2. The essential elements are that *S. cerevisiae* buds from a single site on the surface; that in haploid cells this site is chosen to be adjacent to the site of the previous bud site; and that the actin cytoskeleton is organized in a polarized fashion at this site, ensuring the polarized recruitment and transport of factors required for growth of the bud. Choice of bud site and organization of the cytoskeleton at the bud site both require G-protein signalling. Two types of microfilament structures can be seen in fixed and in living cells: punctate aggregates that accumulate at the bud tip during growth, and filaments that traverse the mother/bud neck and are oriented toward the bud.

## 3.3  Mechanism of localization of Ash1p

How does Ash1p become enriched in daughter cells? The answer is that the *ASH1* mRNA is preferentially localized to the daughter via myosin-mediated movement along actin cables. *In situ* hybridization revealed that the mRNA localizes in particles at the bud tip at the end of anaphase (the period during which the gene is transcribed in both mother and daughter cells) (33, 34). When transcribed from an inducible promoter, the message is found to be capable of being localized to the bud tip throughout the cell cycle. This localization is dependent upon an intact actin cytoskeleton, as demonstrated by drug inhibition (34) and by mutations in actin and actin-binding proteins essential for formation of the actin cables that run from the mother to the bud (33). Mutational disruption of cytoplasmic microtubules and mutational blocking of bud formation reveals that neither microtubules nor bud formation *per se* are required to achieve mRNA localization (33).

Three distinct portions of the *ASH1* mRNA are capable of directing localization of reporter mRNAs to the bud, and removal or replacement of any one does not block localization (33, 35–37). Two of these localization elements are within the coding region, one in the C-terminal region and a bipartite element in the N-terminal region; the third overlaps the coding region near the stop and extends into the 3'UTR. The elements appear to act redundantly rather than additively since localization efficiency does not depend upon the number of elements used (35). It is not known whether the mechanisms used by each element are the same, but there is no obvious sequence similarity among the three or four localization sequences. Almost all the analysis has focused on the 3' element. Although these elements are sufficient to target RNAs to the bud tip, stable anchoring of the message requires translation of the C terminus of the protein (35).

Live imaging of mRNA localization has been possible by constructing fusions of the *ASH1* 3' region with recognition sites for the RNA-binding protein MS2 and then co-expressing this construct with a translational fusion of MS2 and green fluorescent protein (GFP) (36, 38). Most cells that overexpress these two constructs contain a large, single fluorescent particle that localizes in a manner similar to endogenous *ASH1* mRNA. Because behaviour of the particle is similar to the smaller multiple particles of endogenous *ASH1* mRNA seen by *in situ* hybridization, the behaviour of this particle in living cells has been used to investigate mechanisms by which the 3' element mediates localization.

Localization of both RNA and protein is dependent upon the She proteins and Bud6p. All the *she* mutants except *she5/bni1* cause *ASH1* mRNA to be found in both daughter and mother cells and block mating-type switching in both cells; in *she5/bni1* mutants, the mRNA accumulates at the neck joining mother and bud (33, 34). In all *she* mutants, with the possible exception of *she4* for which data are not published, Ash1p is present at similar levels in both mother and daughter and *HO* is not expressed in either cell. She1p, previously identified as Myo4p, is a putative class V minimyosin microfilament motor protein (31). All the data are consistent with a direct role for She1p in transporting *ASH1* mRNA into the bud along the actin cables. GFP reporter particles in *she1* mutants remain in the mother cell; rates of movement of the particles in a wild-type background are consistent with myosin-mediated transport; and She1p co-localizes with the particle in both mother cells and buds (36, 38). Furthermore, *ASH1* mRNA co-localizes with Myo4p and is present in immunoprecipitates of Myo4p (39). She2p and She3p are required for this association. She3p co-localizes with She1p in fixed cells and immunoprecipitates of She3p also contain *ASH1* mRNA. The role of She4p in localization is not known. She4p has been implicated in proper polarization of the actin cytoskeleton (40) and shares sequence similarity with UNC-45, a *C. elegans* protein required for myosin assembly in the thick filaments of body-wall muscles (41). She5p, previously identified as Bni1p, a member of the formin protein family, and Bud6p, an actin-binding protein, are localized to the bud tip (42, 44) and are required to anchor *ASH1* mRNA there (38).

## 3.4 Oriented division in yeast

In yeast, the site of cytokinesis is always at the bud neck. In contrast to flies and worms discussed below, however, the selection of this site is completely independent of the position of the mitotic spindle; even cells with spindles contained completely within the mother cell will undergo cytokinesis at the bud neck. Proper spindle orientation is, however, important for the delivery of a nucleus to the daughter cell and the mechanism of orientation has parallels with the two other systems. Spindle orientation in yeast is dependent upon microtubule-motor-mediated interactions of cytoplasmic microtubules with the cell cortex. (for reviews see refs 45, 46). Although an extensive review of this field is beyond the scope of this chapter, recent results have identified two proteins that link spindle orientation with Ash1p localization. Kar9p is one of a group of proteins that functions in one of two independent but partially redundant pathways that orient spindles along the axis of division (47). It is localized to the bud tip where it appears to play a role in anchoring microtubules. Recent work has shown that this localization is dependent upon microfilaments and upon Bni1p/She5p and Bud6p, two of the proteins required to localize *ASH1* mRNA (48, 49). Thus Bni1p/She5p and Bud6p provide a means to co-ordinate cytoplasmic localization with spindle orientation.

## 3.5 Summary of *S. cerevisiae* asymmetrical cell division

A model for asymmetrical division in *S. cerevisiae* has emerged (Figs 2A and 3A). The process starts in late $G_1$, when the cytoskeleton is polarized using the internal asymmetries from the previous bud scar and a bud begins to form. In anaphase, *ASH1* mRNA is transcribed from both mother and daughter nuclei. The newly transcribed RNA forms a complex with She3p, She1p/Myo4p and perhaps She2p. She1p/Myo4p then transports the *ASH1* mRNA along actin cables to the bud tip, where it is translated and becomes anchored to the actin cytoskeleton under the influence of She5p/Bni1p and Bud6p. As a consequence of this localization, the daughter cell has a high concentration of Ash1p. Ash1p enters the nucleus where it inhibits the activity of Swi5p, thus preventing the expression of *HO*.

# 4. Asymmetrical divisions in the nervous system of *D. melanogaster*

## 4.1 Numb and Prospero are asymmetrically localized and required for differential development of the daughter cells

Prospero was identified by mutations that change gene expression patterns of CNS neurons and in screens for genes expressed in neuroblasts (50, 51). It was sub-

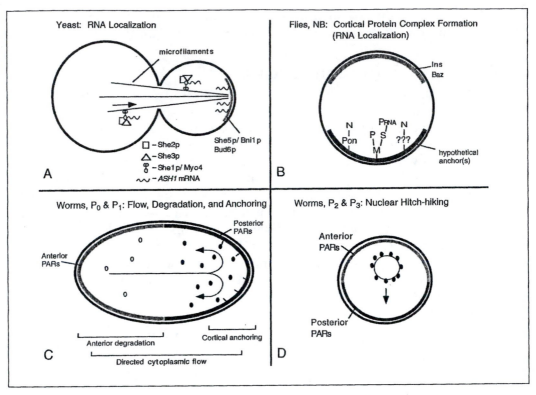

**Fig. 3** Comparison of models of localization mechanisms. (A) Localization of Ash1p by RNA transport in yeast. (B) Localization of Numb and Prospero by formation of cortical protein complexes in fly neuroblasts. (C) Localization of P granules by flow, degradation, and anchoring in the $P_0$ and $P_1$ of worms. (D) Localization of P granules by nuclear hitch-hiking in the $P_2$ and $P_3$ of worms.

sequently found to be asymmetrically localized during the divisions of the NB, embryonic SOP cells (52–54), and cells of the PNR (55). Prospero protein has a divergent homeodomain and is capable of binding DNA and regulating transcription of other genes involved in neuronal identity (50, 51, 56–58). Although Prospero is clearly required for the proper specification of GMC progeny, it does not act as the sole determinant of the differences between NB and GMC (50, 51, 54). Asymmetrical Prospero distribution is essential for its function; mutations that block its localization to the GMC exhibit phenotypes in the CNS identical to weak *prospero* mutations (59). Prospero's role in the embryonic PNS is less clear; *prospero* mutants have no obvious cell-fate transformations in the PNS but neurons that are produced show axon guidance defects. Regardless of its precise role in neuronal development, analysis of Prospero localization has provided information about mechanisms of asymmetrical division.

Numb was identified in screens for P-element-induced lethal mutations with abnormalities in the peripheral nervous system (60) and subsequently shown to be distributed asymmetrically in asymmetrical divisions of PNS, PNR, and CNS (55,

61). In *numb* mutants, most neurons of the PNS are missing and are transformed into support cells. In the simple ES organs of both embryos and adults, both daughters of the SOP behave like IIA. When Numb is expressed ectopically, under heat-shock control, both SOP daughters behave like IIB (61). Numb is also required for the asymmetrical divisions of IIA and IIB (61, 62), fate decisions in other PNS sensory organs (60), asymmetric aldivisions in muscle cell lineages (63, 64), and divisions in the CNS (65–67). Numb mutations, however, have no effect on most NB divisions.

Unlike Ash1p and Prospero, which act in the nucleus, Numb acts in the cytoplasm by inhibiting Notch signalling. *Notch* encodes a large single-pass transmembrane receptor known to act in many tissues and across phyla to determine cell fates (for a recent review on Notch signalling see ref. 68). Reduction of Notch activity causes a fate transformation opposite to loss of Numb: both SOP daughters behave like IIB (69). Expression of constitutively active Notch causes both daughters to behave like IIA (70). Numb binds to Notch *in vitro*, can block Notch signalling when co-expressed with Notch and its ligand, Delta, in tissue culture, and can affect Notch function when ectopically expressed *in vivo* (70, 71). Progress is being made toward understanding the role of Notch signalling in some of these asymmetrical divisions, but will not be discussed here (see ref. 9).

## 4.2  Mechanisms of localization of Numb and Prospero

How are these proteins localized to only one of the daughter cells? The answer is that in many cells of the CNS and PNS both the proteins become localized asymmetrically during mitosis in a crescent at the periphery of the mother cell and are partitioned by cytokinesis into only one daughter (Fig. 2B). In the case of *prospero*, mRNA and protein co-localize whereas only Numb protein is localized. Localization depends upon the actin cytoskeleton and is mediated by a number of proteins, including Miranda, Partner of numb, Staufen, and Inscuteable.

### 4.2.1  Localization of Prospero mRNA and protein in the CNS

Neuroblasts divide in an apical–basal orientation. Prospero mRNA and protein co-localize and change distribution during the cell cycle (52–54, 72, 73). Prospero mRNA and protein initially appear in the cytoplasm of the NB during interphase, where they are localized in an apical cortical crescent. As cells enter mitosis the apical crescent disappears and a basal crescent appears. Evidence from *in vitro* culture of dividing neuroblasts indicates that the apical localization is not a precondition for basal localization (73), so we will focus on the mechanism of basal localization. At cytokinesis, both RNA and protein are partitioned exclusively into the basal daughter, the GMC. RNA localization is dependent upon Staufen, an RNA-binding protein required for asymmetrical localization of *bicoid* and *oskar* RNA in the *D. melanogaster* oocyte (72, 73). Staufen protein co-localizes with *prospero* RNA and protein both apically and basally. In contrast to Ash1p localization, however, localization of *prospero* mRNA is not required for asymmetrical localization of Prospero protein. In *staufen* mutants, *prospero* mRNA fails to localize, but there is no obvious effect on

Prospero distribution or GMC behaviour. Furthermore, in *prospero* mutants Staufen protein localizes normally. Thus, Prospero and Staufen proteins localize independently of each other and localization of *prospero* mRNA is not essential for Prospero localization. Flies carrying a weakly penetrant allele of *prospero*, however, show an enhanced GMC defect in a *staufen* background, indicating that mRNA localization contributes to asymmetrical distribution of Prospero (73).

Prospero protein and mRNA are co-localized in the CNS through direct and indirect interactions with the Miranda protein. Miranda was identified by its ability to bind to a region of Prospero required for its asymmetrical localization and shown to co-localize with Prospero, Staufen, and *prospero* mRNA in NBs (59, 74). After cell division, Staufen protein and *prospero* mRNA are released into the cytoplasm, Prospero protein translocates to the nucleus, and Miranda protein rapidly degrades (59, 74). Whether degradation is a precondition for release of Staufen and Prospero or a consequence of release is not known, but a mutation that replaces the C terminus of Miranda localizes normally but fails to release the protein, indicating that this region is required for release (59). Flies lacking *miranda* function fail to localize Staufen, Prospero protein and mRNA, and produce phenotypes similar to weak *prospero* alleles (59, 74–77). Miranda localizes *prospero* mRNA via a direct interaction with Staufen, which binds to the 3'UTR of *prospero* mRNA (75–77). How Miranda is localized is not clear, but it requires amino acids 1–290 and the Inscuteable protein (discussed below). Miranda is expressed in embryonic SOPs (not adult SOPs), but its role there has yet to be investigated (74, 76).

### 4.2.2 Localization of Numb

Although they are co-localized in many cells, Numb localization has different requirements from Prospero. Miranda binds Numb *in vitro*, but Numb localizes normally *in vivo* in the absence of zygotic Miranda protein (74). Localization of Numb is mediated by sequences in the Numb amino terminus that include a putative *N*-myristoylation signal and a phosphotyrosine-binding domain (PTB) (78). Partner of numb (Pon), a protein that binds to the PTB domain, has a role in Numb localization that is analogous to that of Miranda in Prospero localization. Pon was identified by two-hybrid screening using the PTB domain as bait (79). Pon co-localizes to the crescent with Numb in SOPs, NBs, and in muscle progenitors, and binds to the Numb PTB domain *in vitro* through sequences in its amino terminus. Loss of zygotic Pon affects Numb localization but does not block it. In *pon* mutants, Numb is unlocalized 50% of the time in muscle progenitors, and in the nervous system the timing of Numb localization is delayed (79). Pon is capable of mediating the localization of Numb *in vivo*, however. Numb is present but is not localized in the symmetrically dividing cells of the embryonic epithelium. When Pon is expressed in these cells, however, both it and Numb become localized in a basal crescent (79). It is possible that the mild defect in *pon* mutants is the result of residual maternal product or that another partially redundant mechanism localizes Numb. Consistent with this second possibility, deletion of the PTB domain does not block Numb localization in the CNS (71). Miranda binding does not seem to be the hypothetical second mechanism, because the *miranda;pon* double mutant is no more severe than *pon* alone (cited in ref. 79). Further confusing the issue, however, are the observations that Miranda is present in an immunoprecipitate with Pon and Numb and is reported to bind to Pon in the yeast two-hybrid system (79).

## 4.3 Establishment of cell polarity

Little is known about how polarity is established during these asymmetrical divisions, but different cues appear to operate in NBs, SOPs, and SOP daughters. A major difference between CNS and PNS is that while neuroblasts divide along the apical–basal axis, SOPs and their daughters divide parallel to the plane of the epithelium. SOPs orient along the anterior–posterior axis; IIA and IIB divide in the same plane but IIA divides parallel to the SOP and IIB divides orthogonally. Additional evidence that polarity cues differ comes from the observation that mutations in Inscuteable, the most upstream gene known in organizing cell polarity (see below), affect Numb and Prospero localization in CNS and PNR, but do not obviously affect either in the PNS (55). Evidence that different cues are used even within the SOP lineage comes from a recent study in the adult PNS, indicating that polarity of the SOP is mediated by signalling through the Frizzled receptor, but that polarity in the SOP daughters is not (80).

## 4.4 Co-ordinating cell polarity and spindle orientation: the role of Inscuteable

Inscuteable protein is localized in an apical crescent in NBs, PNR, and the PNS, and is required for asymmetrical localization of all basal crescent components and for proper orientation of the mitotic spindle in PNR and NBs but not SOPs. Inscuteable is a protein of 859 amino acids with a broad expression pattern that includes expression in the PNS, NBs, and PNR (81). In NBs, Inscuteable protein and mRNA appear on the apical side during interphase, and co-localize with Prospero, Staufen, Miranda, and *prospero* mRNAs during prophase (55). Although these other crescent components become localized basally during mitosis, Inscuteable remains apical until anaphase, when it delocalizes and is equally distributed to the daughters. As with *prospero* mRNA, localization of *inscuteable* mRNA is not required to localize Inscuteable protein (82, 83).

The consequences of loss of Inscuteable differ in different parts of the nervous system. Inscuteable appears to be the most upstream gene known to regulate cell polarity in the CNS. *inscuteable* mutants affect distributions of all the other localized components (see already-cited references to Miranda, Pon, Staufen, Prospero) in both PNR and NBs. In the PNR, loss of Inscuteable blocks the formation of basal crescents but does not affect membrane association of Numb and Prospero; in NBs *inscuteable* mutants variably block formation of crescents or mislocalize Prospero and Numb, although mislocalized Numb and Prospero remain together. In the PNS, although Inscuteable is present in apical crescents, there is no detectable defect in *inscuteable* mutants (55).

Inscuteable also influences the orientation of the mitotic apparatus in the CNS and PNR, but not in the PNS. The strongest effect is in the PNR; absence of Inscuteable results in spindle orientations parallel to the epithelium, identical to the divisions in the surrounding ectoderm. A pronounced but weaker effect is seen in the NB divisions that parallels the effect on localization of Numb and Prospero: spindle orientation is somewhat randomized (55). A possible explanation for the difference between PNR and NBs might be different mechanisms of spindle orientation in the two cell types (55). Cells in the PNR remain in the neuroectodermal epithelium and initially orient their centrosomes parallel to the plane of the epithelium during prophase, but then rotate the spindles by 90° early in mitosis. The rotation fails in *inscuteable* mutants. In NBs, one centrosome migrates to the basal pole just after delamination. This movement takes place in the absence of Inscuteable. Thus, in the PNR Inscuteable establishes apical–basal spindle orientation whereas in the CNS Inscuteable maintains or reinforces it. In both cases, however, Inscuteable appears to co-ordinate cytoplasmic localization with spindle orientation.

The fascinating question of how an apically localized protein can mediate the basal localization of other cellular components might be addressed by finding proteins that interact with Inscuteable. Toward this end, functional domains of the protein are being identified. Thus far, a central region required for both asymmetrical Inscuteable distribution and spindle orientation has been identified and that region contains a separable domain for cortical localization. Separate domains for Numb localization and for Miranda/Prospero localization have also been identified (82, 83).

Additional information about regulation of spindle orientation is likely to come from analysis of Bazooka, a protein with sequence similarity to the PAR-3 protein of *C. elegans* (see next section). Bazooka is required for maintaining the integrity of the polarized blastoderm epithelium (84), and is localized apically not only in the epithelium but also in dividing NBs (85). Bazooka mutations cause a low level of misorientation of the mitotic spindle in both epithelium and in NBs (85). The mutations do not uncouple basal crescent formation from spindle orientation, however, suggesting the role of Bazooka is limited to orienting the polar axis of the cell relative to the overlying epithelium. The weak phenotype could be due either to residual maternal product in these zygotic mutants or could reflect partial redundancy of the protein. The co-localization of Bazooka with Inscuteable, and their common affect on spindle orientation, suggest a possible functional relationship between the two proteins.

## 4.5 The role of the cytoskeleton

Although Numb and Prospero are mildly affected by treatments with the microfilament inhibitor cytochalasin (53, 86), treatment *in vivo* with the more potent inhibitor, latrunculin, causes delocalization of Numb and Prospero, but does not block their cortical association (78). Treatments with latrunculin *in vitro*, however, block both cortical association and localization of Prospero (86). Presumably the difference between the *in vivo* and *in vitro* results reflects the better accessibility to the drug in isolated cells; indeed, residual cortical microfilaments were detected after treatments *in vivo* (78). Treatment with colcemid, a microtubule inhibitor, has no effect on protein distribution (53, 54, 86). The effects of latrunculin on Prospero and Numb distributions could reflect indirect effects of the microfilament requirements for localization of Inscuteable (55, 86) or Miranda (75). It is not known whether myosins might be required for asymmetry.

## 4.6 Summary of asymmetrical cell division in *D. melanogaster*

The picture emerging from studies of asymmetrical divisions in the *D. melanogaster* nervous system (Fig. 3B) is more complex and diverges considerably from that of yeast. Two different fate determinants, Prospero and Numb, are co-localized in cortical crescents in many cells of the CNS and PNS. Numb appears to function as a fate determinant in a subset of these cells, while Prospero functions in the others. Their mechanisms of action differ. Prospero, like Ash1p, appears to function as a transcriptional regulator. Numb, however, functions in the cytoplasm by inhibiting an extracellular signalling pathway. Establishment of polarity and localization of the proteins differs among the different cell types as well. Although co-localized, Numb and Prospero do not depend upon each other for localization but rather become localized through 'adaptor proteins', Miranda and Pon, and probably other un-identified proteins. Prospero, but not Numb, requires Miranda protein; and Numb, but not Prospero, requires Pon. The protein Inscuteable plays an important role in the CNS where it appears to be the most upstream, known regulator of asymmetrical cell division. It functions both in directing the formation of the cortical crescent and in orienting the mitotic apparatus, and perhaps acts to co-ordinate the two events. Here again, however, different cell types have different requirements for the protein. In the PNR Inscuteable is absolutely essential, but in NBs it is partially redundant with another uncharacterized polarizing mechanism. Yet another mechanism is likely to function in the PNS to localize Numb and Prospero and to orient spindles, since *inscuteable* mutations have little effect there.

# 5. Asymmetrical divisions in *C. elegans*

As described above and in Fig. 2C, five asymmetrical divisions play an important role in establishing the organization of the early *C. elegans* embryo (87). Other asym-metrical divisions take place during *C. elegans* development, but will not be considered here (see ref. 10).

## 5.1 Determining fates of the founder cells

Three cell-fate determinants have been identified, SKN-1, PIE-1, and PAL-1 (re-viewed in refs 12, 14). SKN-1, PIE-1, and PAL-1 are transcription factors that act

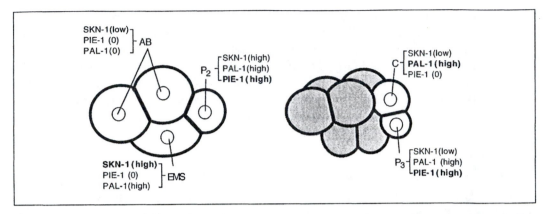

**Fig. 4** Combinatorial model for specification of founder blastomeres in *C. elegans*. Relative nuclear levels of fate determinants SKN-1, PAL-1, and PIE-1 are indicated. (0) indicates no detectable protein. The dominant protein acting in each blastomere is indicated in bold.

combinatorially to determine the fates of the four somatic founder cells arising from $P_1$ (E, MS, C, and D) (88–90). Based on analysis of mutants and protein distributions, a model for how these proteins specify fate has been developed (Fig. 4). The three known transcriptional regulators, SKN-1, PAL-1, and PIE-1 have distributions that overlap both temporally and spatially. The model proposes a dominance relationship between the proteins, such that relative levels at critical times dictate the fate of the cells. SKN-1 promotes the EMS fate and antagonizes PAL-1, PAL-1 promotes the C fate, and PIE-1 antagonizes both. PIE-1 is also required for development of the germ line because in its absence, germ-line cells are missing (89). Although PIE-1 is required, it is not likely to be the sole determinant of germ-line identity. It has been shown to block the transcription of many embryonically expressed genes, raising the possibility that it is a general repressor of transcription and that this repression protects the germ line from the action of early specification genes, such as SKN-1 and PAL-1 (91, 92).

## 5.2 The role of asymmetrical division in distributing cell-fate regulators

SKN-1 and PAL-1 distributions are indirectly dependent upon asymmetrical cell division, but PIE-1 is actively partitioned during the asymmetrical divisions. The mechanisms for restricting distributions of all three are post-transcriptional but, unlike Ash1p and Prospero, do not involve RNA localization; the RNAs for all three are maternally provided and distributed evenly throughout the early embryo. PAL-1 localization depends upon translational control. The MEX-3 protein is required to repress PAL-1 translation in the AB lineage and in the $P_1$ blastomere (90). MEX-3 is an RNA-binding protein present throughout the one- and two-cell embryo but preferentially lost from the $P_1$ daughters at the four-cell stage (93). The mechanism of SKN-

1 localization is unknown, but because little SKN-1 is detectable in $P_0$, its localization to $P_1$ appears to depend on differential regulation of translation or stability in the daughters (94). Thus, cell-specific distribution of SKN-1 and PAL-1 must be regulated indirectly by factors that are localized during the asymmetrical divisions. PIE-1 distribution is more directly dependent upon asymmetrical division and is discussed below.

## 5.3 Establishing polarity

Each of the asymmetrically dividing blastomeres establishes cell polarity anew prior to cytokinesis, but at least three rely on different polarity cues. In $P_0$, polarity arises after fertilization; because sperm position always marks the posterior pole irrespective of egg polarity, it appears that the sperm provides the polarity cue, perhaps via centrosomal activity (95; and see below). In $P_1$, the polarity cue appears to be intrinsic; isolated $P_1$ blastomeres undergo unequal divisions typical of the $P_1$ lineage (96) and produce their normal complement of cell types (97). Presumably some remnant of polarity from the first division persists in the cell but the precise cue or cues are not well understood. The cue for unequal division appears to reside in the posterior (98), but an anterior cue functions in spindle orientation (see below). Although few experiments have focused on $P_2$ and $P_3$ divisions, what is known is consistent with intrinsic polarity cues as well. However, $P_2$ and $P_3$ reverse polarity relative to $P_1$ (96), indicating that the cues are different or are interpreted differently. In EMS, polarity is determined by an interaction between $P_2$ and EMS, which both polarizes the cell and orients the spindle. This interaction involves the Wnt signalling pathway (reviewed in ref. 12).

## 5.4 Localization of P granules and PIE-1

Localization of P granules in $P_0$ occurs during a period of cytoplasmic reorganization in the last third of the cell cycle (16, 99). During this time, internal cytoplasm flows from the anterior to the posterior and peripheral cytoplasm flows from the posterior to the anterior (100). The centrosomes of the sperm are believed to be initiating this flow and it has been proposed that the flow itself establishes the anterior–posterior axis (95). P granules move with this flow, but it appears that posterior localization of the granules results from a combination of directed cytoplasmic flow, differential stability of the granules in posterior versus anterior, and anchoring of granules in the posterior (101).

The localization of P granules is sensitive to cytochalasin treatment and to depletion of a cytoplasmic conventional myosin, suggesting that microfilament-based motility is required (16) (Guo and Kemphues, unpublished). Although it is possible that P granules, like *ASH1* mRNA, are translocated along actin cables, it is likely that sensitivity to cytochalasin reflects a general requirement for microfilaments in polarizing the cell. Pulses of cytochalasin applied during the cytoplasmic reorganization not only block P-granule localization but also cause a variety of polarity defects (102).

Furthermore, oriented cytoplasmic filaments have not been detected with standard methods (103).

Localization of P granules in $P_1$ occurs by a mechanism similar, if not identical, to that used in $P_0$ (101), but in $P_2$ and $P_3$ the mechanism of localization is somewhat different. Although localized degradation or disassembly still plays a role, in these blastomeres P granules associate with the nucleus and move with it as it migrates to the ventral pole, where they are deposited prior to cytokinesis (101). Polarized disappearance of granules and orientation of the mitotic apparatus are sensitive to cytochalasin treatment (101). The existence of a distinct mechanism for localizing P granules in $P_2$ and $P_3$ correlates with the reversal of polarity in these cells; the P granules in $P_2$ localize to the pole opposite that to which they localized in $P_0$ and $P_1$ (96).

PIE-1 localization parallels that of P granules but the distribution is dynamic and more complex (104, 105). Like P granules, PIE-1 is present throughout the cytoplasm in the early $P_0$ and becomes localized to the posterior late in the cell cycle. Some of the PIE-1 protein is associated with P granules but much of the protein remains diffuse in the posterior cytoplasm. A similar polar localization occurs in $P_1$, $P_2$, and $P_3$ but a significant amount of the protein becomes nuclear. In addition, as cells enter mitosis, some PIE-1 appears on centrosomes but, as mitosis progresses, is preferentially lost from the centrosome that will segregate to the somatic daughter. Thus, changes in protein distribution suggest three possible ways that PIE-1 protein is polarized prior to division: co-localization with P granules, differential translation or degradation at one pole, and hitch-hiking on centrosomes. Recent experiments indicate that differential translation is an unlikely means of localization. *Cis*-acting regulatory sequences are neither sufficient nor necessary to localize a PIE-1:GFP fusion; instead, action directly on the PIE-1 protein is suggested by the finding of two protein sequences that mediate localization (K. J. Reese, M. A. Dunn, and G. Seydoux, unpublished results).

## 5.5 Proteins required for localizing P granules

Several proteins required for proper localization of P granules have been identified, but in most cases their mechanism of action is not known. The PAR proteins and PKC-3 have a role in establishing polarity of $P_0$ and may also function in $P_1$, $P_2$, and $P_3$, but do not appear to have a role in EMS. MEX-1 has a weak Par-like phenotype, and the POS-1 and MES-1 proteins appear specific for the asymmetrical divisions of $P_2$ and $P_3$.

### 5.5.1 PAR proteins and PKC-3

The PAR proteins and PKC-3 appear to affect P-granule localization through their role in establishing polarity in the one-cell embryo. Mutations in these genes perturb several aspects of early embryonic polarity, including equal first division, synchronous second division, altered division orientations, and defects in distribution of P granules and other localized molecules, including SKN-1, PAL-1, and PIE-1(105–111; see ref. 13 for a recent review).

In keeping with their proposed role in establishing polarity, PKC-3 and four of the PAR proteins are distributed in a polarized manner at the periphery of $P_0$, $P_1$, $P_2$, and $P_3$ (107, 112–115) (Figs 3C, 3D). Posterior-group proteins (PAR-1, PAR-2) are restricted to the same pole as the P granules (the posterior pole in $P_0$ and $P_1$ and the ventral pole in $P_2$ and $P_3$), and anterior-group proteins (PAR-3, PAR-6, PKC-3) have a reciprocal distribution. PAR-4 and PAR-5 are apolar. The persistence of polarized distributions of the PAR proteins in all asymmetrical divisions of the germ-line lineage suggests that they play a role in each division, but it has not yet been possible to determine whether defects seen beyond the one-cell stage are primary defects or secondary consequences of the abnormal $P_0$ division. The anterior-group proteins are also found at the periphery of all somatic cells in the early embryo. They appear to play an important role in regulating spindle orientation in the AB and $P_1$ (see below), but no role is known in other blastomeres.

The mechanism by which the PAR proteins contribute to polarity establishment is not well understood. The presence in the proteins of conserved motifs consistent with intracellular signalling leads to the hypothesis that the proteins function by localized signalling through kinase cascades. PKC-3, PAR-1, and PAR-4 are serine/threonine kinases (107, 113) (J. Watts, D. Morton, J. Bestman, and K. Kemphues, unpublished results) and PAR-5 is a member of the 14–3–3 protein family (Shakes and Kemphues, unpublished results). The complexity of the effects of mutations in different *par* genes on localized determinants (111) indicates that the *par* genes do not define a single pathway, but that there may be multiple overlapping signals. The targets of the signals are unknown but would be expected to include proteins that regulate the cytoskeleton, translation, cell cycle, and perhaps protein stability.

The three anterior-group genes may be components of a signalling complex. They appear to function together because they give identical mutant phenotypes and are co-dependent for their localization to the cell periphery (106–108, 114). The presence of PDZ domains in PAR-3 and PAR-6 (114–116) is consistent with a role in organizing a signalling complex.

The striking polar distributions of the PAR proteins depends upon microfilaments (B. Etemad-Moghadam and K. Kemphues, unpublished), a cytoplasmic myosin (117), and interactions among the PAR proteins themselves. PAR-2, PAR-4, and PAR-5 are required to exclude the anterior-group genes from the posterior, and the anterior-group genes restrict PAR-2 and PAR-1 to the posterior (112, 114, 115).

## 5.5.2 MEX-1

MEX-1 mutants express phenotypes similar to weak *par-1* alleles, including similar cell-fate transformations and mislocalization of SKN-1, PIE-1, and P granules (89, 94, 105, 118). MEX-1 encodes a protein with two copies of an unusual zinc-finger domain also found in PIE-1 (119). Like PIE-1, MEX-1 is localized to the posterior of $P_0$, segregated to the germ-line lineage, and a fraction of the protein co-localizes with P granules (119). Unlike PIE-1, however, MEX-1 does not associate with centrosomes and is not nuclear. MEX-1 is required to localize PIE-1 to the germ-line lineage; in *mex-1* mutants some PIE-1 protein is present in somatic cells (119). Although the mode of action of MEX-1 remains to be determined, a unifying hypothesis for all of the phenotypes is that MEX-1 functions to anchor regulatory proteins such as PIE-1 and unknown translational regulators to the P granules (119).

### 5.5.3    MES-1 and POS-1

MES-1 and POS-1 affect the $P_2$ and $P_3$ asymmetrical divisions. Loss of *mes-1* function leads to symmetrical divisions of $P_3$, resulting in a transformation of the germ-line cell $P_4$ into muscle, thus maternal effect sterility (120). $P_3$ and, to a lesser extent, $P_2$ blastomeres in *mes-1* embryos exhibit defects in spindle orientation and localization of P granules and PIE-1 (105, 120). $P_0$ and $P_1$, however, are not affected, consistent with other observations that polarity establishment in $P_2$ and $P_3$ is different from that in $P_0$ and $P_1$. MES-1 protein has recently been shown to be restricted to $P_2$ and $P_3$ (L. Berkowitz and S. Strome, unpublished results). Loss of *pos-1* function causes similar phenotypes in $P_2$ and $P_3$, with a transformation of germ line into muscle (121). *pos-1* mutants, however, show additional defects in $P_2$ and $P_3$ not seen in *mes-1* mutants, including an acceleration of division time and a loss of the ability to produce the $P_2$ marker APX-1, suggesting that POS-1 plays a more general role in $P_2$ than does MES-1 (121). POS-1 contains two copies of the unusual zinc finger found in MEX-1 and PIE-1 and, like those two proteins, is localized to the germ-line lineage and partially co-localizes with P granules (121). Unlike PIE-1 and MEX-1, however, little POS-1 is detectable in $P_0$, and POS-1 is not asymmetrically distributed prior to the asymmetrical division. Instead, it is degraded preferentially in the somatic blastomere.

## 5.6    Orientation of the mitotic spindle

Studies of spindle orientation in asymmetrical divisions have focused primarily on the one- and two-cell stages. In $P_0$, the centrosomes, which arise by duplication from the sperm centrosome, orient across the long axis of the fertilized egg. They nucleate long microtubules that mediate pronuclear meeting in the posterior. After meeting, the centrosome/nuclear complex migrates to the centre of the embryo and rotates to align the centrosomes along the long axis of the egg. A slight migration to the posterior coupled with an asymmetrical anaphase results in the unequal first division (106, 122). At the two-cell stage, centrosomes in both cells duplicate and migrate around the nucleus to align orthogonal to the axis of the first division. The AB spindle forms along this axis, but during prophase in the $P_1$ cell, the centrosome/nuclear complex rotates through 90° to align with the axis of polarity in the $P_1$ cell, so that the spindle is positioned appropriately for an asymmetrical division (123).

As with yeast, spindle orientation in *C. elegans* appears to involve microtubule motor-mediated interactions of astral microtubules with the cortex. The centrosome movements in both $P_0$ and $P_1$ have been shown to be dependent upon long microtubules (122, 123) and upon components of the dynactin complex, implicating dynein-based motility (124). The rotation in $P_1$, but not $P_0$ is dependent upon microfilaments (123). The mechanism of orientation in $P_1$ has been studied extensively and involves an interaction between astral microtubules and a region of the anterior cortex that includes the cell-division remnant (125, 126). Antibodies to the dynactin-complex protein p150 (Glued), actin, and actin-capping protein localize to the cell division remnant at the time of rotation in $P_1$ and anti-p150 signal is seen at the site of the polar body extrusion in $P_0$ (124, 127).

The PAR proteins play an important role in spindle orientation at the two-cell stage. Reduction of function of any of the three anterior-group proteins results in ectopic rotation of the centrosome/nuclear complex in AB (107–109). Reduction of function of *par-2* or *par-5* results in failure of rotation in $P_1$ (109, 115). A model to explain these phenotypes is based on distributions of PAR-3 in wild-type and various *par* mutant two-cell embryos (115). In this model, anterior PAR proteins act locally at the cortex to promote interactions of microtubules with the cortical cytoskeleton; if the anterior proteins are uniformly distributed, as occurs in wild-type AB and in both blastomeres of *par-2* and *par-5* mutants, then rotation is blocked. If the protein distribution is asymmetrical, as in $P_1$, or cortical protein is absent, as in *par-3*, *par-6*, and *pkc-3* mutations, rotation is favoured.

The dual role of the anterior-group genes in cytoplasmic localization and in spindle orientation in $P_1$ suggests that they function to co-ordinate these two processes in this blastomere. There is some evidence that they may play a similar role in $P_0$ as well. Mutations in anterior-group genes and mutations in *par-2* and *par-5*, which alter the distribution of anterior-group proteins, have strong effects on asymmetrical placement of spindles in the one-cell embryo (106, 109; D. Morton, D. Shakes, and K. Kemphues, unpublished results). Mutations in *par-2* also lead to a partially penetrant delay in the alignment of the $P_0$ spindle along the long axis (109).

Two additional genes, *let-99* and *gpb-1*, are also involved in controlling spindle orientation, but give phenotypes that are difficult to interpret. In most *let-99* embryos, the rotation in $P_1$ fails, but in half of the embryos ectopic rotation takes place in AB (128). LET-99 appears to act downstream of PAR-3 localization because PAR-3 is localized normally in *let-99* embryos. Depletion of GPB-1, a G protein β subunit, also causes the misorientation of spindles but, in contrast to the other proteins, results in a randomization of the pattern (129).

## 5.7  Summary of asymmetrical cell division in *C. elegans*

Five asymmetrical divisions act to distribute at least three transcriptional regulators, SKN-1, PIE-1, and PAL-1, to blastomeres of the $P_1$ lineage, where they act combinatorially to determine the fates of founder cells MS, E, C, D, and $P_4$. Although dependent upon asymmetrical divisions for their proper distributions, PAL-1 and SKN-1 only become localized after cell division via cell-specific translation or protein stability. Presumably, factors localized during division (e.g. P-granule components or PAR proteins) regulate these processes. P granules, PIE-1, and MEX-1 are localized prior to division and then segregated into one daughter. Establishment of polarity is dependent upon microfilaments but not microtubules. The cues for establishing cell polarity differ among the blastomeres, and include both extrinsic and intrinsic cues. The PAR proteins play a major role in interpreting polarity cues at least in the zygote, where they become distributed in a co-dependent reciprocal fashion at the cell periphery at anterior and posterior poles. How they function remains to be determined, but the presence of protein domains with known roles in intracellular signalling has led to the hypothesis that they provide localized signals for cytoskeletal reorganization, translational regulation, and protein stability. The PAR proteins also function to couple cytoplasmic localization to spindle orientation and positioning.

## 6. Summary: is there a conserved mechanism for asymmetrical cell division?

In the asymmetrical cell divisions we have considered, mechanisms have evolved to accomplish each of the essential steps: establishing polarity, localizing determinants, and orienting cell division. Although many of the components remain to be discovered, enough is known to ask whether there is a common mechanism for asymmetrical cell division. The answer appears to be a fairly compelling 'no' in spite of some superficial similarities. Common to all systems are a dependence upon the microfilament cytoskeleton, restricted cortical localization of essential components, and a means to co-ordinate the axis of asymmetry with the spindle orientation. The molecules that mediate these common processes and the details of their action, however, are fairly diverse.

Diversity appears at every level. Determinants can operate directly at the transcriptional level (Ash1p, Prospero, PAL-1, SKN-1) or can function by antagonizing intercellular signalling (Numb). Mechanisms for localizing determinants are extremely varied. In yeast, protein is localized via mRNA localization. In flies, although localized mRNA plays a supporting role, localization is predominantly mediated through protein–protein interactions. In worms, although some important regulators are localized through protein–protein interactions, cell-specific regulation of translation or protein stability also plays a major role. Even when general mechanisms are common to the different systems, the molecules used vary. For example, co-ordinating the axis of asymmetry with axis of cell division is accomplished by Bin1p/She5p and Bud6p in yeast, by Inscutable in flies, and by the anterior PAR genes in worms, and none of these proteins share significant sequence similarity. With the availability of *S. cerevisiae* and *C. elegans* genome databases, specific comparisons are beginning to be made. Searches for Inscutable, Pon, or Miranda homologues in *C. elegans* have not been successful, and a *C. elegans* Numb-like protein has, surprisingly, no obvious loss-of-function phenotype (A. Schetter, F. Piano, K. Kemphues, unpublished results).

The trend toward diversity is also revealed by a comparison of asymmetrical cell divisions within a single organism. For example, each of the parts of the *D. melanogaster* nervous system differs in its requirements for Inscuteable and Miranda. In *C. elegans*, it appears that $P_2$ and $P_3$ use a different system to establish polarity than do $P_0$ and $P_1$, and that the system in EMS, which depends upon an extrinsic cue, is different from that of the germ line.

It would be surprising, however, not to find any conservation of mechanism beyond cytoskeletal involvement. Perhaps through more detailed studies in flies and worms, and through studies of other organisms, some frequently used 'asymmetry molecules' may be revealed. A hint that this is possible comes from the discovery of conservation of the PAR proteins (85, 113, 114, 130, 131) (F. Piano and K. Kemphues, unpuplished results) and the discovery of asymmetrical Numb in mammals (132).

In conclusion, it appears that evolution has found many ways to accomplish asymmetrical cell division. In the three species reviewed here, cytoskeletal modification and anchoring are common starting points beyond which little conservation is apparent.

# References

1. Horvitz, H. R. and Herskowitz, I. (1992) Mechanisms of asymmetric cell division: two Bs or not two Bs, that is the question. *Cell*, **68**, 237.
2. Strathern, J. N. and Herskowitz, I. (1979) Asymmetry and directionality in production of new cell types during clonal growth: the switching pattern of homothallic yeast. *Cell*, **17**, 371.
3. Nasmyth, K. (1983) Molecular analysis of a cell lineage. *Nature*, **302**, 670.
4. Strathern, J. N., Klar, A. J., Hicks, J. B., Abraham, J. A., Ivy, J. M., Nasmyth, K. A., and McGill, C. (1982) Homothallic switching of yeast mating type cassettes is initiated by a double-stranded cut in the MAT locus. *Cell*, **31**, 183.
5. Nasmyth, K. (1987) The determination of mother cell-specific mating type switching in yeast by a specific regulator of *HO* transcription. *EMBO J.*, **6**, 243.
6. Bobola, N., Jansen, R. P., Shin, T. H., and Nasmyth, K. (1996) Asymmetric accumulation of Ash1p in postanaphase nuclei depends on a myosin and restricts yeast mating-type switching to mother cells. *Cell*, **84**, 699.
7. Sil, A. and Herskowitz, I. (1996) Identification of asymmetrically localized determinant, Ash1p, required for lineage-specific transcription of the yeast HO gene. *Cell*, **84**, 711.
8. Jan, Y. N. and Jan, L. Y. (1998) Asymmetric cell division. *Nature*, **392**, 775.
9. Lu, B., Jan, L. Y., and Jan, Y. N. (1998) Asymmetric cell division: lessons from flies and worms. *Curr. Opin. Genet. Dev.*, **8**, 392.
10. Hawkins, N. and Garriga, G. (1998) Asymmetric cell division: from A to Z. *Genes Dev.*, **12**, 3625.
11. Fuerstenberg, S., Broadus, J., and Doe, C. Q. (1998) Asymmetry and cell fate in the *Drosophila* embryonic CNS. *Int. J. Dev. Biol.*, **42**, 379.
12. Bowerman, B. (1998) Maternal control of pattern formation in early *Caenorhabditis elegans* embryos. *Curr. Top. Dev. Biol.*, **39**, 73.
13. Rose, L. and Kemphues, K. (1998) Early patterning of the *C. elegans* embryos. *Annu. Rev. Genet.*, **32**, 521.
14. Schnabel, R. and Priess, J. R. (1997) Specification of cell fates in the early embryo. In *C. elegans II*, (ed. D. Riddle, T. Blumenthal, B. J. Meyer, and J. R. Priess), p. 361. Cold Spring Harbor Laboratory Press, Cold Spring Harbor.
15. Strome, S. and Wood, W. B. (1982) Immunofluorescence visualization of germ-line-specific cytoplasmic granules in embryos, larvae, and adults of *Caenorhabditis elegans*. *Proc. Natl Acad. Sci., USA*, **79**, 1558.

16. Strome, S. and Wood, W. B. (1983) Generation of asymmetry and segregation of germ-line granules in early *C. elegans* embryos. *Cell*, **35**, 15.

17. Seydoux, G. and Fire, A. (1994) Soma-germline asymmetry in the distributions of embryonic RNAs in *Caenorhabditis elegans*. *Development*, **120**, 2823.

18. Kawasaki, I., Shim, Y. H., Kirchner, J., Kaminker, J., Wood, W. B., and Strome, S. (1998) PGL-1, a predicted RNA-binding component of germ granules, is essential for fertility in *C. elegans*. *Cell*, **94**, 635.

19. Nasmyth, K. (1985) At least 1400 base pairs of 5′-flanking DNA is required for the correct expression of the *HO* gene in yeast. *Cell*, **42**, 213.

20. Andrews, B. J. and Herskowitz, I. (1989) Identification of a DNA binding factor involved in cell-cycle control of the yeast *HO* gene. *Cell*, **57**, 21.

21. Nasmyth, K. and Shore, D. (1987) Transcriptional regulation in the yeast life cycle. *Science*, **237**, 1162.

22. Stillman, D. J., Bankier, A. T., Seddon, A., Groenhout, E. G., and Nasmyth, K. A. (1988) Characterization of a transcription factor involved in mother cell specific transcription of the yeast *HO* gene. *EMBO J.*, **7**, 485.

23. Tebb, G., Moll, T., Dowzer, C., and Nasmyth, K. (1993) SWI5 instability may be necessary but is not sufficient for asymmetric HO expression in yeast. *Genes Dev.*, **7**, 517.

24. Peterson, C. L. and Herskowitz, I. (1992) Characterization of the yeast SWI1, SWI2, and SWI3 genes, which encode a global activator of transcription. *Cell*, **68**, 573.

25. Koch, C. and Nasmyth, K. (1994) Cell cycle regulated transcription in yeast. *Curr. Opin. Cell Biol.*, **6**, 451.

26. Nasmyth, K., Seddon, A., and Ammerer, G. (1987) Cell cycle regulation of SW15 is required for mother-cell-specific HO transcription in yeast. *Cell*, **49**, 549.

27. Nasmyth, K., Adolf, G., Lydall, D., and Seddon, A. (1990) The identification of a second cell cycle control on the HO promoter in yeast: cell cycle regulation of SW15 nuclear entry. *Cell*, **62**, 631.

28. Nasmyth, K., Stillman, D., and Kipling, D. (1987) Both positive and negative regulators of HO transcription are required for mother-cell-specific mating-type switching in yeast. *Cell*, **48**, 579.

29. Sternberg, P. W., Stern, M. J., Clark, I., and Herskowitz, I. (1987) Activation of the yeast HO gene by release from multiple negative controls. *Cell*, **48**, 567.

30. Wang, H., Clark, I., Nicholson, P. R., Herskowitz, I., and Stillman, D. J. (1990) The *Saccharomyces cerevisiae SIN3* gene, a negative regulator of *HO*, contains four paired amphipathic helix motifs. *Mol. Cell Biol.*, **10**, 5927.

31. Jansen, R. P., Dowzer, C., Michaelis, C., Galova, M., and Nasmyth, K. (1996) Mother cell-specific HO expression in budding yeast depends on the unconventional myosin myo4p and other cytoplasmic proteins. *Cell*, **84**, 687.

32. Cosma, M. P., Tanaka, T., and Nasmyth, K. (1999) Ordered recruitment of transcription and chromatin remodeling factors to a cell cycle- and developmentally regulated promoter. *Cell*, **97**, 299.

33. Long, R. M., Singer, R. H., Meng, X., Gonzalez, I., Nasmyth, K., and Jansen, R. P. (1997) Mating type switching in yeast controlled by asymmetric localization of ASH1 mRNA. *Science*, **277**, 383.

34. Takizawa, P. A., Sil, A., Swedlow, J. R., Herskowitz, I., and Vale, R. D. (1997) Actin-dependent localization of an RNA encoding a cell-fate determinant in yeast. *Nature*, **389**, 90.

35. Gonzalez, I., Buonomo, S. B., Nasmyth, K., and von Ahsen, U. (1999) ASH1 mRNA

localization in yeast involves multiple secondary structural elements and Ash1 protein translation. *Curr. Biol.*, **9**, 337.

36. Bertrand, E., Chartrand, P., Schaefer, M., Shenoy, S. M., Singer, R. H., and Long, R. M. (1998) Localization of ASH1 mRNA particles in living yeast. *Mol. Cell*, **2**, 437.

37. Chartrand, P., Meng, X. H., Singer, R. H., and Long, R. M. (1999) Structural elements required for the localization of ASH1 mRNA and of a green fluorescent protein reporter particle *in vivo*. *Curr. Biol.*, **9**, 333.

38. Beach, D. L., Salmon, E. D., and Bloom, K. (1999) Localization and anchoring of mRNA in budding yeast. *Curr. Biol.*, **9**, 569.

39. Munchow, S., Sauter, C., and Jansen, R. P. (1999) Association of the class V myosin myo4p with a localised messenger RNA in budding yeast depends on *she* proteins. *J. Cell Sci.*, **112**, 1511.

40. Wendland, B., McCaffery, J. M., Xiao, Q., and Emr, S. D. (1996) A novel fluorescence-activated cell sorter-based screen for yeast endocytosis mutants identifies a yeast homologue of mammalian eps15. *J. Cell Biol.*, **135**, 1485.

41. Barral, J. M., Bauer, C. C., Ortiz, I., and Epstein, H. F. (1998) Unc-45 mutations in *Caenorhabditis elegans* implicate a CRO1/She4p-like domain in myosin assembly. *J. Cell Biol.*, **143**, 1215.

42. Evangelista, M., Blundell, K., Longtine, M. S., Chow, C. J., Adames, N., Pringle, J. R., Peter, M., and Boone, C. (1997) Bni1p, a yeast formin linking cdc42p and the actin cytoskeleton during polarized morphogenesis. *Science*, **276**, 118.

43. Fujiwara, T., Tanaka, K., Mino, A. M., Kikyo, M., Takahashi, K., Shimizu, K., and Takai, Y. (1998) Rho1p–Bni1p–Spa2p interactions: implication in localization of Bni1p at the bud site and regulation of the actin cytoskeleton in *Saccharomyces cerevisiae*. *Mol. Biol. Cell*, **9**, 1221.

44. Amberg, D. C., Zahner, J. E., Mulholland, J. W., Pringle, J. R., and Botstein, D. (1997) Aip3p/Bud6p, a yeast actin-interacting protein that is involved in morphogenesis and the selection of bipolar budding sites. *Mol. Biol. Cell*, **8**, 729.

45. Stearns, T. (1997) Motoring to the finish: kinesin and dynein work together to orient the yeast mitotic spindle. *J. Cell Biol.*, **138**, 957.

46. Heil-Chapdelaine, R. A., Adames, N. R., and Cooper, J. A. (1999) Formin' the connection between microtubules and the cell cortex. *J. Cell Biol.*, **144**, 809.

47. Miller, R. K. and Rose, M. D. (1998) Kar9p is a novel cortical protein required for cytoplasmic microtubule orientation in yeast. *J. Cell Biol.*, **140**, 377.

48. Miller, R. K., Matheos, D., and Rose, M. D. (1999) The cortical localization of the microtubule orientation protein, Kar9p, is dependent upon actin and proteins required for polarization. *J. Cell Biol.*, **144**, 963.

49. Lee, L., Klee, S. K., Evangelista, M., Boone, C., and Pellman, D. (1999) Control of mitotic spindle position by the *Saccharomyces cerevisiae* formin Bni1p. *J. Cell Biol.*, **144**, 947.

50. Doe, C. Q., Chu-LaGraff, Q., Wright, D. M., and Scott, M. P. (1991) The *prospero* gene specifies cell fates in the *Drosophila* central nervous system. *Cell*, **65**, 451.

51. Vaessin, H., Grell, E., Wolff, E., Bier, E., Jan, L. Y., and Jan, Y. N. (1991) *prospero* is expressed in neuronal precursors and encodes a nuclear protein that is involved in the control of axonal outgrowth in *Drosophila*. *Cell*, **67**, 941.

52. Hirata, J., Nakagoshi, H., Nabeshima, Y., and Matsuzaki, F. (1995) Asymmetric segregation of the homeodomain protein Prospero during *Drosophila* development. *Nature*, **377**, 627.

53. Knoblich, J. A., Jan, L. Y., and Jan, Y. N. (1995) Asymmetric segregation of Numb and Prospero during cell division. *Nature*, **377**, 624.

54. Spana, E. P. and Doe, C. Q. (1995) The *prospero* transcription factor is asymmetrically localized to the cell cortex during neuroblast mitosis in *Drosophila*. *Development*, **121**, 3187.

55. Kraut, R., Chia, W., Jan, L. Y., Jan, Y. N., and Knoblich, J. A. (1996) Role of *inscuteable* in orienting asymmetric cell divisions in *Drosophila*. *Nature*, **383**, 50.

56. Matsuzaki, F., Koizumi, K., Hama, C., Yoshioka, T., and Nabeshima, Y. (1992) Cloning of the *Drosophila prospero* gene and its expression in ganglion mother cells. *Biochem. Biophys. Res. Comm.*, **182**, 1326.

57. Chu-Lagraff, Q., Wright, D. M., McNeil, L. K., and Doe, C. Q. (1991) The *prospero* gene encodes a divergent homeodomain protein that controls neuronal identity in *Drosophila*. *Development, Suppl.*, 79.

58. Hassan, B., Li, L., Bremer, K. A., Chang, W., Pinsonneault, J., and Vaessin, H. (1997) Prospero is a panneural transcription factor that modulates homeodomain protein activity. *Proc. Natl Acad. Sci., USA*, **94**, 10991.

59. Ikeshima-Kataoka, H., Skeath, J. B., Nabeshima, Y., Doe, C. Q., and Matsuzaki, F. (1997) Miranda directs Prospero to a daughter cell during *Drosophila* asymmetric divisions. *Nature*, **390**, 625.

60. Uemura, T., Shepherd, S., Ackerman, L., Jan, L. Y., and Jan, Y. N. (1989) *numb*, a gene required in determination of cell fate during sensory organ formation in *Drosophila* embryos. *Cell*, **58**, 349.

61. Rhyu, M. S., Jan, L. Y., and Jan, Y. N. (1994) Asymmetric distribution of numb protein during division of the sensory organ precursor cell confers distinct fates to daughter cells. *Cell*, **76**, 477.

62. Wang, S., Younger-Shepherd, S., Jan, L. Y., and Jan, Y. N. (1997) Only a subset of the binary cell fate decisions mediated by Numb/Notch signaling in *Drosophila* sensory organ lineage requires *Suppressor of Hairless*. *Development*, **124**, 4435.

63. Ruiz Gomez, M. and Bate, M. (1997) Segregation of myogenic lineages in *Drosophila* requires *numb*. *Development*, **124**, 4857.

64. Carmena, A., Murugasu-Oei, B., Menon, D., Jimenez, F., and Chia, W. (1998) *inscuteable* and *numb* mediate asymmetric muscle progenitor cell divisions during *Drosophila* myogenesis. *Genes Dev.*, **12**, 304.

65. Spana, E. P., Kopczynski, C., Goodman, C. S., and Doe, C. Q. (1995) Asymmetric localization of *numb* autonomously determines sibling neuron identity in the *Drosophila* CNS. *Development*, **121**, 3489.

66. Skeath, J. B. and Doe, C. Q. (1998) *sanpodo* and *Notch* act in opposition to *numb* to distinguish sibling neuron fates in the *Drosophila* CNS. *Development*, **125**, 1857.

67. Buescher, M., Yeo, S. L., Udolph, G., Zavortink, M., Yang, X., Tear, G., and Chia, W. (1998) Binary sibling neuronal cell fate decisions in the *Drosophila* embryonic central nervous system are nonstochastic and require inscuteable-mediated asymmetry of ganglion mother cells. *Genes Dev.*, **12**, 1858.

68. Artavanis-Tsakonas, S., Rand, M. D., and Lake, R. J. (1999) Notch signaling: cell fate control and signal integration in development. *Science*, **284**, 770.

69. Hartenstein, V. and Posakony, J. W. (1990) A dual function of the *Notch* gene in *Drosophila* sensillum development. *Dev. Biol.*, **142**, 13.

70. Guo, M., Jan, L. Y., and Jan, Y. N. (1996) Control of daughter cell fates during asymmetric division: interaction of Numb and Notch. *Neuron*, **17**, 27.

71. Frise, E., Knoblich, J. A., Younger-Shepherd, S., Jan, L. Y., and Jan, Y. N. (1996) The *Drosophila* Numb protein inhibits signaling of the Notch receptor during cell–cell interaction in sensory organ lineage. *Proc. Natl Acad. Sci., USA*, **93**, 11925.

72. Li, P., Yang, X., Wasser, M., Cai, Y., and Chia, W. (1997) Inscuteable and Staufen mediate asymmetric localization and segregation of prospero RNA during *Drosophila* neuroblast cell divisions. *Cell*, **90**, 437.

73. Broadus, J., Fuerstenberg, S., and Doe, C. Q. (1998) Staufen-dependent localization of *prospero* mRNA contributes to neuroblast daughter-cell fate. *Nature*, **391**, 792.

74. Shen, C. P., Jan, L. Y., and Jan, Y. N. (1997) Miranda is required for the asymmetric localization of Prospero during mitosis in *Drosophila*. *Cell*, **90**, 449.

75. Shen, C. P., Knoblich, J. A., Chan, Y. M., Jiang, M. M., Jan, L. Y., and Jan, Y. N. (1998) Miranda as a multidomain adapter linking apically localized Inscuteable and basally localized Staufen and Prospero during asymmetric cell division in *Drosophila*. *Genes Dev.*, **12**, 1837.

76. Fuerstenberg, S., Peng, C. Y., Alvarez-Ortiz, P., Hor, T., and Doe, C. Q. (1998) Identification of Miranda protein domains regulating asymmetric cortical localization, cargo binding, and cortical release. *Mol. Cell. Neurosci.*, **12**, 325.

77. Schuldt, A. J., Adams, J. H., Davidson, C. M., Micklem, D. R., Haseloff, J., Johnston, D. S., and Brand, A. H. (1998) Miranda mediates asymmetric protein and RNA localization in the developing nervous system. *Genes Dev.*, **12**, 1847.

78. Knoblich, J. A., Jan, L. Y., and Jan, Y. N. (1997) The N terminus of the *Drosophila* Numb protein directs membrane association and actin-dependent asymmetric localization. *Proc. Natl Acad. Sci., USA*, **94**, 13005.

79. Lu, B., Rothenberg, M., Jan, L. Y., and Jan, Y. N. (1998) Partner of Numb colocalizes with Numb during mitosis and directs Numb asymmetric localization in *Drosophila* neural and muscle progenitors. *Cell*, **95**, 225.

80. Gho, M. and Schweisguth, F. (1998) Frizzled signalling controls orientation of asymmetric sense organ precursor cell divisions in *Drosophila*. *Nature*, **393**, 178.

81. Kraut, R. and Campos-Ortega, J. A. (1996) *inscuteable*, a neural precursor gene of *Drosophila*, encodes a candidate for a cytoskeleton adaptor protein. *Dev. Biol.*, **174**, 65.

82. Tio, M., Zavortink, M., Yang, X., and Chia, W. (1999) A functional analysis of Inscuteable and its roles during *Drosophila* asymmetric cell divisions. *J. Cell Sci.*, **112**, 1541.

83. Knoblich, J. A., Jan, L. Y., and Jan, Y. N. (1999) Deletion analysis of the *Drosophila* Inscuteable protein reveals domains for cortical localization and asymmetric localization. *Curr. Biol.*, **9**, 155.

84. Muller, H. A. and Wieschaus, E. (1996) *armadillo*, *bazooka*, and *stardust* are critical for early stages in formation of the zonula adherens and maintenance of the polarized blastoderm epithelium in *Drosophila*. *J. Cell Biol.*, **134**, 149.

85. Kuchinke, U., Grawe, F., and Knust, E. (1998) Control of spindle orientation in *Drosophila* by the *par-3*-related PDZ- domain protein Bazooka. *Curr. Biol.*, **8**, 1357.

86. Broadus, J. and Doe, C. Q. (1997) Extrinsic cues, intrinsic cues and microfilaments regulate asymmetric protein localization in *Drosophila* neuroblasts. *Curr. Biol.*, **7**, 827.

87. Sulston, J. E., Schierenberg, E., White, J. G., and Thomson, J. N. (1983) The embryonic cell lineage of the nematode *Caenorhabditis elegans*. *Dev. Biol.*, **100**, 64.

88. Bowerman, B., Eaton, B. A., and Priess, J. R. (1992) *skn-1*, a maternally expressed gene required to specify the fate of ventral blastomeres in the early *C. elegans* embryo. *Cell*, **68**, 1061.

89. Mello, C. C., Draper, B. W., Krause, M., Weintraub, H., and Priess, J. R. (1992) The *pie-1* and *mex-1* genes and maternal control of blastomere identity in early *C. elegans* embryos. *Cell*, **70**, 163.

90. Hunter, C. P. and Kenyon, C. (1996) Spatial and temporal controls target *pal-1* blastomere-specification activity to a single blastomere lineage in *C. elegans* embryos. *Cell*, **87**, 217.

91. Seydoux, G., Mello, C. C., Pettitt, J., Wood, W. B., Priess, J. R., and Fire, A. (1996) Repression of gene expression in the embryonic germ lineage of *C. elegans. Nature*, **382**, 713.

92. Seydoux, G. and Strome, S. (1999) Launching the germline in *Caenorhabditis elegans*: regulation of gene expression in early germ cells. *Development*, **126**, 3275.

93. Draper, B. W., Mello, C. C., Bowerman, B., Hardin, J., and Priess, J. R. (1996) MEX-3 is a KH domain protein that regulates blastomere identity in early *C. elegans* embryos. *Cell*, **87**, 205.

94. Bowerman, B., Draper, B. W., Mello, C. C., and Priess, J. R. (1993) The maternal gene *skn-1* encodes a protein that is distributed unequally in early *C. elegans* embryos. *Cell*, **74**, 443.

95. Goldstein, B. and Hird, S. N. (1996) Specification of the anteroposterior axis in *Caenorhabditis elegans. Development*, **122**, 1467.

96. Schierenberg, E. (1987) Reversal of cellular polarity and early cell–cell interaction in the embryo of *Caenorhabditis elegans. Dev. Biol.*, **122**, 452.

97. Priess, J. R. and Thomson, J. N. (1987) Cellular interactions in early *C. elegans* embryos. *Cell*, **48**, 241.

98. Schierenberg, E. (1988) Localization and segregation of lineage-specific cleavage potential in embryos of *Caenorhabditis elegans. Roux's Arch. Dev. Biol.*, **197**, 282.

99. Rose, L. S., Lamb, M. L., Hird, S. N., and Kemphues, K. J. (1995) Pseudocleavage is dispensable for polarity and development in *C. elegans* embryos. *Dev. Biol.*, **168**, 479.

100. Hird, S. N. and White, J. G. (1993) Cortical and cytoplasmic flow polarity in early embryonic cells of *Caenorhabditis elegans. J. Cell Biol.*, **121**, 1343.

101. Hird, S. N., Paulsen, J. E., and Strome, S. (1996) Segregation of germ granules in living *Caenorhabditis elegans* embryos: cell-type-specific mechanisms for cytoplasmic localisation. *Development*, **122**, 1303.

102. Hill, D. P. and Strome, S. (1988) An analysis of the role of microfilaments in the establishment and maintenance of asymmetry in *Caenorhabditis elegans* zygotes. *Dev. Biol.*, **125**, 75.

103. Strome, S. (1986) Fluorescence visualization of the distribution of microfilaments in gonads and early embryos of the nematode *Caenorhabditis elegans. J. Cell Biol.*, **103**, 2241.

104. Mello, C. C., Schubert, C., Draper, B., Zhang, W., Lobel, R., and Priess, J. R. (1996) The PIE-1 protein and germline specification in *C. elegans* embryos. *Nature*, **382**, 710.

105. Tenenhaus, C., Schubert, C., and Seydoux, G. (1998) Genetic requirements for PIE-1 localization and inhibition of gene expression in the embryonic germ lineage of *Caenorhabditis elegans. Dev. Biol.*, **200**, 212.

106. Kemphues, K. J., Priess, J. R., Morton, D. G., and Cheng, N. S. (1988) Identification of genes required for cytoplasmic localization in early *C. elegans* embryos. *Cell*, **52**, 311.

107. Tabuse, Y., Izumi, Y., Piano, F., Kemphues, K. J., Miwa, J., and Ohno, S. (1998) Atypical protein kinase C cooperates with PAR-3 to establish embryonic polarity in *Caenorhabditis elegans. Development*, **125**, 3607.

108. Watts, J. L., Etemad-Moghadam, B., Guo, S., Boyd, L., Draper, B. W., Mello, C. C., Priess, J. R., and Kemphues, K. J. (1996) *par-6*, a gene involved in the establishment of asymmetry in early *C. elegans* embryos, mediates the asymmetric localization of PAR-3. *Development*, **122**, 3133.

109. Cheng, N. N., Kirby, C. M., and Kemphues, K. J. (1995) Control of cleavage spindle orientation in *Caenorhabditis elegans*: the role of the genes *par-2* and *par-3. Genetics*, **139**, 549.

110. Crittenden, S. L., Rudel, D., Binder, J., Evans, T. C., and Kimble, J. (1997) Genes required for GLP-1 asymmetry in the early *Caenorhabditis elegans* embryo. *Dev. Biol.*, **181**, 36.

111. Bowerman, B., Ingram, M. K., and Hunter, C. P. (1997) The maternal *par* genes and the segregation of cell fate specification activities in early *Caenorhabditis elegans* embryos. *Development*, **124**, 3815.

112. Boyd, L., Guo, S., Levitan, D., Stinchcomb, D. T., and Kemphues, K. J. (1996) PAR-2 is asymmetrically distributed and promotes association of P granules and PAR-1 with the cortex in *C. elegans* embryos. *Development*, **122**, 3075.

113. Guo, S., and Kemphues, K. J. (1995) *par-1*, a gene required for establishing polarity in *C. elegans* embryos, encodes a putative Ser/Thr kinase that is asymmetrically distributed. *Cell*, **81**, 611.

114. Hung, T. J. and Kemphues, K. J. (1999) PAR-6 is a conserved PDZ domain-containing protein that colocalizes with PAR-3 in *Caenorhabditis elegans* embryos. *Development*, **126**, 127.

115. Etemad-Moghadam, B., Guo, S., and Kemphues, K. J. (1995) Asymmetrically distributed PAR-3 protein contributes to cell polarity and spindle alignment in early *C. elegans* embryos. *Cell*, **83**, 743.

116. Kurzchalia, T. and Hartmann, E. (1996) Are there similarities between the polarization of the *C.elegans* embryos and of an epithelial cell? *Trends Cell Biol.*, **6**, 131.

117. Guo, S. and Kemphues, K. J. (1996) A non-muscle myosin required for embryonic polarity in *Caenorhabditis elegans*. *Nature*, **382**, 455.

118. Schnabel, R., Weigner, C., Hutter, H., Feichtinger, R., and Schnabel, H. (1996) *mex-1* and the general partitioning of cell fate in the early *C. elegans* embryo. *Mech. Dev.*, **54**, 133.

119. Guedes, S. and Priess, J. R. (1997) The *C. elegans* MEX-1 protein is present in germline blastomeres and is a P granule component. *Development*, **124**, 731.

120. Strome, S., Martin, P., Schierenberg, E., and Paulsen, J. (1995) Transformation of the germ line into muscle in mes-1 mutant embryos of *C. elegans*. *Development*, **121**, 2961.

121. Tabara, H., Hill, R. J., Mello, C. C., Priess, J. R., and Kohara, Y. (1998) *pos-1* encodes a cytoplasmic zinc-finger protein essential for germline specification in *C. elegans*. *Development*, **126**, 1.

122. Albertson, D. (1984) Formation of the first cleavage spindle in nematode embryos. *Dev. Biol.*, **101**, 61.

123. Hyman, A. A. and White, J. G. (1987) Determination of cell division axes in the early embryogenesis of *Caenorhabditis elegans*. *J. Cell Biol.*, **105**, 2123.

124. Skop, A. R. and White, J. G. (1998) The dynactin complex is required for cleavage plane specification in early *Caenorhabditis elegans* embryos. *Curr. Biol.*, **8**, 1110.

125. Hyman, A. A. (1989) Centrosome movement in the early divisions of *Caenorhabditis elegans*: a cortical site determining centrosome position. *J. Cell Biol.*, **109**, 1185.

126. Keating, H. H. and White, J. G. (1998) Centrosome dynamics in early embryos of *Caenorhabditis elegans*. *J. Cell Sci.*, **111**, 3027.

127. Waddle, J. A., Cooper, J. A., and Waterston, R. H. (1994) Transient localized accumulation of actin in *Caenorhabditis elegans* blastomeres with oriented asymmetric divisions. *Development*, **120**, 2317.

128. Rose, L. S. and Kemphues, K. (1998) The *let-99* gene is required for proper spindle orientation during cleavage of the *C. elegans* embryo. *Development*, **125**, 1337.

129. Zwaal, R. R., Ahringer, J., van Luenen, H. G., Rushforth, A., Anderson, P., and Plasterk, R. H. (1996) G proteins are required for spatial orientation of early cell cleavages in *C. elegans* embryos. *Cell*, **86**, 619.

130. Izumi, Y., Hirose, T., Tamai, Y., Hirai, S., Nagashima, Y., Fujimoto, T., Tabuse, Y., Kemphues, K. J., and Ohno, S. (1998) An atypical PKC directly associates and colocalizes

at the epithelial tight junction with ASIP, a mammalian homologue of *Caenorhabditis elegans* polarity protein PAR-3. *J. Cell Biol.*, **143**, 95.

131. Bohm, H., Brinkmann, V., Drab, M., Henske, A., and Kurzchalia, T. V. (1997) Mammalian homologues of *C. elegans* PAR-1 are asymmetrically localized in epithelial cells and may influence their polarity. *Curr. Biol.*, **7**, 603.

132. Zhong, W., Feder, J. N., Jiang, M. M., Jan, L. Y., and Jan, Y. N. (1996) Asymmetric localization of a mammalian numb homolog during mouse cortical neurogenesis. *Neuron*, **17**, 43.

# 9 | Polarized exocytosis: targeting vesicles to specific domains on the plasma membrane

PATRICK BRENNWALD and JOAN ADAMO

## 1. Introduction

The ability to deliver newly synthesized proteins and lipids to discrete sites on the cell surface allows cells to grow asymmetrically (i.e. tall and thin versus round), to release morphogens on a discrete side of a cell layer during early embryonic development, and to carry out a variety of other processes. The involvement of delivery events is particularly apparent during asymmetrical growth. This is because the ability of the surface to expand requires the addition of new phospholipids and integral membrane proteins, which are the building blocks that allow cell-surface growth. Lipids and integral membrane proteins are extremely hydrophobic in nature and thus they can not reach the plasma membrane by simple diffusion through the cytoplasm. Instead, they are brought to the cell surface in the form of membranous transport vesicles. These transport or secretory vesicles contain materials first synthesized in the endoplasmic reticulum, and then delivered to the Golgi apparatus, within which the secretory vesicles themselves are derived. These post-Golgi secretory vesicles represent the last stage of the exocytic pathway and their fusion with the cell surface is often referred to as exocytosis. In this chapter, we will discuss the mechanisms and machinery that regulate the delivery and fusion of secretory vesicles with specific sites on the plasma membrane, the process of polarized exocytosis.

## 2. Epithelial cells, neurons, and yeast as models for polarized exocytosis

As with much of modern biological research, most of what we know about cell polarity and polarized exocytosis comes from a wealth of research on a relatively

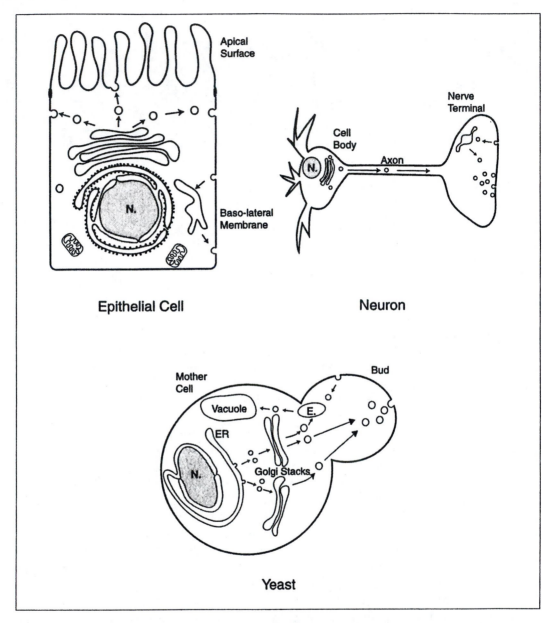

**Fig. 1** Three major cell types used to study polarized exocytosis. The exocytic pathway by which newly synthesized proteins and lipids are targeted to the cell surface is outlined for each cell type. In each cell type, the major site of exocytic release is at very precise points on the plasma membrane. In epithelial cells, basolateral proteins appear to be delivered to the lateral face just below the point of cell–cell contact. The release of neurotransmitters in neurons by exocytosis occurs at the active zone of the nerve terminal to release cargo into the synapse. In yeast, during much of the growth cycle, cell-surface growth is restricted to the daughter cell. One difference between cell types is that in mature neurons and epithelial cells, exocytic release is balanced by membrane retrieval via endocytosis so that no net growth in the surface occurs.

small number of model systems. In particular, for the analysis of trafficking, polarized epithelial cells, neurons and budding yeast are the most prominent systems (Fig. 1). Each of these systems has distinct and shared features in the characteristics of its polarity. For example, the polarity of membrane proteins in epithelial cells is maintained due to the presence of tight junctions which serve as diffusion barriers, separating the apical and basolateral domains (see also Chapter 4). In neurons, diffusion barriers are located between the cell body and the axon. No such diffusion barriers have been demonstrated in yeast. However, in all three systems, the exocytic release of proteins is largely constrained to highly defined sites on the plasma membrane.

While each of these cell types has its own unique features, it is likely that beneath the superficial differences in each cell type, there is a common blueprint for the molecular mechanisms governing this polarity (1). In this chapter, we will focus primarily on studies of polarized exocytosis in budding yeast (see also Chapter 2). This organism has provided what is currently the most complete description of the factors involved in polarized exocytosis, although even in this system much remains to be elucidated. While some mechanistic details will clearly differ in each system, it is likely that the principles learned from yeast will have widespread applicability to the mechanism of polarized exocytosis in mammalian cells.

## 3. Polarized exocytosis involves two distinct stages: polarized delivery of secretory vesicles and the docking/fusion of secretory vesicles with the plasma membrane

In all three of the systems discussed above, the targeting of vesicles to discrete sites on the plasma membrane can be divided into two major events. The first event is the directed movement of secretory vesicles derived from Golgi apparatus to their site of fusion at the cell periphery. In some cells, such as in neurons (Fig. 1, middle panel), where vesicles carrying biosynthetic products to a synapse may travel as much as 1 metre from the cell body where they are produced, movement occurs over great distances. In contrast, the distance of vesicle movement in budding yeast is on the order of a few microns (Fig. 1, right panel). Nevertheless, in both cell types, the physical movement of the vesicle is not left to passive diffusion, but rather involves directed movement along elements of either the microtubule-based (neurons) or actin-based (budding yeast) cytoskeleton.

The second stage involves molecular recognition events mediated by Rab-GTPases, the exocyst complex, SNARE proteins, and other factors that ultimately lead to fusion of the vesicle membrane with the plasma membrane. Interestingly, a few factors appear to function at both vesicle movement and vesicle docking/fusion steps. In summary, the process of vesicle targeting is a product of both the directed movement of vesicles along cytoskeletal tracks to the correct region of the cell, and the precise targeting of vesicles to discrete regions of the plasma membrane by the localized action of components of the docking and fusion machinery.

## 4. Polarized exocytosis can be a dynamic process: in budding yeast the site of cell-surface delivery changes during the cell cycle

In mature neurons or epithelial cells, cell polarity and sites of polarized exocytosis are stably maintained. In contrast, sites of polarized exocytosis can change dramatically in other cells types. In budding yeast, for example, the site of cell-surface exocytosis, and hence growth, changes during the cell cycle. A schematic diagram of this cycle of localization of the site of exocytosis is shown in Fig. 2. Early in the development of the bud (cells with incipient bud sites and with very small buds) exocytosis is highly polarized to the bud tip. This correlates precisely with the region of cell surface expansion and the localization of several components of the exocytic machinery (2–6). As the bud enlarges, the site of exocytosis becomes less tightly localized. Around the time of nuclear division and migration into the bud, there is a short period of isotropic growth, during which the bud appears to grow

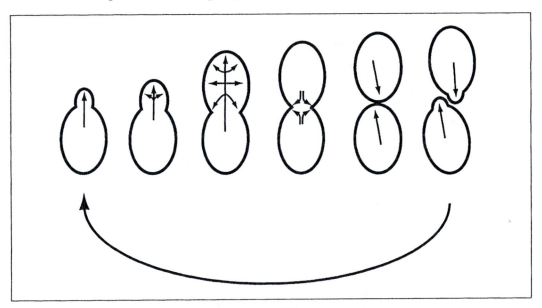

**Fig. 2** Changes in polarity of exocytosis during the yeast cell-division cycle. The direction of polarized secretion changes during the growth cycle of a haploid yeast cell. Growth at incipient bud sites and in small buds is highly polarized to the bud tip. As the size of the daughter cell increases, there is a delocalization of growth around the bud tip. A switch to isotropic growth precedes the final enlargement of the bud. Polarized secretion is reoriented to the mother–bud neck just prior to cytokinesis. Haploid cells have axial bud-site selection, where the new bud occurs at a site adjacent to the previous bud scar. In contrast, diploid cells can make a 180° switch and form the next bud at the opposite pole of the cell. The localization of markers of the exocytic machinery, such as Sec4, Sec3, or Sec8, mirrors this polarity.

uniformly and exocytic markers show no polarization. Immediately following this period, the site of exocytosis and the exocytic apparatus reorient to the neck or mother–bud junction, in preparation for cytokinesis. In haploid cells, this orientation is roughly maintained as the next bud site is adjacent to the previous mother–bud junction.

## 5. Polarized vesicle delivery: secretory vesicles derived from the Golgi apparatus must be physically transported from the mother cell into the bud

In mammalian cells, there is generally a single, well-organized Golgi apparatus in each cell, consisting of several sausage-shaped cisternae or stacks of elongated membranes. Distinct compartments within the Golgi are often distinguished by the presence of different processing enzymes residing within each compartment. The compartment that is most proximal to the endoplasmic reticulum is called the *cis*-Golgi. The next compartment is the medial, then the *trans*, and the last compartment of the Golgi is usually referred to as the *trans*-Golgi network (TGN), from which exocytic secretory vesicles are derived. In many mammalian cell types, the Golgi apparatus often appears as a large 'cap' on one side of the nucleus. In the yeast *S. cerevisiae*, the Golgi apparatus is less organized, and immunofluorescent staining for a number of yeast Golgi markers reveal 8–14 distinct punctae which appear to be randomly distributed in the cell. Since the mother cell is much larger than the daughter cell during most of the cell cycle, most of the Golgi elements giving rise to secretory vesicles are present in the mother cell. Therefore, most of the Golgi-derived secretory vesicles originate in the mother cell and must be delivered to the bud. As we will see below, mutations in the genes encoding the vesicle delivery machinery cause cells to accumulate vesicles primarily in the mother cell and to a lesser extent in the bud. In contrast, mutations in genes encoding proteins of the docking and fusion apparati accumulate vesicles almost exclusively in the bud.

## 5.1 Polarized delivery: the role of the actin cytoskeleton and type V myosins

In mammalian cells, both the microtubule and actin cytoskeletons play a prominent role in delivery of vesicles to the cell perimeter, while in yeast, vesicle delivery appears to be dependent primarily on the actin cytoskeleton. Temperature-sensitive mutants in the *ACT1* gene, coding for yeast actin, show a number of defects in polarized exocytosis: the mother cells are enlarged and rounded, the cells show a partial defect in secretion, and they accumulate post-Golgi secretory vesicles in both the mother cell and the bud (7).

The actin cytoskeleton in yeast is highly polarized. Cortical actin patches are present almost exclusively in the bud of pre-mitotic cells and actin cables run parallel to the mother–bud axis from the bud tip into the mother cell (8). The importance of actin

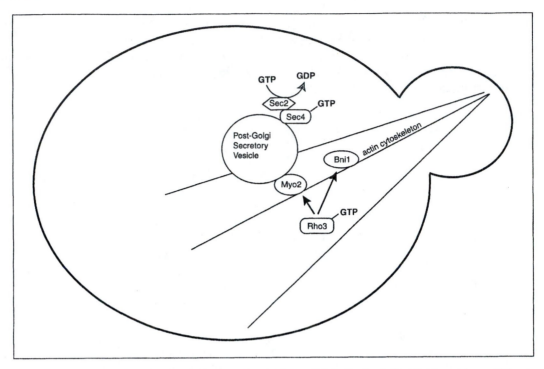

**Fig. 3** Factors involved in vesicle delivery from the mother cell into the bud. Post-Golgi vesicles, which are thought to be produced largely in the mother cell, must be transported along actin cables into the bud. The unconventional type V myosin, Myo2, has been implicated as a likely motor protein for this directed movement. The exchange of GDP for GTP on Sec4, mediated by the exchange factor Sec2, appears to be a necessary prerequisite for this delivery event. Rho3 appears to be a key player in that it may co-ordinate both the polarization of the actin cytoskeleton (proposed to be through an interaction with Bni1) as well as regulate the function of Myo2 in this process (43). The precise effect of Rho3 on Myo2 is unknown at the present time.

cables in polarized exocytosis was demonstrated recently using a temperature-sensitive mutant in a gene encoding tropomyosin in yeast. Yeast cells containing this mutation show a rapid loss of actin cables within minutes of a shift to the non-permissive temperature. However, the cortical actin patches remain largely un-affected in the course of these rapid shifts. Consistent with a loss of polarity in exocytic transport, the cells form large, round mother cells immediately following the temperature shift, but this effect is blocked by a second mutation in *SEC6*, an essential component of the docking fusion apparatus (9). In addition, a number of components of the exocytic fusion machinery that are normally concentrated at the site of exocytic fusion, such as the Rab-GTPase Sec4, present on secretory vesicles, and Sec8, a component of the plasma membrane exocyst complex, are dispersed in response to the loss of actin cables (12). Taken together, these observations suggest a prominent role for actin cables in the polarized delivery of secretory vesicles to the bud (Fig. 3).

Just as kinesin and dynein motor proteins are thought to move 'cargo' along microtubules, myosin proteins can act as motors to move associated cargo along actin filaments (Fig. 3). The best candidates for the motor proteins that move secretory vesicles along actin cables to their site of fusion are the unconventional type V myosins. The yeast type V myosin, Myo2, has been implicated in this process by a number of observations:

1. A temperature-sensitive mutant in the *MYO2* gene, *myo2–66*, was found to accumulate large numbers of secretory vesicles following a shift to the non-permissive temperature (10).

2. Myo2 shows extensive genetic interactions with many of the components of the exocytic machinery, including the Rab-GTPase, Sec4 (11).

3. Myo2 localizes to regions of polarized growth and co-localizes with Sec4 and the exocyst subunit Sec8 (9, 12, 13).

4. While *sec6–4* mutants normally accumulate secretory vesicles in the bud, in a *myo2–66*, *sec6–4* double mutant, secretory vesicles accumulate primarily in the mother cell, indicating a defect in delivery of the secretory vesicles from the mother cell into the bud (11).

5. The accumulation of secretory vesicles at the bud tip, observed by immuno-fluorescent staining with α-Sec4 antibodies, is absent in a *myo2–66* mutant following a shift to the non-permissive temperature (14).

Taken together, there is strong evidence that the type V myosin, Myo2, has a role in delivering post-Golgi secretory vesicles from their site of synthesis at Golgi stacks, which are predominantly localized in the mother cell, to their site of fusion in the bud.

A homologue of Myo2, named *dilute*, has been identified in mice as a gene which, when mutated, causes an alteration in coat colour. Like Myo2, *dilute* is a type V myosin. The coat colour phenotype appears to result from a defect in the movement of membrane-bound pigment granules, or melanosomes, to the dendritic processes of melanocytes (15). Therefore, it is likely that at least a subset of the type V myosins present in mammalian cells perform a function similar to that of Myo2 in yeast—that is delivery of vesicular cargo to precise sites within the cell.

## 5.2 Marking vesicles for polarized delivery: the role of the Rab-GTPase Sec4, and its exchange factor, Sec2

Rab proteins, like SNARE proteins described in more detail below, are a highly conserved family of proteins. A distinct member of the Rab family appears to function at each stage of intracellular membrane trafficking (4). The first evidence that this family of proteins functions in vesicular transport came from the analysis of Sec4, a Rab-GTPase required for Golgi to cell-surface transport (16). Sec4 is found on the surface of secretory vesicles and its function is required for some step (see below) in the docking and fusion of these vesicles with the plasma membrane (17). Interestingly, Sec4 itself is highly polarized in wild-type yeast cells, demonstrating a bright staining of sites of polarized growth—the bud tip of pre-mitotic buds and the mother–neck junction of cells nearing cytokinesis (4).

While the major function of Sec4 appears to be mediation of the docking and fusion of vesicles with the plasma membrane, recent evidence suggests that Sec4 may also play a role in marking vesicles for delivery from the mother to the bud. These data come from a study identifying Sec2 as a protein involved in the exchange of GTP for GDP on Sec4 (Fig. 3). This is a necessary event since loss of this exchange function results in a complete block in exocytosis. The localization of Sec4 to sites of polarized growth was examined in the 10 late-acting (post-Golgi) *sec* mutants. Interestingly, *sec2* mutants showed a dramatic loss of the Sec4 patch at sites of polarized growth. Only diffuse staining was observed. Immunoelectron microscopy of the Sec4 staining suggests that the patch of Sec4 observed by immunofluorescence at sites of growth represents a small cluster of vesicles in the process of fusing. In *sec2* mutants, this cluster fails to form, and instead the vesicles appear dispersed in the mother cell (14). This suggests that the presence of GTP-bound Sec4 on post-Golgi vesicles is necessary for some step of the delivery of the vesicles to sites of fusion. Thus, GTP-bound Sec4 appears to 'mark' the vesicles for delivery by the actinomyosin system. The mechanism by which this recognition occurs is not at all understood. Apparently, it is independent of the recognition of GTP-bound Sec4 by Sec15, a component of the exocyst complex which functions in the docking and fusion reaction (see below). *sec15* mutants have no effect on vesicle delivery or Sec4 staining (11, 14).

## 5.3 Co-ordinating polarized exocytosis and actin polarity: the role of the Rho-GTPase, Rho3

Recent studies from several laboratories, including our own, have implicated a member of the Rho-GTPase family as an essential regulator of both actin polarity and polarized exocytosis. The initial studies on Rho3 demonstrated a role for this GTPase in regulating the polarity and integrity of the actin cytoskeleton in yeast. Mutations or loss of Rho3 cause the normally polarized cortical actin patches to become randomly distributed between the mother and bud, and actin cables to become either very short or undetectable (18).

*RHO3* shows a number of strong genetic interactions with components of the exocytic apparatus. *SEC4* was previously identified in a screen for high-copy suppressors of a *rho3* deletion (18) and we have found that *RHO3* itself is a potent suppressor of a cold-sensitive effector mutation of Sec4 (J. Adamo, G. Rossi, and P. Brennwald, submitted). In addition we have both shown that the genes encoding the Sec9-binding proteins, *SRO7/77*, and *SEC9* itself can suppress the severe growth defect associated with loss of Rho3 (19). This suggests that in addition to its role in regulating actin polarity, Rho3 may play a direct role in exocytosis.

Mutagenesis of the effector domain of Rho3 has confirmed this prediction. From an exhaustive collection of point mutations within the effector domain of *RHO3*, several mutants were recovered with cold-sensitive growth defects. Of the four mutants examined, three had clear defects in actin polarity, the expected phenotype based on previous studies. One of the mutants, however, had no detectable defects in actin polarity but demonstrated a significant defect in secretion of several peri-plasmic proteins at both permissive and non-permissive conditions. Consistent with the constitutive secretion defect associated with this mutant, in cells examined by electron microscopy, a significant number of 80–100 nm secretory vesicles were found to accumulate at both permissive and non-permissive conditions. However, at permissive conditions the vesicles were found almost exclusively in the bud, while after a short shift to the non-permissive conditions a dramatic accumulation of vesicles appeared in the mother cell (J. Adamo, G. Rossi, and P. Brennwald, sub-mitted). As discussed above for Myo2 and Sec2 mutants, this phenotype strongly suggests a role for Rho3 in vesicle delivery from the mother into the bud.

# 6. Polarized fusion: defining sites of docking and fusion of secretory vesicles on the plasma membrane

The second stage of polarized exocytosis involves the recruitment and assembly of factors necessary for the docking and fusion of the vesicles with the plasma membrane. Exactly what the docking step entails has been somewhat controversial, and more recently the idea has surfaced that there may be an initial interaction of the vesicle with the target membrane, known as tethering. This term has been used to separate the initial association of the vesicle with the target membrane from the intermembrane associations which occur later and which appear to be required for the final fusion event. Rab-GTPases and other peripherally associated membrane proteins appear to be important in the tethering reaction (20, 21). The interactions between SNARE proteins present on the vesicle and target membrane appear to be important at a later stage that is either just upstream of, or concomitant with, the fusion of the vesicle with the target membrane (22–24).

## 6.1 The Rab-GTPase, Sec4, and the exocyst complex may act in the initial docking or 'tethering' of vesicles to the plasma membrane

Genetic dissection of the exocytic pathway led to the identification of 10 *sec* mutants which are required for transport from the Golgi apparatus to the cell surface. Of the original 10 'late-acting' *sec* genes, six have been found to reside in a single high-molecular-weight complex (5, 25–27). This multimeric complex, recently termed the 'exocyst' complex, is composed of at least six subunits: Sec3, Sec5, Sec6, Sec8, Sec10, and Sec15 proteins, identified in a screen for secretory defective mutants (28), plus at least one additional subunit, Exo70, which was identified from protein sequencing of the purified complex (27). The exocyst complex is peripherally associated with the plasma membrane in yeast (Fig. 4) and immunofluorescence experiments reveal that

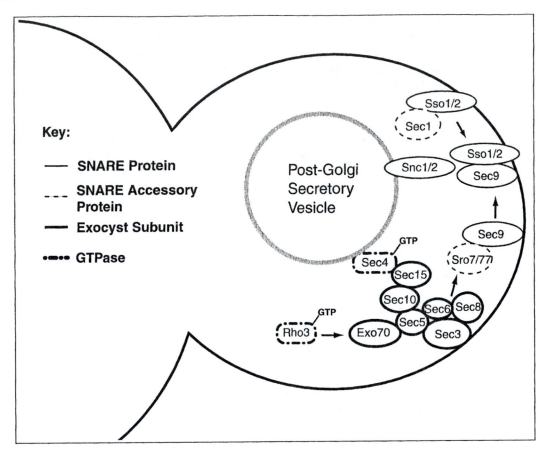

**Fig. 4** A model of the bud tip with the SNAREs, exocyst complex, GTPases, and other known SNARE accessory proteins. The docking and fusion machinery at the bud tip includes a Rab-GTPase (Sec4), a Rho-GTPase (Rho3), the exocyst complex (Exo70/Sec3/5/6/8/10/15), the SNARE proteins (Snc1/2, Sso1/2, and Sec9) and the SNARE accessory proteins (Sec1 and Sro7/77). The bringing together of each of these components is thought to result in the tethering, docking, and fusion of the vesicle with the correct site at the bud tip. The assembly of Sec9, Sso1/2 heterodimers is likely to be regulated at least in part by the interaction of the accessory proteins Sro7/77 and Sec1. Genetic evidence suggests that the SNARE proteins act downstream of the Rab-GTPase, Rho3, and the exocyst complex. The proposed interaction of Rho3 with the Exo70 component of the exocyst (42), along with Myo2 on the vesicles, implies a mode by which the vesicles can be tethered prior to fusion. Rho3 may regulate the exocyst complex during this late stage of polarized exocytosis, and play a role in vesicle delivery (see Fig. 3). One possibility is that Sro7/77 co-ordinates exocyst function with SNARE assembly; this would be consistent with the strong genetic interactions observed between these components (19).

it localizes to sites of polarized growth (5, 6, 12, 25). Interestingly, the polarized localization of most of the members of the complex appears to depend on both an intact actin cytoskeleton and exocytic trafficking to the cell surface.

The localization of one subunit of this complex appears to be distinct from the rest. Analysis of a Sec3p fused to GFP has shown that, like other components of the exocyst, it is localized to sites of polarized secretion. However, Sec3, unlike Sec8, which is thought to be representative of the complex as a whole, localizes to the plasma membrane independent of the status of the actin cytoskeleton or the secretory pathway (6). Sec3 is also unaffected by mutations in the septins—*cdc3*, *10*, *11*, and *12* (6). This suggests that the localization of Sec3 is a primary event in the polarized localization of the exocyst complex on the plasma membrane in yeast. Consequently, it may serve as a landmark for this entire polarized secretory process in yeast.

How might the exocyst complex be involved in targeting vesicles? At least one possible answer to this question comes from the recent studies of another subunit of the exocyst, Sec15. Guo *et al.* (29) detected a two-hybrid interaction between Sec15 and the Rab-GTPase Sec4. This interaction appears to be with the GTP-bound form of Sec4 because it is only seen when a GTP-hydrolysis-deficient mutant is used. This connection may serve as a bridge between the vesicle and the initial docking event at the plasma membrane.

Mammalian homologues of the components of the exocyst complex—Sec5, Sec6, Sec8, Sec10, Sec15, and Exo70—have all been identified in neurons (30–32). Similar to the yeast complex, these proteins are associated with each other in a high-molecular-weight complex that is peripherally associated with the neuronal plasma membrane. Antibodies raised against the neuronal Sec6 and Sec8 complexes detected related proteins in MDCK polarized epithelial cells (see also Chapter 4). Upon initiation of calcium-dependent cell–cell adhesion, this 17S complex is recruited to the lateral plasma membrane at sites of cell–cell contact. In a permeabilized cell exocytosis assay, antibodies against Sec8 of this complex inhibit delivery of basolateral proteins, but not delivery of apical proteins. This suggests that the complex has a specific function in the targeting of vesicles to the lateral membrane (33). Therefore, targeting of basolateral proteins to sites of cell–cell contact is analogous to targeting of post-Golgi vesicles in yeast to the bud tip (see also Chapter 4).

## 6.2 SNARE proteins: a highly conserved protein family required for a late step in the docking and fusion of vesicles with the plasma membrane

The role of SNARE proteins in exocytosis was first revealed by work on exocytosis of neurotransmitters in nerve terminals. A complex was found which was composed of one integral membrane protein on synaptic vesicles, synaptobrevin (also known as VAMP), and two integral plasma membrane proteins, syntaxin and SNAP-25. In 1993 Sollner *et al.* (34) demonstrated that these proteins form a stable stoichiometric complex with each other. However, the localization of these integral membrane proteins in two distinct membranes suggested that the formation of the complex might be involved in a critical step in the docking and fusion machinery. This led to the SNARE hypothesis, which suggested that the specificity of the vesicle target membrane docking and fusion is mediated by the selective association of these proteins with each other (34, 35). In yeast, homologues of syntaxin, synaptobrevin, and SNAP-25 exist that are functionally and structurally conserved (24, 36–38).

## 6.3 The SNARE regulatory proteins Sro7/77 and Sec1 may play a role in defining the sites of vesicle fusion at the plasma membrane by regulating localized assembly of t-SNARE complexes

Sec9 and the closely related proteins Sso1 and Sso2 are normally found distributed evenly along the plasma membrane. The lack of any polarity in the localization of the plasma membrane SNARE proteins Sec9 and Sso1/2 in yeast, and SNAP-25 and syntaxin in neurons, would appear to preclude any direct role for these proteins in targeting vesicles to particular sites on the plasma membrane (38, 39). However, only the heterodimer of Sec9 and Sso1/2 (or SNAP-25 and syntaxin in the neuron) is functional in binding to the vesicle SNARE, Snc1/2 (synaptobrevin or VAMP in the neuron). In fact, only a tiny fraction of the Sso1/2 and Sec9 present on the plasma membrane is in a complex (19). This is likely due in part to the fact that Sso1 and Sso2 form an intermolecular interaction. This interaction potently inhibits the ability to form t-SNARE complexes (40, 41).

It is likely that SNARE regulatory proteins mediate the assembly of active t-SNAREs (Fig. 4). The best-characterized accessory factor is Sec1, a member of a highly conserved family of proteins which, like SNARE proteins, acts at a number of different steps in the exocytic and endocytic pathway. Sec1 family members are thought to act as positive regulators of syntaxin (i.e. Sso1/2) family members. One possibility is that Sec1 binds to Sso1/2 in a conformation that increases its ability to form t-SNARE complexes with Sec9 by reducing the strength of the intermolecular inhibition by the N-terminal domain of Sso1/2 discussed above. Consistent with the possibility that Sec1 plays a role in polarized fusion, recent work has shown that Sec1 has a highly polarized distribution on the plasma membrane, being concentrated to the site of growth (42). Therefore, it is possible that while the major pools of plasma membrane SNARE proteins are not polarized in their distribution, the active subcomplexes of the proteins may be formed in a spatially restricted manner by the polarized action of the Sec1 and Sro7/77 proteins, which might control this assembly process (see Fig. 4).

## 6.4 The Rho-GTPase, Rho3, may act to regulate exocyst function through an interaction with the Exo70 subunit

In addition to its possible role in vesicle delivery in conjunction with Myo2, there is significant evidence that Rho3 may play a role in docking and fusion as well. Rho3 shows strong genetic interactions with a number of components that appear to act in the second stage of this pathway—the tethering, docking, and fusion step. In particular, we have identified Rho3 as a potent suppressor of an effector mutant in *sec4* (J. Adamo, G. Rossi, and P. Brennwald, submitted), and previously Sec4 has also been isolated as a suppressor of a *rho3* mutant (18). In addition, we have recently shown that overexpression of several other components, including the t-SNAREs Sec9 and

Sso1/2, can strongly suppress the loss of Rho3 function (19). Taken together, these observations suggest that Rho3 may have functions both in the vesicle delivery and in the docking and fusion step. Consistent with this idea, Rho3 has been shown by two-hybrid analysis to have strong interactions with a portion of Exo70, a component of the exocyst (43). This conclusion has been further supported by the fact that mutations in *rho3* show clear defects in exocytosis which are independent of defects in the actin cytoskeleton. Under certain conditions these defects appear to be at the level of efficient docking and fusion of secretory vesicles with the plasma membrane and not at the level of the delivery step (J. Adamo, G. Rossi, and P. Brennwald, submitted). Since the SNARE protein, Sec9, and SNARE-binding protein, Sro7/77, have strong genetic interactions with components of the exocyst complex, it is likely that these proteins all function together at sites of vesicle docking and fusion (19). Furthermore, the pattern of suppression would suggest that Sro7/77 acts downstream of the Rab-GTPase, Sec4, and the exocyst complex in polarized exocytosis (see Fig. 4).

# 7. Conclusions

Yeast genetics and studies in other cell types have given us a skeleton view of how the complex process of polarized exocytosis might work. However, there is still much to be learned about how this system functions, and how variations in the arrangement and interactions of these proteins might be exploited for spatial and temporal regulation and to mediate specific processes relevant in other cell types. It is not at all understood how the Rab interaction with the exocyst might be intertwined with the SNAREs. While it might be through the Sro7/77/tomosyn family, there is no evidence at this point to support such a possibility.

The analysis of polarized exocytosis in yeast and in other cell types has begun to delineate many of the essential components. In broad terms, these components appear to act in one of two major events. The first is the delivery of vesicles to the correct region of the cell with which they will fuse. The second step is the recruitment of the exocytic machinery to specific sites on the plasma membrane where these fusion events are allowed to occur. The sites of fusion in eukaryotic cells are often not static and therefore must be re-specified during a number of different processes. In yeast, the rearrangements occur under regulation of the cell-cycle control machinery, while in epithelial cells rearrangements occur as cells form contacts with other cells or respond to environmental cues (see Chapters 4 and 7). This exquisite regulation that underlies many aspects of morphogenesis and motility requires co-ordination between rearrangements of the cytoskeleton and the recruitment/activation of the exocytic machinery.

The functions of Rho3 appear to be particularly interesting because it has the properties of a molecule that can potentially co-ordinate three different steps necessary for polarized exocytosis. Rho3 has a role in polarization of the actin cytoskeleton, in the transport of vesicles from the mother cell into the bud, and in the docking and fusion of secretory vesicles with the plasma membrane. This co-ordination might be especially important to yeast cells, because the site of fusion is constantly changing throughout the cell cycle. At present, no Rho family member associating with or regulating the exocyst complex has been defined in mammalian cells. However, the region of yeast Exo70 that interacts with Rho3 is reasonably well conserved with its mammalian counterpart, and therefore such an interaction may indeed exist.

Future work will undoubtedly begin to unravel the precise details of the effect of activated Rho3 on Myo2 and Exo70, and the details of how the Rab/exocyst complex is functionally tied to SNARE complex assembly at sites of fusion. Furthermore, understanding the specific details of these interactions in epithelial cells and neurons will give us a much better understanding of the core mechanisms that underlie polarized exocytosis.

# Acknowledgements

We would like to thank our colleagues for scientific advice and for sharing unpublished information, and Luba Katz for critical reading of the manuscript.

# References

1. Drubin, D. G. and Nelson, W. J. (1996) Origins of cell polarity. *Cell*, **84**, 335.
2. Tkacz, J. S. and Lampen, J. O. (1973) Surface distributon of invertase on growing *Saccharomyces* cells. *J. Bacteriol.*, **113**, 1073.
3. Field, C. and Schekman, R. (1980) Localized secretion of acid phosphatase reflects the pattern of cell surface growth in *Saccharomyces cerevisiae*. *J. Cell Biol.*, **86**, 123.
4. Novick, P. and Brennwald, P. (1993) Friends and family: the role of the Rab GTPases in vesicular traffic. *Cell*, **75**, 597.
5. TerBush, D. R. and Novick, P. (1995) Sec6, Sec8, and Sec15 are components of a multi-subunit complex which localizes to small bud tips in *Saccharomyces cerevisiae*. *J. Cell Biol.*, **130**, 299.
6. Finger, F. P., Hughes, T. E., and Novick, P. (1998) Sec3p is a spatial landmark for polarized secretion in budding yeast. *Cell*, **92**, 559.
7. Novick, P. and Botstein, D. (1985) Phenotypic analysis of temperature-sensitive yeast actin mutants. *Cell*, **40**, 405.
8. Amberg, D. C. (1998) Three-dimensional imaging of the yeast actin cytoskeleton through the budding cell cycle. *Mol. Biol. Cell*, **9**, 3259.
9. Pruyne, D. W., Schott, D. H., and Bretscher, A. (1998) Tropomyosin-containing actin cables direct the Myo2p-dependent polarized delivery of secretory vesicles in budding yeast. *J. Cell Biol.*, **143**, 1931.
10. Johnston, G. C., Prendergast, J. A., and Singer, R. A. (1991) The *Saccharomyces cerevisiae* MYO2 gene encodes an essential myosin for vectorial transport of vesicles. *J .Cell Biol.*, **113**, 539.
11. Govindan, B., Bowser, R., and Novick, P. (1995) The role of Myo2, a yeast class V myosin, in vesicular transport. *J. Cell Biol.*, **128**, 1055.
12. Ayscough, K. R., Stryker, J., Pokala, N., Sanders, M., Crews, P., and Drubin, D. G. (1997) High rates of actin filament turnover in budding yeast and roles for actin in establishment and maintenance of cell polarity revealed using the actin inhibitor latrunculin-A. *J. Cell Biol.*, **137**, 399.

13. Lillie, S. H. and Brown, S. S. (1994) Immunofluorescence localization of the unconventional myosin, Myo2p, and the putative kinesin-related protein, Smy1p, to the same regions of polarized growth in *Saccharomyces cerevisiae. J. Cell Biol.*, **125**, 825.

14. Walch-Solimena, C., Collins, R. N., and Novick, P. J. (1997) Sec2p mediates nucleotide exchange on Sec4p and is involved in polarized delivery of post-Golgi vesicles. *J. Cell Biol.*, **137**, 1495.

15. Wu, X., Bowers, B., Rao, K., Wei, Q., and Hammer, J. A. R. (1998) Visualization of melanosome dynamics within wild-type and dilute melanocytes suggests a paradigm for myosin V function *in vivo. J. Cell Biol.*, **143**, 1899.

16. Salminen, A. and Novick, P. J. (1987) A ras-like protein is required for a post-Golgi event in yeast secretion. *Cell*, **49**, 527.

17. Goud, B., Salminen, A., Walworth, N. C., and Novick, P. J. (1988) A GTP-binding protein required for secretion rapidly associates with secretory vesicles and the plasma membrane in yeast. *Cell*, **53**, 753.

18. Imai, J., Toh-e, A., and Matsui, Y. (1996) Genetic analysis of the *Saccharomyces cerevisiae RHO3* gene, encoding a rho-type small GTPase, provides evidence for a role in bud formation. *Genetics*, **142**, 359.

19. Lehman, K., Rossi, G., Adamo, J. E., and Brennwald, P. (1999) Yeast homologs of tomosyn and *lethal giant larvae* function in exocytosis and are associated with the plasma membrane SNARE, Sec9. *J. Cell Biol.*, **146**, 125.

20. Pfeffer, S. (1999) Transport-vesicle targeting: tethers before SNAREs. *Nature Cell Biology*, **1**, 17.

21. Cao, X., Ballew, N., and Barlowe, C. (1998) Initial docking of ER-derived vesicles requires Uso1p and Ypt1p but is independent of SNARE proteins. *EMBO J.*, **17**, 2156.

22. Ungermann, C., Sato, K., and Wickner, W. (1998) Defining the functions of trans-SNARE pairs [see comments]. *Nature*, **396**, 543.

23. Weber, T., Zemelman, B. V., McNew, J. A., Westermann, B., Gmachl, M., Parlati, F., Sollner, T. H., and Rothman, J. E. (1998) SNAREpins: minimal machinery for membrane fusion. *Cell*, **92**, 759.

24. Katz, L., Hanson, P. I., Heuser, J. E., and Brennwald, P. (1998) Genetic and morphological analyses reveal a critical interaction between the C-termini of two SNARE proteins and a parallel four helical arrangement for the exocytic SNARE complex. *EMBO J.*, **17**, 6200.

25. Bowser, R. and Novick, P. (1991) Sec15 protein, an essential component of the exocytotic apparatus, is associated with the plasma membrane and with a soluble 19.5S particle. *J. Cell Biol.*, **112**, 1117.

26. Bowser, R., Muller, H., Govindan, B., and Novick, P. (1992) Sec8p and Sec15p are components of a plasma membrane-associated 19.5S particle that may function downstream of Sec4p to control exocytosis. *J. Cell Biol.*, **118**, 1041.

27. TerBush, D. R., Maurice, T., Roth, D., and Novick, P. (1996) The Exocyst is a multiprotein complex required for exocytosis in *Saccharomyces cerevisiae. EMBO J.*, **15**, 6483.

28. Novick, P., Field, C., and Schekman, R. (1980) Identification of 23 complementation groups required for post-translational events in the yeast secretory pathway. *Cell*, **21**, 205.

29. Guo W., Roth, D., Walch-Solimena, C., and Novick, P. (1999) The exocyst is an effector for Sec4p, targeting secretory vesicles to sites of exocytosis. *EMBO J.*, **18**, (4), 1071.

30. Ting, A. E., Hazuka, C. D., Hsu, S. C., Kirk, M. D., Bean, A. J., and Scheller, R. H. (1995) rSec6 and rSec8, mammalian homologs of yeast proteins essential for secretion. *Proc. Natl Acad. Sci., USA*, **92**, 9613.

31. Hazuka, C. D., Hsu, S. C., and Scheller, R.H. (1997) Characterization of a cDNA encoding a subunit of the rat brain rsec6/8 complex. *Gene*, **187**, 67.

32. Kee, Y., Yoo, J. S., Hazuka, C. D., Peterson, K. E., Hsu, S. C., and Scheller, R. H. (1997) Subunit structure of the mammalian exocyst complex. *Proc. Natl Acad. Sci., USA*, **94**, 14438.

33. Grindstaff, K. K., Yeaman, C., Anandasabapathy, N., Hsu, S. C., Rodriguez-Boulan, E., Scheller, R. H., and Nelson, W. J. (1998) Sec6/8 complex is recruited to cell–cell contacts and specifies transport vesicle delivery to the basal-lateral membrane in epithelial cells. *Cell*, **93**, 731.

34. Sollner, T., Whiteheart, S. W., Brunner, M., Erdjument-Bromage, H., Geromanos, S., Tempst, P., and Rothman, J. E. (1993) SNAP receptors implicated in vesicle targeting and fusion. *Nature*, **362**, 318.

35. Warren, G. (1993) Cell biology. Bridging the gap. *Nature*, **362**, 297.

36. Protopopov, V., Govindan, B., Novick, P., and Gerst, J. E. (1993) Homologs of the synaptobrevin/VAMP family of synaptic vesicle proteins function on the late secretory pathway in S. cerevisiae. *Cell*, **74**, 855.

37. Aalto, M. K., Ronne, H., and Keranen, S. (1993) Yeast syntaxins Sso1p and Sso2p belong to a family of related membrane proteins that function in vesicular transport. *EMBO J.*, **12**, 4095.

38. Brennwald, P., Kearns, B., Champion, K., Keranen, S., Bankaitis, V., and Novick, P. (1994) Sec9 is a SNAP-25-like component of a yeast SNARE complex that may be the effector of Sec4 function in exocytosis. *Cell*, **79**, 245.

39. Bennett, M. K., Calakos, N., and Scheller, R. H. (1992) Syntaxin: a synaptic protein implicated in docking of synaptic vesicles at presynaptic active zones. *Science*, **257**, 255.

40. Rossi, G., Salminen, A., Rice, L. M., Brunger, A. T., and Brennwald, P. (1997) Analysis of a yeast SNARE complex reveals remarkable similarity to the neuronal SNARE complex and a novel function for the C terminus of the SNAP-25 homolog, Sec9. *J. Biol. Chem.*, **272**, 16610.

41. Nicholson, K. L., Munson, M., Miller, R. B., Filip, T. J., Fairman, R., and Hughson, F.M. (1998) Regulation of SNARE complex assembly by an N-terminal domain of the t- SNARE Sso1p. *Nature Struct. Biol.*, **5**, 793.

42. Carr, C. M., Grote, E., Munson, M., Hughson, F. M., and Novick, P. J. (1999) Sec1p binds to SNARE complexes and concentrates at sites of secretion. *J. Cell Biol.*, in press.

43. Robinson, N. G., Guo, L., Imai, J., Toh-e, A., Matsui, Y., and Tamanoi, F. (1999) Rho3 of *Saccharomyces cerevisiae*, which regulates the actin cytoskeleton and exocytosis, is a GTPase which interacts with myo2 and exo70. *Mol. Cell Biol.*, **19**, 3580.

# 10 | Morphogenesis of skin epithelia

PIERRE A. COULOMBE and KEVIN McGOWAN

## 1. Introduction

By virtue of its strategic location at the surface of the body, the skin has multiple functions that define in many ways the relationship that the whole organism has with its external environment (1–4). The epithelial compartment of the skin acts as a barrier that restricts the exchange of water-soluble compounds, prevents the entry of infectious micro-organisms, filters out harmful ultraviolet irradiation, and contributes to immunological surveillance. In addition, epithelial specializations such as the hair and nail contribute significantly to thermal insulation, tactile sensation, sexual behaviours, and, in some cases, flight (feathers) and natural defence (claws). These various epithelial compartments lie over, or in some cases are embedded within, a connective tissue known as the dermis. Through its rich vascularization and innervation and a complex extracellular matrix, the dermis makes essential contributions to the properties and functions of skin. The regulation of cutaneous blood flow represents an important homeostatic mechanism for the regulation of body heat, and the afferent and efferent innervation of the skin are an integral part of sensory perception (3, 4). More relevant to this text, however, the dermis is the source of instructions that play essential roles in the morphogenesis and the homeostasis of mature skin tissue (5–13).

This impressive array of functions is made possible by a unique organization reflected in the histological appearance of the skin (Fig. 1). Excluding surface specializations such as hair and nail (and their functional equivalents in vertebrates other than mammals), the skin is generally considered to extend from the epidermis, a stratified squamous epithelium at its surface, to the hypodermis, an adipose-tissue-rich compartment found at its bottom (3, 4). Between the epidermis and the adipose tissue lies the dermis. Compared to the epithelial compartment, the dermis is fairly acellular (see Fig. 1), a notion that should not be taken as an indication of a poor organization. In addition to an extensive network of nerves and blood vessels, the dermis features a complex assortment of cell types and extracellular polymers. The latter includes several types of fibrillar and non-fibrillar collagens (which together represent 75% of the skin's dry weight), elastin, and various types of proteoglycans

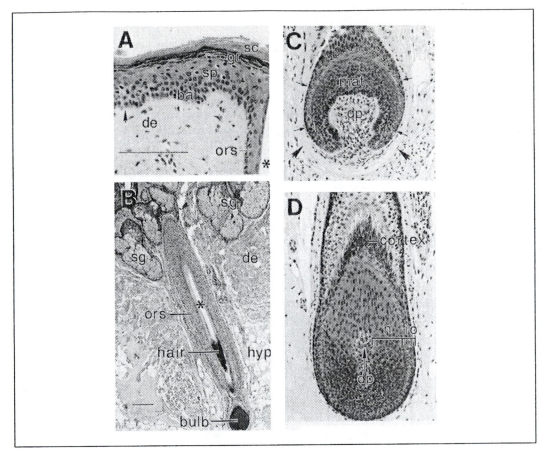

**Fig. 1** Histology of human skin. Human skin tissues were fixed in Bouin's fixative, embedded in paraffin, and 5 μm-sections were stained with haematoxylin and eosin. (A) Epidermis and part of a hair-follicle unit. The arrowhead at left depicts the interface between the dermis (de) and the epidermis. The main compartments of epidermis are labelled as follows: ba, basal layer; sp, spinous layer; gr, granular layer; sc, stratum corneum layer. The asterisk marks a hair-follicle canal. (B) Pilosebaceous unit. A late anagen-stage hair follicle is seen to project down into the hypodermis (hyp). The various compartments of this unit as follows: bulb, hair bulb; ors, outer root sheath; sg, sebaceous gland; de, dermis. The asterisk depicts the hair canal. (C) and (D) show two examples of anagen-stage hair bulb at a higher magnification. In (C), the arrowheads depict the outer boundary of the specialized extracellular matrix that surrounds the hair follicle, while the short arrows depict the outer boundary of the outer root sheath. dp, dermal papillae; mat, matrix epithelial cells. In (D), additional compartments of the hair bulb are highlighted, owing to a favourable sectioning plane. h, cellular precursors for the hair shaft; i, inner root sheath; o, outer root sheath. Bar in (A) equals 200 μm and applies for (C) and (D); bar in (B) equals 200 μm.

and glycosaminoglycans (2–4). A delicate balance exists between the cellular and the acellular elements found in the dermis, and its maintenance is critical to the wide range of functions performed by the skin as an organ (14, 15).

The concept of polarity, the major theme of this volume, applies particularly well to the skin. Polarity, or asymmetry, is indeed an intrinsic and omnipresent determinant of the biology of this fascinating organ. This chapter is focused nearly exclusively on the epithelial compartment of the skin, which provides ample material to illustrate some of the mechanisms responsible for the acquisition of the functional asymmetry that characterizes complex epithelia. Since it has been competently covered elsewhere in this volume (see Chapter 4), we will not deal with the topic of epithelial cell polarity in this chapter. Rather, the emphasis is placed on how the polarity of skin epithelia as a tissue is acquired during development, and maintained throughout adult life. For practical reasons our discussion is restricted to the epidermis and the hair follicle, since they represent two of the best-understood complex epithelia.

## 2. Skin epithelia are polar and asymmetrical in many respects

### 2.1 The epidermis as an example of stratified squamous epithelium

All complex epithelia are markedly asymmetrical in their basic organization, and this is an important aspect of their properties and function. Because they often occur at tissue surfaces, these epithelial sheets are subjected to significant frictional forces that result in desquamation and call for continuous self-renewal. Even under baseline conditions, therefore, complex epithelia tend to turn over at a significant rate. Complex epithelia such as the epidermis maintain a constant thickness and architecture because cell loss at their surface is precisely compensated by a steady production of new cells at the base of the tissue (16–18). In between the base and the apex of the tissue is a sizeable compartment featuring cells that are progressively acquiring the molecular traits of terminally differentiated squames. The features of the differentiation programme specifically executed in the epidermis are such that four horizontal compartments can be readily recognized in histological cross-sections (Fig. 1A). They consist of:

(1) the basal layer, which features relatively undifferentiated, mitotically active cells lying on a basal lamina;

(2) the spinous layer, which contains early differentiating cells that show spine-like radial projections at their periphery (hence the name);

(3) the granular layer, made of cells that are further differentiated and which display characteristic basophilic granules in their cytoplasm; and

(4) the stratum corneum layer, which consists of stacks of terminally differentiated, flattened cells (squames) that have lost their nucleus and all of their organelles (2–4, 18).

This programme of differentiation is carried out by keratinocytes, the major cell type within the epidermis, over a period of 2–4 weeks in human skin (3, 17). The discovery and characterization of molecules expressed as part of this programme established that, as expected, terminal differentiation is incredibly more complex than the four successive steps suggested by histology. This notion will be further discussed in Section 4, below.

Other than through its programme of terminal differentiation, the epidermis is also made polar by virtue of the distribution and the attributes of the non-keratinocyte cell types it contains (3, 4, 18). The neural-crest-derived melanocytes (~2–3% of total epidermal cell number) are the exclusive source of the melanin and related pigments that mediate protection against harmful ultraviolet radiations and impart the hair and epidermis with their colour. These dendritic-shaped cells are located within the basal layer and send several long and thin cytoplasmic processes between neighbouring keratinocytes. This cytoarchitecture facilitates the transfer of the bulk of the photoprotective melanin granules produced by melanocytes to the cytoplasm of these keratinocytes. Merkel cells are slow-adapting type I mechanoreceptors that also occur within the basal layer. They feature typical secretory granules in their cytoplasm and are associated with the termini of intraepidermal neurites that extend from sensory neurons innervating the skin. The frequency of Merkel cells is variable in skin epithelia, with the highest numbers found at sites of high tactile sensitivity. Finally, Langerhans cells (3–8% of cells) are a population of bone-marrow-derived cells that migrate into and establish themselves within the spinous layer of epidermis. These dendritic-shaped cells are involved in a variety of T-cell responses, and thus contribute to the skin's role of immunological surveillance. The distribution of each of these cell types within the epidermis thus appears to best serve its primary role(s) in skin tissue as a whole.

## 2.2 The hair as an example of an epithelial appendage

Whether it is hair, nail, tooth, or glands, all major types of epithelial appendages are markedly asymmetrical. Throughout this chapter the mammalian hair (and to a lesser extent, the early stage bird feather) will be used as an example. The hair typically occurs in the context of a pilosebaceous unit (Fig. 1B), which consists of the hair-producing tissue itself, one or several sebaceous glands that funnel their secretions into the hair canal, an arrector pili muscle that inserts in a specific location below the sebaceous gland, and a pocket of specialized mesenchymal cells known as the dermal papilla at the base of the unit, in a region known as the hair bulb (1–4, 19). This papilla (see Fig. 1C, 1D) is the source of essential instructions during both development and homeostatis in the adult (20; see below). Particularly in mammals, several types of hair can be recognized, based on criteria such as size, curl (shape), colour, anatomical location, sensitivity to androgenic hormones, features of the growth cycle, and more (1, 10, 19, 21).

Histologically, the cylinder-shaped hair is comprised of three major epithelial compartments, organized in concentric layers (1–4, 19). From outside in, these are the outer root sheath, inner root sheath, and hair shaft (Fig. 1B, 1D). The outer root sheath is a stratified epithelium that resembles and is contiguous with the epidermis (Fig. 1A, 1B). It has its own pool of progenitor cells located within the outermost (basal) layer that contacts the basal lamina. Within the outer root sheath, differentiation proceeds along an axis that is parallel to the surface of the skin (22 and references therein). A local swelling of the outer root sheath occurs at the point of insertion of the arrector pili muscle. This region is known as the bulge, and is believed to contain the pool of stem cells for all the epithelia found in hairy skin tissue (13, 23, 24). This pool of epithelial cells plays a key regulatory role during the adult hair cycle (see below). The inner root sheath (Fig. 1D) features three distinct layers of epithelial cells known as (from outside in) Henle's, Huxley's, and a cuticle. The hair shaft itself is also comprised of three layers, known as the hair cuticle, the hair cortex, and the medulla (again, from outside in). A single type of progenitor cell, the matrix epithelial cells (Fig. 1C, 1D) gives rise to all three layers of the inner root sheath and of the hair shaft, a remarkable occurrence given that they are each the result of a distinct programme of differentiation. Presumably, the position of a matrix cell relative to the dermal papilla plays a key role in its choice of a specific differentiation programme (3, 4, 19). The growing hair is therefore a complex epithelial mosaic involving multiple programmes of terminal differentiation, each characterized by unique biochemical markers (for a recent review see ref. 19).

In mature skin tissue, the hair undergoes a cycle, with phases of active growth interspersed with phases of rest (3, 10, 13, 19, 21). Three main stages are recognized: *anagen*, during which the follicular unit is maximally elongated, features a prominent bulb structure, and actively produces new hair (Fig. 1B); *catagen*, a transient phase during which the germinative (lower) segment of the follicle is destroyed by a massive wave of apoptosis (with the notable exception of the dermal papilla), thereby drastically shortening the follicle; and *telogen*, a resting phase during which only the permanent part of the follicle persists. Resting hairs extend down to a region just below the bulge, the postulated reservoir of stem cells. The dermal papilla remains physically proximal to the epithelium at the base of the follicle throughout the hair cycle (10, 12, 13, 21). The size of the dermal papilla, which is largest in full anagen and becomes small and compact in telogen, is the primary determinant of the width of the hair being produced. The length of the hair, on the other hand, is primarily determined by the relative length of the anagen and telogen phases (10, 12, 21). Cycle-related variations also occur in the vasculature, innervation, and in the composition of the extracellular matrix that surround the hair tissue (21). In wild animals, seasonal moulting represents a manifestation of synchrony in the progression of large groups of hair follicles through the cycle, and is under endocrine control (10, 21). In the skin of laboratory rodents, as in humans, only the first and second hair cycles are believed to be truly synchronized (9, 10, 11). The existence of this fascinating growth cycle gives the hair a unique status in mammalian biology, and accordingly many of the signalling pathways that play fundamental roles in development and morphogenesis are also at play in adult skin.

Beyond its histological make-up and its growth cycle, additional attributes make the hair interesting from the standpoint of polarity. Mammals feature several types of hair (cf. above) that are organized according to intricate patterns of density, spacing, and orientation relative to the main axes of the body, limbs, etc. (2, 7, 10, 11). In rodents, for instance, the highly specialized vibrissae follicles are restricted to the anterior segment of the head, whereas foot-pad skin features no hair follicles. In addition, most hairs lie at a characteristic angle relative to the skin surface (10, 11, 25). These characteristics are reflected in the organization of every one of the components making up a pilosebaceous unit, and are established at an early stage during embryonic development.

To summarize, our description of skin epithelia in this section was meant to emphasize the existence of polarity or asymmetry at three distinct levels of anatomy and/or organization:

(1) that which is intrinsic to any stratified epithelium (of which the skin has several);

(2) that exemplified by appendages, which are mosaics of multiple cell types integrated to form a complex functional unit; and, finally,

(3) that associated with variations in the spatial arrangement of a set of basic modules (e.g. epidermis, hair, gland) to generate a pattern that is adapted to fit the regional needs of the body in terms of the density, orientation, growth characteristics, etc., of each module.

This remarkably complex architecture arises during development, starting from a monolayer of unpolarized epithelial cells known as the embryonic ectoderm. The events responsible for the morphogenesis of skin epithelia are incredibly interesting, and are discussed in the next section.

# 3. Morphogenesis of skin epithelia

## 3.1 Generalities

The morphogenesis of skin epithelia is like that of any other tissue, in that it occurs as a continuum of cellular and molecular events during which cell fates become progressively restricted (26–28). Such events typically proceed according to spatially and temporally defined gradients in the embryo, so that at any given time of development, different body sites may be at a different stage of their progression through a fixed programme. Extensive studies in amphibians, chicken, and rodents have shown that the major features of this morphogenetic programme are conserved among vertebrates (7, 10–12, 19, 29). This said, the onset of the events leading to the morphogenesis of skin epithelia relative to development as a global phenomenon may vary significantly between species, as may progression through the different steps. Within mammals, for instance, hair follicle formation is initiated during the first half of pregnancy in humans and during the second half in rats and mice (4, 10). The description given in this section emphasizes skin epithelia, and applies to vertebrates in a general fashion. The reader interested in the development of the dermis, a derivative of the mesoderm and neural crest, is referred to other texts (30).

Dating back several decades, classical experiments involving the recombination and transplantation of tissues dissected from various species have led to the identification of the cellular origin of the key molecular cues that preside over the morphogenesis of skin epithelia. These pioneering efforts, largely ignored in this text (reviewed in 5–8, 10, 12, 20), yielded (at least) two major concepts that paved the way for the identification, in recent years, of a plethora of molecules and pathways involved in these events. The first concept is that morphogenetic events involve interactions between different embryonic tissue layers (28, 31, 32). Skin epithelia are no exception, as their development involves reciprocal interactions between ectoderm and mesoderm (see below). The second concept is that the adoption, on the part of a precursor cell, of a given fate among a choice of two or more fates is controlled by a combination of long-range and short-range interactions (recently reviewed in 25, 27, 28). Long-range signalling occurs in the form of gradients of secreted molecules having the classically defined properties of morphogens. Accordingly, the fate of a precursor cell depends upon the precise concentration of morphogen to which it is exposed. These choices are locally reinforced in the context of short-range interactions that require physical proximity of the participating cells. The superimposition of multiple signals having different operational ranges, as well as synergistic or antagonistic influences, underlies both global patterning and the establishment of local boundaries during development.

For pragmatic reasons, it is useful to arbitrarily partition the development of the tissue of interest into a limited number of discrete steps. Here, we segment the morphogenesis of skin epithelia as a succession of three steps:

(1) commitment of embryonic precursor cells towards a 'general' epithelial fate;

(2) commitment of these cells towards a specific type of epithelium; and

(3) morphogenesis and differentiation within that specified epithelium.

## 3.2 Commitment of embryonic precursor cells towards a 'general' epithelial fate

The precursor cells that will give rise to all skin epithelia have their origin in the embryonic ectoderm (10, 33). Depending on the signalling cues received, the ectoderm will become either the neural plate (on the dorsal side of the vertebrate embryo), epithelium (lateral and ventral sides of the embryo), or neural crest (interface between neural and epithelial fates). A series of elegant molecular studies, conducted mostly in *Xenopus*, has led to the discovery that a long-range signalling gradient, generated along the dorsal–ventral axis of the embryo, underlies the specification of these three fates (for a recent review see ref. 33). The key protagonist(s) in this gradient are secreted proteins known as bone morphogenetic proteins (BMPs), which are members of the TGF-β superfamily (34), and a variety of associated regulators. When unopposed, BMP-2 and/or BMP-4 signals the ectoderm towards an epithelial fate via an action on specific cell-surface receptors (35). Interference with the BMP-2/4 signal allows the ectoderm to proceed along a default pathway, which is to give rise to the neural plate (36). Several secreted molecules, including chordin, follistatin, and noggin, can antagonize the BMP-2/4 signal by direct binding to the protein (37, 38). These proteins, chordin in particular, have been shown to occur as a gradient, with peak levels on the dorsal side of the embryo and gradually decreasing levels along the dorsal–ventral axis.

Yet another type of protein, personified by tolloid, a member of the astacin family of metalloproteases closely related to BMP-1 (39), is involved in regulating these fate specifications. Indeed, tolloid and related proteins are capable of cleaving chordin and its relatives when bound to BMP-2/4, thereby releasing the latter and enabling it to bind its receptor (40; see ref. 41 for a recent review). The biochemical interplay between BMP-2/4, the protagonist, chordin and related proteins, the antagonists, and tolloid and related proteins, which 'antagonize the antagonists', therefore creates a sophisticated signalling gradient that underlies the commitment of embryonic ectoderm cells towards either a neural, neural crest, or an epithelial fate (33, 41).

## 3.3  Patterning the skin, or the rise of the epithelial appendages

A series of inductive interactions between the ectoderm and the mesoderm lying immediately beneath it leads to the formation of all types of skin appendages and thus sets up the pattern characteristic of the mature tissue. Pioneering tissue recombination experiments, in which embryonic epithelium and mesenchyme harvested from various body sites at different times of development were recombined and transplanted into suitable hosts, have collectively shown that these interactions are reciprocal, operate over relatively short distances, and can be broken down into three discrete steps (10). A *first message* originates from the mesenchyme and instructs the ectoderm directly above it to form an appendage. The induced epithelium will, in turn, elaborate a *second message*, instructing local mesenchymal cells to aggregate and form a papilla. The *third message* emanates from the dermal papilla and causes the induced epithelium to form a specific type of appendage (e.g. hair, nail, tooth, gland, etc.).

The available evidence indicates that the first message is universal, in that it can be interpreted across species boundaries, whereas the second and third messages function in a species-specific manner (10, and references therein). Major advances have been made in the past 10 years in our understanding of the molecules and pathways involved in these tissue interactions. Here again, the net balance between opposing influences plays a key role in directing cells towards one among a limited number of fates. Established signalling molecules, such as fibroblast growth factors (FGFs) and the BMPs, have been implicated in these events. Newer ones, such as Wnts, Notch, and sonic hedgehog have made the transition from being known as segment polarity genes in *Drosophila* to being key players in vertebrate tissue morphogenesis. Most remarkably, however, many of these molecules have been genetically and mechanistically implicated in the generation of a variety of solid tumours, underscoring the intimate relationship between neoplastic transformation and normal development.

Studies performed primarily in chick embryos identified an FGF molecule as a candidate effector for the *first message* (25, 29), which is issued by the mesenchyme and targets the epithelium. FGFs are a family of cytokines that impact profoundly on proliferation, differentiation, migration, and other basic cellular activities (42). Accordingly, FGF-mediated signalling is prevalent during development as well as in mature tissues undergoing remodelling. Secreted FGF signals are transduced by a small family of cell-surface receptors, and ligand activity can be significantly modulated by high-affinity interactions with the heparan sulphate proteoglycans that occur in extracellular matrices (42). A specific FGF family member, FGF-4, is expressed in a patchy, grid-like fashion in the embryonic ectoderm just prior to its patterning. Consistent with the proposed role, ectopic delivery of FGF-2 or FGF-4 via different experimental means (retroviruses, coated beads) is sufficient to initiate new placode formation in chick embryonic ectoderm (43, 44). The available evidence suggests that reception of the FGF signal antagonizes the BMP-2/4 signal, which pushes the ectoderm towards the formation of epidermis, the default pathway. In support of this competition model, an artificial increase in the levels of BMP-2/4 inhibit placode formation in chick embryonic skin (45). That BMP-2/4 promotes the default pathway in this context is interesting, given that at an earlier stage of development it causes the pre-neurulation ectoderm to adopt an epithelial fate at the expense of a neural fate (26, 33).

The Notch signalling pathway functions to refine the emerging pattern promoted by the competition between FGF-4 and BMP-2/4 (46). Notch is another example of a pathway discovered in *Drosophila* and subsequently found to be conserved throughout metazoans (see ref. 47 for a recent review). Notch and its vertebrate orthologues are large integral membrane proteins exposed at the cell surface, where they acts as receptors in the context of a short-range signalling pathway. The main ligands for Notch, Delta (chicken), and Jagged/Serrate (*Drosophila*), are transmembrane proteins themselves, and are believed to act exclusively on adjacent Notch-receptor-bearing cells (47). Activation of either one of two known Notch receptors, Notch-1 and Notch-2, through ligand binding results in the transcription of distinct sets of target genes. The transduction of the signal likely begins with the cleavage of the intracellular domain of the Notch receptor, which may itself be part of the complex imported into the nucleus (47, 48). Because of its unique features, the Notch pathway functions to sharpen, at a local level, the boundaries between adjacent multicellular cellular domains.

In chick embryonic skin, Delta-1 is present in the mesenchymal condensate during early placode formation (49). The emerging evidence supports a model in which the Notch pathway helps to fortify the boundaries between the 'ectodermal domains' created as a result of the antagonism between FGF and BMP signals (25, 49–51). According to this model, Delta-1 expression in mesenchymal cells is stimulated by an ectodermally derived FGF signal. Delta-1 induces Notch-1 expression in the ectoderm directly above it and Notch-2 expression in adjacent mesenchymal cells. Notch-1 activation would stimulate placode growth, whereas Notch-2 expression would inhibit it (49–51). One can therefore envision a scenario whereby iterations through this pathway would act locally to consolidate the boundary between 'appendage' and 'non-appendage' domains in the ectoderm (47). Although much remains to be resolved, the emerging picture for the molecular events surrounding the *first message* is already seen to involve complex interactions between different tissue layers, as well as a plethora of molecular signals.

The *second message* in the trilogy of epithelial–mesenchymal interactions originates from the induced epithelium and instructs the mesenchymal cells underneath it to condense into a papilla. The significance of this event extends beyond development, as the papilla will retain its inductive capabilities throughout the life of the organism. There is increasingly strong experimental evidence that the Wnt and sonic hedgehog signalling pathways may be part of the *second message*. Both these pathways impact profoundly on appendageal growth at an early stage of development, and may well do so by promoting the formation or stimulating the function of the dermal papilla.

Wnt genes encode a large family of evolutionarily conserved, secreted proteins that play critical roles as intercellular signalling molecules during development (52). Studies on *Wingless*, a segment polarity gene in *Drosophila* that is orthologous to mouse's Wnt-1, contributed significantly to our understanding of this pathway. A key effector of Wnt signals is β-catenin, a cytoplasmic protein that is structurally and functionally related to the protein Armadillo in *Drosophila*, and which had long been know to be a structural component of adherens junctions (53, 54). The latter is a cell–cell junction device that features classical cadherins as its transmembrane adhesive entity and which is linked, through a protein complex containing β-catenin, to the actin cytoskeleton on the intracellular side (53). β-Catenin was subsequently also found to occur in the nucleus under specific conditions, where is it is part of a complex with lef-1 (lymphoid enhancer factor 1), a member of the HMG box-containing T-cell factor (TCF) family of architectural transcription factors (55, 56). Binding of *Wingless*/Wnt ligands to the *Frizzled* family of cell-surface receptors results in the stabilization of cytoplasmic β-catenin and enables its translocation to the nucleus, where it impacts on the transcription of several genes as part of a DNA-binding complex with a TCF family member (55–57). This effector role of β-catenin in the Wnt signalling pathway is distinct from its structural role in the adherens junction. Genetic screens and positional cloning have together identified several key players upstream from the formation of the β-catenin/TCF complex, including glycogen synthase kinase-3β, the adenomatous polyposis colon (APC) protein, and axin (56, 58, 59). Collectively these cytoplasmic proteins act to sequester β-catenin and target it for rapid degradation in a phosphorylation-regulated manner.

The critical importance of the Wnt pathway for the growth and morphogenesis of epithelial appendages has been demonstrated mostly through gene manipulation experiments in mice. Inactivation of the mouse *lef-1* gene, which is expressed in both the ectodermal and mesenchymal compartments at an early stage of placode form-ation (60–62), results in the absence of vibrissae, a severe reduction in the number of pelage hair follicles, and alterations in other types of appendages (63). In contrast, targeted expression of *lef-1* in the progenitor (basal) layer of skin epithelia in trans-genic mice results in anomalies in the positioning and orientation of hair follicles, and interestingly, in the formation of ectopic appendages (60). On the other hand, inactivation of the β-catenin gene has not yet provided much useful information about its role in appendageal morphogenesis, and it is lethal at a very early stage of mouse development (64). Here again, however, a straight transgenesis strategy has paid dividends, as the targeted expression of a constitutively active form of β-catenin in the progenitor layer of mouse skin epithelia was found to cause a massive stimulation of hair follicle morphogenesis at a postnatal stage (65). This experiment demonstrated that the artificial stabilization of β-catenin can, under specific conditions met specifically in postnatal skin (even though the promoter used becomes active at mid-gestation), initiate the formation of a new and complete pilosebaceous unit, including a dermal papilla. Collectively, these and other studies point to a Wnt molecule(s) as a potential component of the second message. Candidate molecules are Wnt10b, which is present in the early mouse hair placode (66), and Wnt7a, expressed in the early chicken feather placode (29). Of interest, other Wnt family members have recently been implicated at a later stage of hair growth and differentiation (127).

Sonic hedgehog (Shh) is yet another example of a signal that plays an important role at this juncture. As for *Wingless–Armadillo* (cf. above), the *Hedgehog* pathway was discovered while screening for genes affecting segment polarity in *Drosophila*, and a vertebrate counterpart, Shh, was subsequently identified in vertebrates and associated with key patterning events during development (67–70). The Shh signal is secreted, can operate over short as well as long distances, and induces transcription of specific target genes (25, 69, 70). The mechanism by which this pathway operates is original in several respects. Hedgehog proteins undergo an autocleavage reaction during their biosynthesis that produces an amino-terminal fragment which becomes modified with a cholesterol adduct (68). Subsequently, activation of the pathway by the processed and secreted ligand requires two sequential inhibitory steps at the plasma membrane, as follows. The gene *Patched* encodes a large integral membrane protein that acts as the sole Shh receptor at the surface of its target cells. In that absence of the Shh ligand, Patched inhibits the function of Smoothened, an integral membrane protein related to G-protein-coupled receptors. Binding of Shh to Patched relieves this inhibition, and enables Smoothened to activate the downstream effectors of this pathway (69, 70). In *Drosophila*, this results in the transcription of several genes, including *Wingless* (Wnt), *Decapentaplegic* (a BMP-like molecule), and *Patched* itself (implying self-regulation via a negative feedback loop). The identity of these target genes underscores the existence of cross-talk between the major signalling pathways operating during skin morphogenesis.

In developing mouse skin, *Shh* mRNA is first detected in the ectoderm, where it appears in the emerging placode prior to the condensation of the underlying mesenchyme (71; see ref. 72 for similar findings in chick embryo skin). As placodes expand into the mesenchyme and become elongated, *Shh* expression is maintained at the distal tip, proximal to the dermal papilla (71). *Patched*, on the other hand, is expressed in both the ectoderm and the underlying mesenchyme (66, 73). Introduction of a null mutation in the mouse *Shh* gene causes embryolethality, owing to multiple developmental defects (74). Subsequent studies involving the grafting of newborn skin from *Shh* null mice in a suitable recipient host mouse strain showed that the growth and morphogenesis of hair follicles is severely impaired at a postplacode stage, correlating with a poorly developed dermal papilla (66, 73). These interesting findings demonstrate that the Shh pathway generally functions to promote and regulate the growth of appendages. Additionally, they may implicate Shh in the elaboration of a functional mesenchymal papillae, which corresponds to the endpoint of the second message. Consistent with a role in this context, studies in the chick embryo showed that FGFs stimulate the expression of *Shh*, while BMPs suppress it (44, 45).

As mentioned above, the *third and final message* emanates from the dermal papilla and instructs the induced ectoderm to give rise to a specific type of appendage (hair, gland, nail; tooth, thymus, etc.). This requires the onset of differentiation-specific gene expression, a rich topic deserving its own chapter. Selective aspects of differentiation-specific gene expression will be addressed in Section 4 below.

Also germane to hair-patterning events are the determination of the type, size, shape, and orientation of the follicles, and we will limit our discussion of these to a few comments. The major determinant of hair calibre is the size of the dermal papillae (10), and this may presumably be under the control of the *second message*. With regards to hair shape, transgenic mouse studies have implicated transforming growth factor-α (TGF-α), a ligand of the epidermal growth factor (EGF) receptor, in this process (75, 76). Finally, recent studies in chick embryos suggest that Hox genes (77) may contribute to define the angle and orientation of the hair follicle during skin morphogenesis. Indeed, studies by Chuong and colleagues have shown that the non-cluster Hox family genes *msx-1* and *msx-2* are each expressed in the anterior segment of the emerging placodes (29, 78 and refs therein). Much remains to be learned, however, about how these attributes of asymmetry are set during embryogenesis.

## 3.4 The adult hair cycle likely recapitulates key steps of appendageal morphogenesis

In adult skin, re-entry of a resting (telogen) hair follicle into a growth (anagen) mode involves interactions between the dermal papilla and a special population of keratinocytes located in the bulge (recently reviewed in ref. 13). The stimulus (stimuli) that trigger(s) bulge keratinocytes to proliferate, migrate downward, and re-form a hair-producing bulb structure originate(s) from the dermal papilla. The bulge is appropriately located at the distal end of the permanent segment of the pilosebaceous unit, and contains a population of slowly cycling cells believed to be epithelial stem cells (23, 24; for a slightly different view see refs 79, 80). The molecular events surrounding re-entry into anagen are likely to be analogous to those surrounding the inception and execution of the *third message* during the embryonic phase of hair-follicle development. Depending on the source and the nature of the signal(s) that cause(s) the activation of the dormant dermal papilla during re-entry into anagen, in fact, it is even possible that the similarities between adult hair cycling and embryonic hair morphogenesis partially extends to the *second message*. However related they may be, the two series of events can not be entirely the same, as appendages are not regenerated following full-thickness injury to Adult skin (15, 81).

Some of the signals that have been implicated directly in regulating transitions between the various phases of the adult hair cycle are well known to developmental biologists. Among them, for instance, is a FGF family member. Mice carrying a null mutation in the FGF-5 gene, which is normally expressed in the outer root sheath at a late stage of anaphase, have abnormally long hair owing to a significant delay of entry into catagen (82). Another is the EGF receptor signalling pathway. Several groups have shown that the inactivation of the mouse EGF receptor gene is lethal, with the time of death ranging from mid-gestation to 3 weeks postnatally, depending upon the genetic background (83–85). Follow-up studies have shown, again through skin grafting experiments, that EGF receptor null hair follicles can not transit into telogen and exhibit a prolonged inflammatory reaction (extended catagen ?) that coincides with hair loss (86). Targeted expression of a dominant-negative EGF receptor mutant to the skin of transgenic mice yielded very similar findings (87).

Additional signals promoting specific transitions between phases of the hair cycle have been identified, but their mode of action remains unclear. The *hairless* mutation is a recessive trait that arose as a result of the insertion of a leukaemia virus in the HRS/J mouse strain (88 and refs therein). These mice initially grow a normal coat of hair, but lose all pelage hair in a head-to-tail fashion due to a failure of re-entry into anagen after completion of the first hair cycle. The cloning and subsequent character-ization of the gene disrupted by this insertion revealed that it produces, through alternative splicing, two mRNAs encoding a short and a long form of a protein that contains a potential zinc-finger domain (88, 89). The short form of *hairless* mRNA occurs only in the skin (89), where intriguingly it is found in epidermis as well as hair (88). Consistent with the mouse phenotype, mutations were recently discovered in the *hairless* gene of individuals suffering from an inherited, severe form of alopecia (90). The function of the *hairless* gene products is not yet known; intriguingly, *hairless* was rediscovered in the context of a screen for thyroid-hormone-responsive genes in developing rat brain (91).

Neurotrophins are a small family of secreted proteins related to nerve growth factor that are expressed in target tissues that regulate the survival, differentiation, and function of neurons (92). Gene knock-out and transgenic mouse experiments showed that neurotrophin-3 (NT-3) is essential for the survival of most sensory neurons before they reach their target early in development, and that skin-derived NT-3 is essential for the development of mechanoreceptors associated with hair follicles and Merkel cells (93–95). Recent studies have shown that NT-3 is also expressed in a nerve-tissue-independent fashion at sites of epithelial–mesenchymal interactions (including hair placodes) during mouse development (96). Early during hair morphogenesis, interestingly, NT-3 expression is restricted to the mesenchyme and appears to depend upon an ectoderm-derived Wnt signal (96). In adult skin, expression of NT-3 mRNA and of its high-affinity receptor, TrkC, occurs in the outer root sheath, matrix, and dermal papilla, with peak levels reached during late anagen and early catagen (95, 97). Constitutive expression of NT-3 in transgenic mouse skin epithelia causes precocious entry of hair follicles into catagen. Conversely, mice hemizygous for a null mutation in the NT-3 gene (93) show a delay in the entry of hair follicles into catagen (97). These findings strongly suggest that a NT-3 signal plays a critical role in the transition into catagen for the mature hair follicle, a notion that is consistent with its co-expression with several markers of apoptosis (97).

## 3.5 The periderm layer represents an enigma

Coinciding with the determination of appendages, the surface ectoderm begins to stratify in vertebrate embryos (see Fig. 2). A new cell type, the periderm, appears at the surface of the two-layer ectoderm (98 and refs therein). Periderm epithelial cells are mitotically active (99, 100 and refs therein) and, other than their location, they can be easily discriminated by their flat shape (Fig. 2A) along with molecular markers (see below). The appearance of the periderm layer follows a cranio-caudal pattern similar to that which characterizes appendages, leading to the speculation that the same signals may be involved (100). An important study in which the fate of precursor cells was mapped in developing mice directly supports this notion, in that the periderm and appendages were found to share a cellular precursor that is distinct from that giving rise to epidermis (99). It is generally believed that the periderm layer is shed prior to birth and has no known equivalent in mature skin epithelia. While its function remains to be established (98), it is clear that the periderm does not mediate a barrier function akin to the differentiated layers of adult epidermis (101).

## 3.6 A link between developmental pathways and skin cancer

Given their prominent role in the control of both the growth and differentiation of precursor cells, it should come as no surprise that some of the signalling pathways operating during skin morphogenesis have recently been implicated in cancer. Thus, mutations in components of the Wnt signalling pathway have been discovered to occur in many cancers, including hair-follicle tumours (102,103). A priori, this is not surprising, as the first Wnt gene, *wnt-1*, was discovered as a proto-oncogene in a virus-induced mammary tumour (104). Moreover, the APC gene was discovered by positional cloning in the context of a search for genes responsible for colon cancer (105). Since then it has been clearly established that most cases of colorectal cancer are associated with an activation of the Wnt signalling pathway via stabilization of the cytoplasmic pool of free β-catenin (102, 106). In most instances of this cancer, this stabilization event results from inactivating (somatic) mutations in APC, but a small fraction of cases is associated with activating mutations in β-catenin (102). Pilomatricomas, on the other hand, are hair-follicle tumours that are believed to originate from hair matrix epithelial cells. Fuchs and her colleagues (103) recently discovered that most cases of pilomatricomas involve the activation of Wnt signalling as a consequence of somatic mutations in β-catenin as opposed to defects in APC. A mechanistic link between mutational activation of this pathway and uncontrolled cell growth has recently been elucidated by the discovery that TCF/β-catenin complexes directly regulate the c-MYC and cyclin D1 genes (107–109).

Likewise, mutations in various components of the Shh pathway have been discovered in another skin tumour, basal cell carcinoma (BCC). Basal cell carcinoma is the most common tumour affecting light-skinned people (110) and is thought to arise from the uncontrolled proliferation of hair-follicle precursor epithelial cells (25). A number of studies have shown that the uncontrolled stimulation of the Shh pathway by various experimental means causes BCC-like lesions in chick and mouse skin (29, 72, 111, 112). In instances of human BCC, inactivating somatic mutations have been discovered in Patched (113,114,115), which normally acts to repress the pathway, while activating mutations were found in Smoothened (116), which is an activator of Shh signalling. That key regulators of the Wnt and the Shh pathways are frequently targeted in different types of human epithelial cancers underscores their importance in the regulation of cell growth in these systems (25).

# 4. Polarity as reflected through differentiation-specific gene expression

## 4.1 Generalities

It is easily understood that the structural and functional polarity of stratified epithelia goes hand-in-hand with the stage-specific expression of several regulatory genes and an even greater number of structural genes. Through the application of the up-and-coming DNA chip technology (117), it will soon be possible to determine, for instance, what fraction of the human genome is expressed in a given stratified epithelium alone (likely to be an impressive number), and get a better appreciation for the timing and extent of the changes in gene expression that occur as epithelial cells progress through differentiation. One of the exciting prospects of this technology is the ability to identify groups of genes that show co-ordinate regulation at the mRNA level in the absence of any obvious biochemical or functional relationship between their products (see ref. 118 for an elegant example involving yeast cells). Such large gene groupings will provide a glimpse of how basic activities such as migration, compaction, or differentiation are accomplished at the whole cell and even tissue levels.

However appealing the prospect of applying DNA chip technology may be, a great deal of information has already been gathered about the molecular events that underlie morphogenesis, differentiation, and homeostasis in complex epithelia. As discussed throughout this chapter, genetic screens involving *Drosophila* and other model organisms have contributed immensely to this. Another significant source that has not been discussed yet is the study of multigene families whose individual members show differentiation-related expression. There are many such families whose individual members show an interesting regulation in complex epithelia like epidermis, and we will name only a few. The retinoic acid receptors and related transcription factors represent a relevant example, in part because they contribute to regulate this differentiation programme (119–121). The desmocollin and desmoglein classes of desmosomal cadherins, involved in cell–cell adhesion, each feature several genes and alternatively spliced mRNAs that are regulated in a stratification-related manner in the epidermis (122–124). Connexin genes, which encode gap-junction-forming proteins enabling metabolic coupling between participating cells, also show intriguing spatial patterns in the epidermis (125,126). Relevant to this discussion as well is the epidermal differentiation complex, which is located on human chromosome 1q21 and consists of at least 28 genes expressed during epidermal differentiation (128). Of these, at least 13 genes encode precursors of the cornified envelope, while another 13 genes encode S100A calcium-binding proteins (129). It is probably fair to say, however, that no group of genes has contributed more to our understanding of terminal differentiation in complex epithelia than those encoding keratins.

## 4.2  Introduction to keratins

In spite of a recent revolution in gene discovery, the single molecular criterion that most successfully defines the type and differentiation status of an epithelial cell remains keratin gene expression. Keratins are structural proteins encoded by two groups of genes, type I and type II, which are distinct by their genomic structure and sequence relatedness. Each sequence type is represented by more than 20 genes that are clustered within separate loci in mammalian genomes (for recent reviews see refs 130,131). Keratin proteins have the intrinsic ability to polymerize into 10 nm wide filaments, and, accordingly, they belong to the superfamily of intermediate filament proteins. Keratin filaments occur in most, if not all, epithelial cells (132, 133), where they are particularly abundant. These several micrometre long filaments are attached at the surface of the nucleus as well as at junctional complexes in the plasma membrane, and typically span the entire cytoplasm in between (134). A major function of these keratin filaments is to provide mechanical resilience to epithelial cells and tissues. Accordingly, functional alterations in keratin proteins (e.g. via mutations) result in fragility syndromes whereby the epithelial cell type(s) affected rupture when subjected to trivial mechanical trauma (134–137).

At a biochemical level, a keratin filament is an obligatory heteropolymer involving type I and type II proteins (138) in a 1 : 1 molar ratio (139). As a result of this requirement, epithelial cells must co-ordinately regulate the expression of at least one type I and one type II gene. As it turns out, type I and type II keratin genes are regulated in a pairwise, epithelial tissue-type, and differentiation-specific manner (see below), creating patterns that have been conserved among vertebrates (131–133). Each type of keratin genes can be further partitioned into two subgroups based upon expression in soft or hard epithelia (the latter includes hair, nail, and subsets of the dorsal tongue epithelium). The type I keratins expressed in soft and hard epithelia are designated as K9–K20 and Ha1–Ha7, respectively. The type II keratins expressed in soft and hard epithelia are designated as K1–K8 and Hb1–Hb7, respectively (131, 133). Generally speaking, type I keratins tend to be smaller and more acidic proteins compared to type II keratins (132, 133). Whereas early *in vitro* studies showed that any combination of type I and type II keratins can produce a fibrous polymer (140), the notion that these genes have been highly conserved in both their sequence and their regulation among higher vertebrates suggests a important functional relationship between keratin pairs and the structure and function of epithelial cells (133). Consistent with this notion, gene-replacement studies in transgenic mice recently showed that skin epithelia do not tolerate substitutions in type I keratin sequences, even when they involve related proteins (141–143).

## 4.3  Keratin gene expression in mature skin epithelia

The profound asymmetry that characterizes skin epithelia is mirrored by their patterns of keratin gene expression. Starting with epidermis, as many as 11 keratin genes can be expressed in keratinocytes alone, depending upon their differentiation status and regional tissue specifications. Mitotically active basal cells express the type II *K5* and the type I *K14* and *K15* genes (133, 144, 145). A small subset of basal cells may also express *K19* (146, 147; see below). Coinciding with the onset of differentiation in interfollicular epidermis, transcription of the basal-layer-specific genes ceases, while that of *K1* (type II) and *K10* (type I) is induced (131, 133). Accordingly, the presence of K1 and K10 proteins can be easily detected in the lowermost suprabasal layers, and occasionally, in postmitotic cells that have yet to 'leave' the basal layer (148). Yet another type II keratin gene, *K2e*, is expressed at a later stage of epidermal differentiation (i.e. in the granular layer) (149). In the thicker epidermis of human palm and sole skin, the major differentiation-specific type I gene is *K9* (at the expense of *K10*; 150), while the type II K6 and type I K16 and K17 occur in a patchy fashion (151; P. A. Coulombe and K. McGowan, unpublished data). Such differentiation-related patterns in keratin gene expression are very typical of complex epithelia (131, 133). A great deal can be said, therefore, about progress through epidermal differentiation from the standpoint of keratin gene expression. However, much remains to be learned about the molecular mechanisms that regulate the spatial and temporal expression of keratin genes in this context (129,131,152).

In agreement with its much greater complexity, a larger number of keratin genes (more than 25) is expressed in the pilosebaceous unit. Starting with the hair follicle, the outer root sheath resembles epidermis, in that K5, K14, K15, and K19 are preferentially expressed in the outermost, progenitor layer. The case of K19 is an intriguing one, as this keratin is preferentially associated with the least differentiated keratinocytes (stem cells?) of the bulge (145, 146). At least two K6 isoforms (153) and K16 are expressed in the postmitotic layers of the outer root sheath (154). A recently discovered type II keratin, designated K6hf, occurs specifically in the companion layer (the innermost of the outer root sheath (ORS) layers; 155). Finally, the *K17* gene is expressed throughout the ORS layers (62, 156). Likewise, the various cell layers of the inner root sheath also feature a well-developed array of keratin filaments (157), but the identity of the proteins making them up remains unclear (158). In the hair shaft proper, the cortical epithelial cells express several type I and type II hard keratin genes, depending on their progression through differentiation (19). Finally, at least the type I keratin K19 (159) and K17 (62; P. A. Coulombe and K. McGowan, unpublished data) are expressed in the medulla, but we don't yet know which is their type II partner(s) in these cells. Sebaceous glands are complex epithelia as well, and accordingly they express several type I and type II keratin genes (not to be discussed here; 160). Here again, therefore, the profoundly asymmetrical organization of a complex epithelial unit, namely the pilosebaceous apparatus, is partly reflected through its pattern of keratin gene expression.

## 4.4   Keratin gene expression during skin morphogenesis

Not surprisingly, the major steps taking place during the morphogenesis of mammalian skin epithelia are paralleled by changes in the regulation of keratin gene expression (19, 62, 100, 161–163 and refs therein). The description given here applies to mouse fetal development, which lasts between 19 and 20 days (164). The morphological events that surround hair morphogenesis in this species are illustrated in Fig. 2. Up to embryonic day 9.5 (abbreviated as ed9.5), the single-layered post-neurulation ectoderm covering the embryo expresses the keratins characteristic of adult simple epithelia, K8 and K18 (165–167). Expression of the transcription factor AP2 begins at ed9.5, soon followed by that of its target genes K5 and K14 (100). Similarly to the signalling events described in Section 3 above, the expression of these genes occurs first in the head region, and spreads rostrally in a wave-like spatiotemporal pattern. At least initially, these cells are uniformly immunopositive for K8/K18 and K5/K14 proteins. The notion that onset of *K5* and *K14* gene expression would reflect the decision to form a complex epithelium in embryonic ectoderm is an attractive and interesting one, given that their expression persists in the (least differentiated) progenitor cells in adult complex epithelia. Similar observations have been reported in human fetal skin (161,162).

The next fate-determination event that ectoderm cells undergo is whether to give rise to epidermis or non-epidermis (i.e. appendages or periderm) (99). We recently showed that onset of K17 protein expression likely reflects this event (62). Immunoreactivity for this keratin is first detected in a subset of individual ectodermal cells in the one- or two-layer stage ectoderm in ed10.5 embryos. Over the next 48 hours, the K17-positive single cells evolve into small cell clusters distributed periodically over the surface of the embryo, and then into morphologically distinct placodes and periderm (Fig. 3A–3D). Coincident with stratification, K8/K18 become polarized to the periderm. Whether a distinct type II keratin gene is co-induced along with the type I K17 is not clear. The expression of K6, a type II keratin that is frequently co-regulated with K17 in adult epithelia, begins at a later stage and is initially restricted to the periderm (62). In the early stage placodes, therefore, the only known type II keratin is K5. Recent studies have established can K17 can be promiscuous in its choice of a type II partner (168).

As we have discussed throughout this chapter, the placodes (and possibly the periderm as well; 98) form directly above sites of mesenchymal cell condensates (Fig. 2). It is in that context that expression of the transcription factor lef-1, a late-stage effector in Wnt signalling pathways, is first detected (60, 61). In fact, there is a striking correspondence between cytoplasmic staining for K17 (Fig. 3B and 3C) and nuclear staining for lef-1 in placode epithelial cells, while the mesenchymal cells located

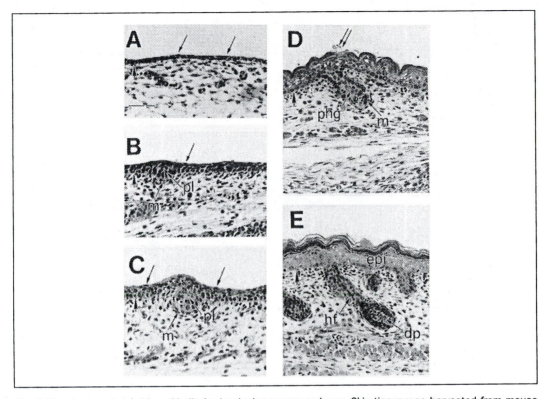

**Fig. 2** Morphogenesis of skin epithelia in developing mouse embryos. Skin tissue was harvested from mouse embryos at specific times of development, fixed in Bouin's, embedded in paraffin, and 5 μm sections were stained with haematoxylin and eosin (62). All samples shown are from the dorsal side of the trunk, and in all frames the thick arrowhead at left depicts the interface between the epithelium and mesenchyme. (A) Surface ectoderm from an ed12.5 embryo. Two layers can be recognized. The upper one is flat and consists of the periderm. The short arrows point to periderm cell nuclei. (B) and (C) Two examples of ed14.5 ectoderm. The epithelium has begun to stratify, and placodes (pl) are forming above mesenchymal cell condensates (m). The placode shown in C is larger than that shown in B, reflecting a later stage of development. Again, the short arrows point to periderm cell nuclei. (D) ed16.5 ectoderm. Many of the developing follicles have reached the primary hair germ (phg) stage, and feature a larger and well-defined condensate of mesenchymal cells (m). The epidermis has stratified further and begins to show signs of differentiation. The double arrows point to a possible example of periderm shedding. (E) ed18.5 ectoderm. The morphology of mature epidermis (epi) can now be recognized (compare with Fig. 1A). The hair follicle (hf) has grown further, and it too shows a more mature morphology. dp, dermal papillae. Bar in (A) equals 50 μm and applies for all frames.

underneath contain only lef-1 (62). The notion that the K17 gene may be at least partly regulated by lef-1 (and/or related TCF family members) in this context is supported by two facts: the presence of consensus binding sites for lef-1 in its 5′ upstream sequence, and the observation that K17 expression is induced in transgenic mice that ectopically express lef-1 under the control of a K14 gene promoter (62). If proven, this relationship would establish K17 as a target gene for the Wnt signal(s) postulated to function at an early stage in the morphogenesis of all appendages (see above).

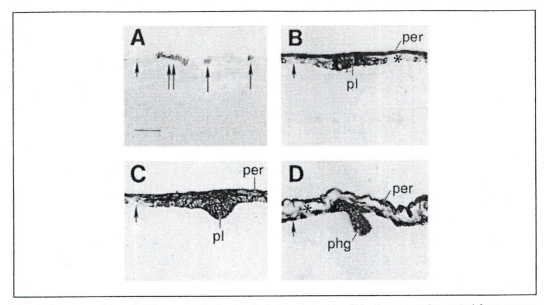

**Fig. 3** Ontogeny of keratin 17 expression in developing mouse skin. Skin tissue was harvested from mouse embryos at specific times of development, snap frozen, and embedded in OCT compound. Sections (thickness 5 μm) were immunostained with an antiserum to K17 followed by detection by horseradish peroxidase (62). All tissue samples shown are from the dorsal side of the trunk, and in all frames the thick arrowhead at left depicts the interface between the epithelium and mesenchyme. (A) ed12.5 skin in which either single cells (arrows) or small clusters of cells (double arrow) are positive for the K17 antigen in the ectoderm. (B) and (C) Two examples of hair placodes in ed14.5 skin. The placodes (pl) and periderm (per) epithelial cells are strongly positive for the K17 antigen, while the embryonic epidermis (see asterisk in B) is negative. (D) ed16.5 skin showing a primary hair germ (phg) that is strongly positive for K17. The periderm is uniformly labelled as well, while the interfollicular epidermis is mostly negative. Bar in (A) equals 50 μm and applies for all frames.

The next occurrence of new keratin gene expression occurs at ed14.5–15.5, coinciding with the beginning of differentiation within each of the major lineages in developing skin epithelia (Fig. 2D). In the epidermis, for instance, expression of *K1/K10* begins in the suprabasal layers (100 and refs therein), while several hard keratin genes are expressed in a subset of cells within primary hair germs (19 and refs therein). Pursuing the study of the mechanisms regulating the expression of differentiation-specific structural genes, such as those encoding keratins, represents a sound strategy to bridge the knowledge gap that currently exists between the molecular cues responsible for tissue patterning, as defined in Section 3 of this text, and the actual effectors of polarity at a cellular and multicellular levels (129, 131, 152). Alternatively, the study of developmental molecules may unexpectedly provide insights into the mechanisms of differentiation-specific gene expression. Consider, for instance, the recent discovery that the *Hox13c* gene is involved in the regulation of hard keratin gene expression in developing hair follicles (169). It is to be hoped that the progress made to that end will be as spectacular as that which we have recently witnessed in our understanding of the early steps of skin patterning.

# Acknowledgements

We sincerely apologize to those authors whose work was not directly recognized here. Due to the breadth of the subject, together with space limitations, we chose to refer to review articles wherever possible. The work in the authors' laboratory is supported by grants AR42047 and AR44232 from the National Institutes of Health. K.M. is supported by a NRSA Fellowship grant CA-67513 from the National Cancer Institute.

# References

1. Pinkus, H. (1958) Embryology of the hair. In *The biology of hair growth*, (ed. W. Montagna and R. A. Ellis), pp. 1–32. Academic Press, New York.

2. Montagna, W. and Parakkal, P. F. (1974) *The structure and function of skin*, (3rd edn). Academic Press, New-York.

3. Stenn, K. S. (1983) The skin. In *Histology: Cell and tissue biology*, (ed. L. Weiss), pp. 569–605. Elsevier Biomedical, New York.

4. Holbrook, K. A. and Wolff, K. (1993) The structure and development of skin. In *Dermatology in general medicine*, (ed. T. B. Fitzpatrick, A. Z. Eisen, K. Wolff, I. M. Freedberg, and M. D. Austen), pp. 97–144. McGraw-Hill, New-York.

5. Sengel, P. (1976) *The morphogenesis of skin*. Cambridge University Press, Cambridge.

6. Sengel, P. (1983) Epidermal–dermal interactions during formation of skin and cutaneous appendages. In *Biochemistry and physiology of skin*, (ed. A. L. Goldsmith). Oxford University Press, New York.

7. Dhouailly, D. (1977) Regional specification of cutaneous appendages in mammals. *Wilhem. Roux's Archives*, **181**, 3–10.

8. Sawyer, R. H. (1983) The role of epithelial–mesenchymal interactions in regulating gene expression during avian scale morphogenesis. In *Epithelial–mesenchymal interactions in development*, (ed. R. H. Sawyer and J. F. Fallon), Prager, New-York.

9. Messenger, A. G. (1993) The control of hair growth: an overview. *J. Invest. Dermatol.*, **101**, 4S–9S.

10. Hardy, M. H. (1992) The secret life of the hair follicle. *Trends Genet.*, **8**, 55–61.

11. Fisher, C. (1994) The cellular basis of development and differentiation in mammalian keratinizing epithelia. In *The keratinocyte handbook*, (ed. I. M. Leigh, B. Lane, F. Watt), pp. 131–52. Cambridge University Press, Cambridge.

12. MacKenzie, I. C. (1994) Epithelial–mesenchymal interactions in the development and maintenance of epithelial tissues. In *The keratinocyte handbook*, (ed. I. M. Leigh, B. Lane, and F. Watt), pp. 243–58. Cambridge University Press, Cambridge.

13. Cotsarelis, G. (1997) The hair follicle: dying for attention, *Am. J. Pathol.*, **151**, 1505–9.

14. Gilchrest, B. A. (1993) Aging of the skin. In *Dermatology in general medicine*, (ed. T. B. Fitzpatrick, A. Z. Eisen, K. Wolff, I. M. Freedberg, M. D. Austen), pp. 150–8. McGraw-Hill, New-York.

15. Martin, P. (1997) Wound healing—aiming for the perfect skin regeneration. *Science*, **276**, 75–81.

16. Hall, P. A. and Watt, F. M. (1989) Stem cells: the generation and maintenance of cellular diversity. *Development*, **106**, 619–33.

17. Dover, R. and Wright, N. A. (1993) Epidermal cell kinetics, In *Dermatology in general*

*medicine*, (ed. T. B. Fitzpatrick, A. Z. Eisen, K. Wolff, I. M. Freedberg, and M. D. Austen), pp. 159–71. McGraw-Hill, New-York.

18. Holbrook, K. A. (1994) Ultrastructure of the epidermis. In *The keratinocyte handbook*, (ed. I. M. Leigh, B. Lane, and F. Watt), pp. 1–42. Cambridge University Press, Cambridge.

19. Powell, B. C. and Rogers, G. E. (1997) The role of keratin proteins and their genes in the growth, structure and properties of hair. *EXS*, **78**, 59–148.

20. Jahoda, C. A., Horne, K. A., and Oliver, R. F. (1984) Induction of hair growth by implantation of cultured dermal papilla cells. *Nature*, **311**, 560–2.

21. Stenn, K. S., Combates, N. J., Eilertsen, K. J., Gordon, J. S., Pardinas, J. R., Parimoo, S., and Prouty, S. M. (1996) Hair follicle growth controls. *Dermatol. Clin.*, **14**, 543–58.

22. Coulombe, P. A., Kopan, R., and Fuchs, E. (1989) Expression of keratin K14 in the epidermis and hair follicle: insights into complex programs of differentiation. *J. Cell Biol.*, **109**, 2295–312.

23. Cotsarelis, G., Sun, T. T., and Lavker, R. M. (1990) Label-retaining cells reside in the bulge area of pilosebaceous unit: implications for follicular stem cells, hair cycle, and skin carcinogenesis. *Cell*, **61**, 1329–37.

24. Lavker, R. M., Miller, S., Wilson, C., Cotsarelis, G., Wei, Z. G., Yang, J. S., and Sun, T. T. (1993) Hair follicle stem cells: their location, role in hair cycle, and involvement in skin tumor formation. *J. Invest. Dermatol.*, **101**, 16S–26S.

25. Oro, A. E. and Scott, M. P. (1998) Splitting hairs: dissecting roles of signaling systems in epidermal development. *Cell*, **95**, 575–8.

26. Sasai, Y. and De Robertis, E. M. (1997) Ectodermal patterning in vertebrate embryos. *Dev. Biol.*, **182**, 5–20.

27. Gurdon, J. B., Dyson, S., and St Johnston, D. (1998) Cells' perception of position in a concentration gradient. *Cell*, **95**, 159–62.

28. Hogan, B. L. (1999) Morphogenesis. *Cell*, **96**, 225–33.

29. Chuong, C. M. (1998) *Molecular basis of epithelial appendage morphogenesis*. R. G. Landes Co., Austin, TX.

30. Le Douarin, N. M., Ziller, C., and Couly, G. F. (1993) Patterning of neural crest derivatives in the avian embryo: *in vivo* and *in vitro* studies. *Dev. Biol.*, **159**, 24–49.

31. Wessels, N. K. (1977) *Tissue interactions and development*. W. A. Benjamin, Menlo Park, CA.

32. Gurdon, J. B. (1992) The generation of diversity and pattern in animal development. *Cell*, **68**, 185–199.

33. Hemmati-Brivanlou, A. and Melton, D. (1997) Vertebrate embryonic cells will become nerve cells unless told otherwise. *Cell*, **88**, 13–17.

34. Hogan, B. L. (1996) Bone morphogenetic proteins: multifunctional regulators of vertebrate development. *Genes Dev.*, **10**, 1580–94.

35. Wilson, P. A. and Hemmati-Brivanlou, A. (1995) Induction of epidermis and inhibition of neural fate by Bmp-4. *Nature*, **376**, 331–3.

36. Hawley, S. H., Wunnenberg-Stapleton, K., Hashimoto, C., Laurent, M. N., Watabe, T., Blumberg, B. W., and Cho, K. W. (1995) Disruption of BMP signals in embryonic *Xenopus* ectoderm leads to direct neural induction. *Genes Dev.*, **9**, 2923–35.

37. Piccolo, S., Sasai, Y., Lu, B., and De Robertis, E. M. (1996) Dorsoventral patterning in *Xenopus*: inhibition of ventral signals by direct binding of chordin to BMP-4. *Cell*, **86**, 589–98.

38. Zimmerman, L. B., De Jesus-Escobar, J. M., and Harland, R. M. (1996) The Spemann organizer signal noggin binds and inactivates bone morphogenetic protein 4. *Cell*, **86**, 599–606.

39. Shimell, M. J., Ferguson, E. L., Childs, S. R., and O'Connor, M. B. (1991) The *Drosophila* dorsal–ventral patterning gene tolloid is related to human bone morphogenetic protein 1. *Cell*, **67**, 469–81.

40. Piccolo, S., Agius, E., Lu, B., Goodman, S., Dale, L., and De Robertis, E. M. (1997) Cleavage of Chordin by Xolloid metalloprotease suggests a role for proteolytic processing in the regulation of Spemann organizer activity. *Cell*, **91**, 407–16.

41. Mullins, M. C. (1998) Holy tolloido: Tolloid cleaves SOG/Chordin to free DPP/BMPs. *Trends Genet.*, **14**, 127–9.

42. Szebenyi, G. and Fallon, J. F. (1999) Fibroblast growth factors as multifunctional signaling factors. *Int. Rev. Cytol.*, **185**, 45–106.

43. Song, H., Wang, Y., and Goetinck, P. F. (1996) Fibroblast growth factor 2 can replace ectodermal signaling for feather development. *Proc. Natl Acad. Sci., USA*, **93**, 10246–9.

44. Jung, H. S., Francis-West, P. H., Widelitz, R. B., Jiang, T. X., Ting-Berreth, S., Tickle, C., *et al.* (1998) Local inhibitory action of BMPs and their relationships with activators in feather formation: implications for periodic patterning. *Dev. Biol.*, **196**, 11–23.

45. Noramly, S. and Morgan, B. A. (1998) BMPs mediate lateral inhibition at successive stages in feather tract development. *Development*, **125**, 3775–87.

46. Kopan, R. and Weintraub, H. (1993) Mouse notch: expression in hair follicles correlates with cell fate determination. *J. Cell Biol.*, **121**, 631–41.

47. Artavanis-Tsakonas, S., Rand, M. D., and Lake, R. J. (1999) Notch signaling: Cell fate control and signal integration in development. *Science*, **284**, 770–6.

48. Schroeter, E. H., Kisslinger, J. A., and Kopan, R. (1998) Notch-1 signalling requires ligand-induced proteolytic release of intracellular domain. *Nature*, **393**, 382–6.

49. Crowe, R., Henrique, D., Ish-Horowicz, D., and Niswander, L. (1998) A new role for Notch and Delta in cell fate decisions: patterning the feather array. *Development*, **125**, 767–75.

50. Crowe, R. and Niswander, L. (1998) Disruption of scale development by Delta-1 misexpression. *Dev. Biol.*, **195**, 70–4.

51. Viallet, J. P., Prin, F., Olivera-Martinez, I., Hirsinger, E., Pourquie, O., and Dhouailly, D. (1998) Chick Delta-1 gene expression and the formation of the feather primordia. *Mech. Dev.*, **72**, 159–68.

52. Wodarz, A. and Nusse, R. (1998) Mechanism of Wnt signaling in development. *Annu. Rev. Cell Dev. Biol.*, **14**, 59–88.

53. Takeichi, M. (1991) Cadherin cell adhesion receptors as a morphogenetic regulator. *Science*, **251**, 1451–5.

54. Barth, A. I., Nathke, I. S., and Nelson, W. J. (1997) Cadherins, catenins and APC protein: interplay between cytoskeletal complexes and signaling pathways. *Curr. Opin. Cell Biol.*, **9**, 683–90.

55. Clevers, H. and van de Wetering, M. (1997) TCF/LEF factor earn their wings. *Trends Genet.*, **13**, 485–9.

56. Eastman, Q. and Grosschedl, R. (1999) Regulation of lef-1/TCF transcription factors by Wnt and other signals. *Curr. Opin. Cell Biol.*, **11**, 233–40.

57. Orsulic, S. and Peifer, M. (1996) Cell–cell signalling: Wingless lands at last. *Curr. Biol.*, **6**, 1363–7.

58. Zeng, L., Fagotto, F., Zhang, T., Hsu, W., Vasicek, T. J., Perry, W. L., *et al.* (1997) The mouse Fused locus encodes Axin, an inhibitor of the Wnt signaling pathway that regulates embryonic axis formation. *Cell*, **90**, 181–92.

59. Behrens, J., Jerchow, B. A., Wurtele, M., Grimm, J., Asbrand, C., Wirtz, R., *et al.* (1998)

Functional interaction of an axin homolog, conductin, with beta-catenin, APC, and GSK3beta. *Science*, **280**, 596–9.

60. Zhou, P., Byrne, C., Jacobs, J., and Fuchs, E. (1995) Lymphoid enhancer factor 1 directs hair follicle patterning and epithelial cell fate. *Genes Dev.*, **9**, 700–13.

61. Kratochwil, K., Dull., M., Farinas, I., Galceran, J., Grosscheld, R. (1996) Lef1 expression is activated by BMP-4 and regulates inductive tissue interactions in tooth and hair development. *Cell*, **10**, 1382–94.

62. McGowan, K. and Coulombe, P. A. (1998) Onset of keratin 17 expression coincides with the definition of major epithelial lineages during mouse skin development, **143**: 469–86.

63. van Genderen, C., Okamura, R. M., Farinas, I., Quo, R.-G., Parslow, T. G., Bruhn, L., and Grosscheld, R. (1994) Development of several organs that require inductive epithelial-mesenchymal interactions is imparied in Lef-1-deficient mice. *Genes Dev.*, **8**, 2691–703.

64. Haegel, H., Larue, L., Ohsugi, M., Fedorov, L., Herrenknecht, K., and Kemler, R. (1995) Lack of beta-catenin affects mouse development at gastrulation. *Development*, **121**, 3529–37.

65. Gat, U., DasGupta, R., Degenstein, L., and Fuchs, E. (1998) *De novo* hair follicle morphogenesis and hair tumors in mice expressing a truncated beta-catenin in skin. *Cell*, **95**, 605–14.

66. St-Jacques, B., Dassule, H. R., Karavanova, I., Botchkarev, V. A., Li, J., Danielian, P. S., *et al.* (1998) Sonic hedgehog signaling is essential for hair development. *Curr. Biol.*, **8**, 1058–68.

67. Goodrich, L. V., Johnson, R. L., Milenkovic, L., McMahon, J. A., and Scott, M. P. (1995) Conservation of the hedgehog/patched signaling pathway from flies to mice: Induction of a mouse patched gene by hedgehog. *Genes Dev.*, **10**, 301–12.

68. Beachy, P. A., Cooper, M. K., Young, K. E., von Kessler, D. P., Park, W. J., Hall, T. M., *et al.* (1997) Multiple roles of cholesterol in hedgehog protein biogenesis and signaling. *Cold Spring Harbor Symp. Quant. Biol.*, **62**, 191–204.

69. Johnston, R. L. and Scott, M. P. (1998) New players and puzzles in the Hedgehog signaling pathway. *Curr. Opin. Genet. Dev.*, **8**, 450–6.

70. Inham, P. W. (1998) Transducing hedgehog: The story so far. *EMBO J.* **17**, 3505–11.

71. Iseki, S., Araga, A., Ohuchi, H., Nohno, T., Yoshioka, H., Hayashi, F., and Noji, S. (1996) Sonic hedgehog is expressed in epithelial cells during development of whisker, hair, and tooth. *Biochem. Biophys. Res. Comm.* **218**, 688–93.

72. Morgan, B. A., Orkin, R. W., Noramly, S., and Perez, A. (1998) Stage-specific effects of sonic hedgehog expression in the epidermis. *Dev. Biol.*, **201**, 1–12.

73. Chiang, C., Swan, R. Z., Grachtchouk, M., Bolinger, M., Litingtung, Y., Robertson, E. K., *et al.*, (1999) Essential role for Sonic hedgehog during hair follicle morphogenesis. *Dev. Biol.*, **205**, 1–9.

74. Chiang, C., Litingtung, Y., Lee, E., Young, K. E., Corden, J. L., Westphal, H., and Beachy, P. A. (1996) Cyclopia and defective axial patterning in mice lacking Sonic hedgehog gene function. *Nature*, **383**, 407–13.

75. Luetteke, N. C., Qiu, T. H., Peiffer, R. L., Oliver, P., Smithies, O., and Lee, D. C. (1993) TGF alpha deficiency results in hair follicle and eye abnormalities in targeted and waved-1 mice. *Cell*, **73**, 263–78.

76. Mann, G. B., Fowler, K. J., Gabriel, A., Nice, E. C., Williams, R. L., and Dunn, A. R. (1993) Mice with a null mutation of the TGF alpha gene have abnormal skin architecture, wavy hair, and curly whiskers and often develop corneal inflammation. *Cell*, **73**, 249–61.

77. Capecchi, M. R. (1997) Hox genes and mammalian development. *Cold Spring Harbor Symp. Quant. Biol.*, **62**, 273–81.

78. Chuong, C. M., Widelitz, R. B., Ting-Berreth, S., and Jiang, T. X. (1996) Early events during avian skin appendage regeneration: dependence on epithelial–mesenchymal interaction and order of molecular reappearance. *J. Invest. Dermatol.*, **107**, 639–46.

79. Jones, P. H., Harper, S., and Watt, F. M. (1995) Stem cell patterning and fate in human epidermis. *Cell*, **80**, 83–93.

80. Rochat, A., Kobayashi, K., and Barrandon, Y. (1994) Location of stem cells of human hair follicles by clonal analysis. *Cell*, **76**, 1063–73.

81. Coulombe, P. A. (1997) Towards a molecular definition of keratinocyte activation after acute injury to stratified epithelia. *Biochem. Biophys. Res. Comm.*, **236**, 231–8.

82. Hebert, J. M., Rosenquist, T., Gotz, J., and Martin, G. R. (1994) FGF5 as a regulator of the hair growth cycle: evidence from targeted and spontaneous mutations. *Cell*, **78**, 1017–25.

83. Luetteke, N. C., Phillips, H. K., Qiu, T. H., Copeland, N. G., Earp, H. S., Jenkins, N. A., and Lee, D. C. (1994) The mouse waved-2 phenotype results from a point mutation in the EGF receptor tyrosine kinase. *Genes Dev.*, **8**, 399–413.

84. Threadgill, D. W., Dlugosz, A. A., Hansen, L. A., Tennebaum, T., Lichti, U., Yee, D., *et al.* (1995). Targeted disruption of mouse EGF receptor: Effect of genetic background on mutant phenotype. *Science*, **269**, 230–4.

85. Sibilia, M. and Wagner, E. R. (1995) Strain-dependent epithelial defects in mice lacking the EGF receptor. *Science*, **269**, 234–7.

86. Hansen, L. A., Alexander, N., Hogan, M. E., Sundberg, J. P., Dlugosz, A., Threadgill, D. W., *et al.* (1997) Genetically null mice reveal a central role for epidermal growth factor recpetor in the differentiation of the hair follicle and normal hair development. *Am. J. Pathol.*, **150**, 1959–75.

87. Murillas, R., Larcher, F., Conti, C., Santos, M., Ullrich, A., and Jorcano, J. (1995) Expression of a dominant negative mutant of epidermal growth factor receptor in the epidermis of transgenic mice elicits striking alterations in hair follicle development and skin structure. *EMBO J.*, **14**, 5216–23.

88. Cachon-Gonzalez, M. B., Fenner, S., Coffin, J. M., Moran, C., Best, S., and Stoye, J. P. (1994) Structure and expression of the hairless gene in mice. *Proc. Natl Acad. Sci., USA*, **91**, 7717–21.

89. Cichon, S., Anker, M., Vogt, I. R., Rohleder, H., Putzstuck, M., Hillmer, A., *et al.* (1998) Cloning, genomic organization, alternative transcripts and mutational analysis of the gene responsible for autosomal recessive universal congenital alopecia. *Hum. Mol. Genet.*, **7**, 1671–9.

90. Ahmad, W., Faiyaz ul Haque, M., Brancolini, V., Tsou, H. C., ul Haque, S., Lam, H., *et al.* (1998) Alopecia universalis associated with a mutation in the human hairless gene. *Science*, **279**, 720–4.

91. Thompson, C. C. (1996) Thyroid hormone-responsive genes in developing cerebellum include a novel synaptotagmin and a hairless homolog. *J. Neurosci.*, **16**, 7832–40.

92. Lewin, G. R. and Barde, Y. A. (1996) Physiology of the neurotrophins. *Annu. Rev. Neurosci.*, **19**, 289–317.

93. Ernfors, P., Lee, K. F., Kucera, J., and Jaenisch, R. (1994) Lack of neurotrophin-3 leads to deficiencies in the peripheral nervous system and loss of limb proprioceptive afferents. *Cell*, **77**, 503–12.

94. Airaksinen, M. S., Koltzenburg, M., Lewin, G. R., Masu, Y., Helbig, C., Wolf, E., *et al.* (1996) Specific subtypes of cutaneous mechano-receptors require neurotrophin-3 following peripheral target innervation. *Neuron*, **16**, 287–95.

95. Albers, K. M., Perrone, T. N., Goodness, T. P., Jones, M. E., Green, M. A., and Davis, B. M. (1996) Cutaneous overexpression of NT-3 increases sensory and sympathetic neuron number and enhances touch dome and hair follicle innervation. *J. Cell Biol.*, **134**, 487–97.

96. Patapoutian, A., Backus, C., Kispert, A., and Reichardt, L. F. (1999) Regulation of neurotrophin-3 expression by epithelial–mesenchymal interactions: The role of Wnt factors. *Science*, **283**, 1180–3.

97. Botchkarev, V. A., Welker, P., Albers, K. M., Botchkareva, N. V., Metz, M., Lewin, G. R., *et al.* (1998) A new role for neurotrophin-3: involvement in the regulation of hair follicle regression (catagen). *Am. J. Pathol.*, **153**, 785–99.

98. M'Boneko, V. and Merker, H.-J. (1988) Development and morphology of the periderm of mouse embryos (Days 9–12 of gestation). *Acta Anat.*, **133**, 325–36.

99. Sanes, J. R., Rubenstein, J. L. R., and Nicolas, J.-F. (1986) Use of a recombinant retrovirus to study post-implantation cell lineage in mouse embryos. *EMBO J.*, **5**, 3133–42.

100. Byrne, C., Tainsky, M., and Fuchs, E. (1994) Programming gene expression in developing epidermis. *Development*, **120**, 2369–83.

101. Hardman, M. J., Sisi, P., Banbury, D. N., and Byrne, C. (1998) Patterned acquisition of skin barrier function during development. *Development*, **125**, 1541–52.

102. Sparks, A. B., Morin, P. J., Vogelstein, B., and Kinzler, K. W. (1998) Mutational analysis of the APC/beta-catenin/Tcf pathway in colorectal cancer. *Cancer Res.*, **58**, 1130–4.

103. Chan, E. F., Gat, U., McNiff, J. M., and Fuchs, E. (1999) A common human skin tumour is caused by activating mutations in β-catenin. *Nature Genet.*, **21**, 410–13.

104. Nusse, R. and Varmus, H. E. (1982) Many tumors induced by the mouse mammary tumor virus contain a provirus integrated in the same region of the host genome. *Cell*, **31**, 99–109.

105. Kinzler, K. W., Nilbert, M. C., Su, L. K., Vogelstein, B., Bryan, T. M., Levy, D. B., *et al.* (1991) Identification of FAP locus genes from chromosome 5q21. *Science*, **253**, 661–5.

106. Morin, P. J., Sparks, A. B., Korinek, V., Barker, N., Clevers, H., Vogelstein, B., and Kinzler, K. W. (1997) Activation of beta-catenin-Tcf signaling in colon cancer by mutations in beta-catenin or APC. *Science*, **275**, 1787–90.

107. Tetsu, O. and McCormick, F. (1999) Beta-catenin regulates expression of cyclin D1 in colon carcinoma cells. *Nature*, **398**, 422–6.

108. Shtutman, M., Zhurinsky, J., Simcha, I., Albanese, C., D'Amico, M., Pestell, R., BenZe'ev, A. (1999) The cyclin D1 gene is a target of the β-catenin/lef-1 pathway. *Proc. Natl Acad. Sci., USA*, **96**, 5522–7.

109. He, T. C., Sparks, A. B., Rago, C., Hermeking, H., Zawel, L., da Costa, L. T., Morin, P. J., Vogelstein, B., and Kinzler, K. W. (1998) Identification of c-MYC as a target of the APC pathway. *Science*, **281**, 1509–12.

110. Carter, D. M. and Lin, A. N. (1993) Basal cell carcinoma. In *Dermatology in general medicine*, (ed. T. B. Fitzpatrick, A. Z. Eisen, K. Wolff, I. M. Freedberg, and M. D. Austen), pp. 840–7. McGraw-Hill, New-York.

111. Oro, A. E., Higgins, K. M., Hu, Z., Bonifas, J. M., Epstein, E. H. Jr, and Scott, M. P. (1997) Basal cell carcinomas in mice overexpressing sonic hedgehog. *Science*, **276**, 817–21.

112. Dahmane, N., Lee, J., Robins, P., Heller, P., and Ruiz i Altaba, A. (1997) Activation of the transcription factor Gli1 and the Sonic hedgehog signalling pathway in skin tumours. *Nature*, **389**, 876–81.

113. Johnson, R. L., Rothman, A. L., Xie, J., Goodrich, L. V., Bare, J. W., Bonifas, J. M., *et al.* (1996) Human homolog of patched, a candidate gene for the basal cell nevus syndrome. *Science*, **272**, 1668–71.

114. Hahn, H., Wicking, C., Zaphiropoulous, P., Gailani, M., Shanley, S., Chidambaram, A., *et al*. (1996) Mutations of the human homolog of *Drosophila* patched in the nevoid basal cell carcinoma syndrome. *Cell*, **85**, 841–51.

115. Gailani, M. R., Stahle-Backdahl, M., Leffell, D., Glynn, M., Zaphiropoulos, P. G., Pressman, C., *et al*. (1996) The role of the human homologue of *Drosophila* patched in sporadic basal cell carcinomas. *Nature Genetics*, **14**, 78–81.

116. Xie, J., Murone, M., Luoh, S. M., Ryan, A., Gu, Q., Zhang, C., *et al*. (1998) Activating Smoothened mutations in sporadic basal-cell carcinoma. *Nature*, **391**, 90–2.

117. Johnston, M. (1998) Gene chips: Array of hope for understanding gene regulation. *Curr. Biol.*, **8**, R171–R174.

118. DeRisi, J. L., Iyer, V. R., and Brown, P. O. (1997) Exploring the metabolic and genetic control of gene expression on a genomic scale. *Science*, **278**, 680–6.

119. Rees, J. (1992) The molecular biology of retinoic acid receptors: orphan from good family seeks home. *Br. J. Dermatol.*, **126**, 97–104.

120. Saitou, M., Sugai, S., Tanaka, T., Shimouchi, K., Fuchs, E., Narumiya, S., and Kakizuka, A. (1995) Inhibition of skin development by targeted expression of a dominant-negative retinoic acid receptor. *Nature*, **374**, 159–62.

121. Attar, P. S., Wertz, P. W., McArthur, M., Imakado, S., Bickenbach, J. R., and Roop, D. R. (1997) Inhibition of retinoid signaling in transgenic mice alters lipid processing and disrupts epidermal barrier function. *Mol. Endocrinol.*, **11**, 792–800.

122. Garrod, D. R. (1993) Desmosomes and hemidesmosomes. *Curr. Opin. Cell Biol.*, **5**, 30–40.

123. Koch, P. J. and Franke, W. W. (1994) Desmosomal cadherins: another growing multigene family of adhesion molecules. *Curr. Opin. Cell Biol.*, **6**, 682–7.

124. Kowalczyk, A. P., Bornslaeger, E. A., Norvell, S. M., Palka, H. L., and Green, K. J. (1999) Desmosomes: intercellular adhesive junctions specialized for attachment of intermediate filaments. *Int. Rev. Cytol.*, **185**, 237–302.

125. Salomon, D., Masgrau, E., Vischer, S., Ullrich, S., Dupont, E., Sappino, P., Saurat, J., and Meda, P. (1994) Topography of mammalian connexins in human skin. *J. Invest. Dermatol.*, **103**, 240–7.

126. Risek, B., Klier, F. G., and Gilula, N. B. (1992) Multiple gap junction genes are utilized during rat skin and hair development. *Development*, **116**, 639–51.

127. Millar, S., Willert, K., Salinas, P. C., Roelink, H., Nusee, R., Sussman, D., and Barsh, G. S. (1999) Wnt signaling in the control of hair growth and structure. *Dev. Biol.*, **207**, 133–49.

128. Mischke, D. (1998) The complexity of gene families involved in epithelial differentiation: Keratin genes and the epidermal differentiation complex. In *Subcellular biochemistry: intermediate filaments*, (ed. J. R. Harris and H. Herrmann), pp. 71–95. Plenum Publishing, London.

129. Eckert, R. L., Crish, J. F., and Robinson, N. A. (1997) The epidermal keratinocyte as a model for the study of gene regulation and cell differentiation. *Physiol. Rev.*, **77**, 397–424.

130. Fuchs, E. and Weber, K. (1994) Intermediate filaments: structure, dynamics, function, and disease. *Annu. Rev. Biochem.*, **63**, 345–82.

131. Fuchs, E. (1995) Keratins and the skin. *Annu. Rev. Cell Dev. Biol.*, **11**, 123–53.

132. Moll, R., Franke, W. W., Schiller, D. L., Geiger, B., and Krepler, R. (1982) The catalog of human cytokeratins: patterns of expression in normal epithelia, tumors and cultured cells. *Cell*, **31**, 11–24.

133. O'Guin, W. M., Schermer, A., Lynch, M., Sun, T.-T. (1990) Differentiation-specific expression of keratin pairs. In *Cellular and molecular biology of intermediate filaments*, (ed. R. D. Goldman and P. M Steinert), pp. 301–34. Plenum Publishing, London.

134. Coulombe, P. A. and Fuchs, E. (1994) Molecular mechanisms of keratin gene disorders and other bullous diseases of the skin. In *Molecular mechanisms in epithelial cell junctions: From development to disease,* (ed. S. Citi), pp. 259–85. R. G. Landes, Austin, TX.

135. McLean, W. H. I. and Lane, E. B. (1995) Intermediate filaments in diseases. *Curr. Opin. Cell Biol.,* **7**, 118–25.

136. Korge, B. P. and Krieg, T. (1996) The molecular basis for inherited bullous diseases. *J. Mol. Med.,* **74**, 59–70.

137. Fuchs, E. and Cleveland, D. W. (1998) A structural scaffolding of intermediate filaments in health and disease. *Science,* **279**, 514–19.

138. Steinert, P. M., Idler, W. W., and Zimmerman, S. B. (1976) Self-assembly of bovine epidermal keratin filaments *in vitro. J. Mol. Biol.,* **108**, 547–67.

139. Coulombe, P. A. (1993) The cellular and molecular biology of keratins: beginning a new era. *Curr. Opin. Cell Biol.,* **5**, 17–29.

140. Hatzfeld, M. and Franke, W. W. (1985) Pair formation and promiscuity of cytokeratins: formation *in vitro* of heterotypic complexes and intermediate-sized filaments by homologous and heterologous recombinations of purified polypeptides. *J. Cell Biol.,* **101**, 1826–41.

141. Hutton, E., Paladini, R. D., Yu, Q. C., Yen, M.-Y., Coulombe, P. A., and Fuchs, E. (1998) Functional differences between keratins of stratified and simple epithelia. *J. Cell Biol.,* **143**, 487–99.

142. Paladini, R. D. and Coulombe, P. A. (1998) Directed expression of keratin 16 to the progenitor basal cells of transgenic mouse skin delays skin maturation. *J. Cell Biol.,* **142**, 1035–51.

143. Paladini, R. D. and Coulombe, P. A. (1999) The functional diversity of epidermal keratins revealed by the partial rescue of the keratin 14 null phenotype by keratin 16. *J. Cell Biol.,* **146**, 1185–1201.

144. Purkis, P. E., Steel, J. B., Mackenzie, I. C., Nathrath, W. B., Leigh, I. M., and Lane, E. B. (1990) Antibody markers of basal cells in complex epithelia. *J. Cell Sci.,* **97**, 39–50.

145. Lloyd, C., Yu, Q. C., Cheng, J., Turksen, K., Degenstein, L., Hutton, E., and Fuchs, E. (1995) The basal keratin network of stratified squamous epithelia: Defining K15 function in the absence of K14. *J. Cell Biol.,* **129**, 1329–44.

146. Stasiak, P. C., Purkis, P. E., Leigh, I. M., and Lane, E. B. (1989) Keratin 19: predicted amino acid sequence and broad tissue distribution suggest it evolved from keratinocyte keratins. *J. Invest. Dermatol.,* **92**, 707–16.

147. Michel, M., Torok, N., Godbout, M. J., Lussier, M., Gaudreau, P., Royal, A., and Germain, L. (1996) Keratin 19 as a biochemical marker of skin stem cells *in vivo* and *in vitro*: keratin 19 expressing cells are differentially localized in function of anatomic sites, and their number varies with donor age and culture stage. *J. Cell Sci.,* **109**, 1017–28.

148. Schweizer, J., Kinjo, M., Furstenberger, G., and Winter, H. (1984) Sequential expression of mRNA-encoded keratin sets in neonatal mouse epidermis: basal cells with properties of terminally differentiating cells. *Cell,* **37**, 159–70.

149. Collin, C., Moll, R., Kubicka, S., Ouhayoun, J. P., and Franke, W. W. (1992) Characterization of human cytokeratin 2, an epidermal cytoskeletal protein synthesized late during differentiation. *Exp. Cell Res.,* **202**, 132–41.

150. Langbein, L., Heid, H. W., Moll, I., and Franke, W. W. (1993) Molecular characterization of the body site-specific human epidermal cytokeratin 9: cDNA cloning, amino acid sequence, and tissue specificity of gene expression. *Differentiation,* **55**, 57–71.

151. Swensson, O., Langbein, L., McMillan, J. R., Stevens, H. P., Leigh, I. M., McLean, W. H.,

*et al.* (1998) Specialized keratin expression pattern in human ridged skin as an adaptation to high physical stress. *Br. J. Dermatol.*, **139**, 767–75.

152. Byrne, C. (1997) Regulation of gene expression in developing epidermal epithelia. *Bioessays*, **19**, 691–8.

153. Takahashi, K., Paladini, R., and Coulombe, P. A. (1995) Cloning and characterization of multiple human genes and cDNAs encoding highly related type II keratin 6 isoforms. *J. Biol. Chem.*, **270**, 18581–92.

154. Stark, H. J., Breikreutz, D., Limat, A., Bowden, P., and Fusenig, N. E. (1987) Keratins of the human hair follicle: 'Hyperproliferative' keratins consistently expressed in outer rrot sheat cells *in vivo* and *in vitro*. *Differentiation*, **35**, 236–48.

155. Winter, H., Langbein, L., Praetzel, S., Jacobs, M., Rogers, M. A., Leigh, I. M., *et al.* (1998) A novel human type II cytokeratin, K6hf, specifically expressed in the companion layer of the hair follicle. *J. Invest. Dermatol.*, **111**, 955–62.

156. Troyanovsky, S. M., Leube, R. E., and Franke, W. W. (1992) Characterization of the human gene encoding cytokeratin 17 and its expression pattern. *Eur. J. Cell Biol.*, **59**, 127–37.

157. Steinert, P. M. (1978) Structural features of the alpha-type filaments of the inner root sheath cells of the guinea pig hair follicle. *Biochemistry*, **17**, 5045–52.

158. Stark, H. J., Breitkreutz, D., Limat, A., Ryle, C. M., Roop, D., Leigh, I., and Fusenig, N. (1990) Keratins 1 and 10 or homologues as regular constituents of inner root sheath and cuticle cells in the human hair follicle. *Eur. J. Cell Biol.*, **52**, 359–72.

159. Heid, H. W., Moll, I., and Franke, W. W. (1988) Patterns of expression of trichocytic and epithelial cytokeratins in mammalian tissues. I. Human and bovine hair follicles. *Differentiation*, **37**, 137–57.

160. Hughes, B. R., Morris, C., Cunliffe, W. J., and Leigh, I. M. (1996) Keratin expression in pilosebaceous epithelia in truncal skin of acne patients. *Br. J. Dermatol.*, **134**, 247–56.

161. Moll, R., Moll, I., and Wiest, W. (1982b) Changes in the pattern of cytokeratin polypeptides in epidermis and hair follicles during skin development in human fetuses. *Differentiation*, **23**, 170–8.

162. Dale, B. A., Holbrook, K. A., Kimball, J. R., Hoff, M., and Sun, T. T. (1985) Expression of epidermal keratins and filaggrin during human fetal skin development. *J. Cell Biol.*, **101**, 1257–69.

163. Kopan, R. and Fuchs, E. (1989) A new look into an old problem: Keratins as tools to investigate determination, morphogenesis, and differentiation in skin. *Genes Dev.*, **3**, 1–15.

164. Kaufman, M. H. (1992) *The atlas of mouse development*. Academic Press, London.

165. Jackson, B. W., Grund, C., Winter, S., Franke, W. W., and Illmensee, K. (1981) Formation of cytoskeletal elements during mouse embryogenesis II. Epithelial differentiation and intermediate filaments in early post-implantation embryos. *Differentiation*, **20**, 203–16.

166. Thorey, I. S., Meneses, J. J., Neznanov, N., Kulesh, D. A., Pedersen, R. A., and Oshima, R. G. (1993) Embryonic expression of human keratin 18 and K18-beta-galactosidase fusion genes in transgenic mice. *Dev. Biol.*, **160**, 519–34.

167. Baribault, H., Price, J., Miyai, K., and Oshima, R. G. (1993) Mid-gestational lethality in mice lacking keratin 8. *Genes Dev.*, **7**, 1191–202.

168. Wawersik, M., Paladini, R. D., Noensie, E., and Coulombe, P. A. (1997) A proline residue in the α-helical rod domain of type I keratin 16 destabilizes keratin heterotetramers and influences incorporation into filaments. *J. Biol. Chem.*, **272**, 32557–65.

169. Godwin, A. R. and Capecchi, M. R. (1998) Hoxc13 mutant mice lack external hair. *Genes Dev.*, **12**, 11–20.

# Index

Printed in the United States
1412300002B/69